Physical Activity and Bone Health

Karim Khan
University of British Columbia
Heather McKay
University of British Columbia
Pekka Kannus
UKK Institute, Tampere
Don Bailey
University of Saskatchewan
John Wark
University of Melbourne
Kim Bennell
University of Melbourne

Human Kinetics

Library of Congress Cataloging-in-Publication Data

Physical activity and bone health / Karim Khan . . . [et al.].
p. ; cm.
Includes bibliographical references and index.
ISBN 0-88011-968-3
1. Bones--Physiology. 2. Bones--Anatomy. 3. Exercise--Physiological aspects. 4. Bones--Diseases--Exercise therapy. I. Khan, Karim.
[DNLM: 1. Bone and Bones--physiology. 2. Bone Development. 3. Exercise. WE 200 P5775 2001]
QP88.2 .P47 2001
612.7'5--dc21 00-054606

ISBN: 0-88011-968-3

Acquisitions Editor: Loarn D. Robertson, PhD; **Developmental Editor:** Renee T. Thomas; **Assistant Editor:** Amanda S. Ewing; **Copyeditor:** Patsy Fortney; **Proofreader:** Jim Burns; **Indexer:** Marie Rizzo; **Permission Managers:** Courtney Astle, Dalene Reeder; **Graphic Designer:** Robert Reuther; **Graphic Artist:** Kathleen Boudreau-Fuoss; **Photo Manager:** Clark Brooks; **Cover Designer:** Jack W. Davis; **Art Manager:** Craig Newsom; **Illustrators:** Argosy, Tom Roberts; **Printer:** Sheridan Books

Printed in the United States of America 10 9 8 7 6 5 4 3 2 1

Human Kinetics
Web site: www.humankinetics.com

United States: Human Kinetics
P.O. Box 5076
Champaign, IL 61825-5076
800-747-4457
e-mail: humank@hkusa.com

Canada: Human Kinetics
475 Devonshire Road Unit 100
Windsor, ON N8Y 2L5
800-465-7301 (in Canada only)
e-mail: orders@hkcanada.com

Europe: Human Kinetics
P.O. Box IW14
Leeds LS16 6TR, United Kingdom
+44 (0) 113 278 1708
e-mail: humank@hkeurope.com

Australia: Human Kinetics
57A Price Avenue
Lower Mitcham, South Australia 5062
08 8277 1555
e-mail: liahka@senet.com.au

New Zealand: Human Kinetics
P.O. Box 105-231
Auckland Central
09-523-3462
e-mail: hkp@ihug.co.nz

Contents

PART III Evidence and Prescription: A Life Span Approach 101

Contributors

Nick Carter
Defence Forces Medical Rehabilitation Centre, Headley Court, Epsom, England

Ari Heinonen
The Bone Research Group, UKK Institute for Health Promotion Research, Tampere, Finland

Mark Forwood
Department of Anatomical Sciences, University of Queensland, Brisbane, Australia

Kerry MacKelvie
Bone Health Research Group, School of Human Kinetics, University of British Columbia, Vancouver, Canada

Moira Petit
M S Hershey Medical Center, Pennsylvania State University, Hershey, Pennsylvania, United States

Preface

Never before has the relationship between physical activity and bone health received so much attention as it does today. This can be gauged by public awareness of conditions such as osteoporosis, by the ever-increasing number of journal articles published on the subject, and by the increasing emphasis on this field in health-related university courses.

As the information available on the relationship between physical activity and bone health burgeons, the need for clear, concise, and correct synthesis becomes increasingly important. No longer are we able to get up to speed by spending a night perusing the journals, or even a weekend on the Web. For these reasons we have written *Physical Activity and Bone Health*—to provide a handy, trusted guidebook for those who choose to explore the frontiers of bone health.

If you saw the title of this book and thought, Now that's what I've been looking for, then this book is for you. You may be an exercise specialist, a health provider, or perhaps a member of the American College of Sports Medicine (ACSM) or the American Society of Bone and Mineral Research (ASBMR). Perhaps you work or study in a human movement science program and have an interest in bone.

This book is also aimed at people working in health-related fields. If you are a personal trainer or fitness instructor, you will have been asked all the questions about bone health that are raised in this book, and we know that clients have been asking you for bone health programs. If you are a clinician wanting to understand more about conditions that affect bone such as osteoporosis and stress fractures, or if you are a health worker and your friends expect you to advise them about optimal exercise to maintain a healthy skeleton, then you will appreciate *Physical Activity and Bone Health*.

Physical Activity and Bone Health has five parts that can be read in any order so that you can happily jump from section to section (after all, we ought to encourage jumping, in a book about bone health!). Part I provides a clear outline of the fundamental aspects of the structure and function of bone. This includes a handy digest of anatomy, physiology, biomechanics, and how bone properties can be measured.

Part II provides essential background information about factors other than physical activity that influence bone: age, genetics, soft tissue factors, the endocrine system, nutrition, and lifestyle. We encourage awareness of these determinants so that the role of physical activity can be seen in perspective.

Part III is a walk, if you will, through the life span. In this section we critically review the literature that reports the effect of targeted bone loading on the healthy skeleton during childhood and adolescence, during adult life, and in the advancing years. This part provides the evidence that your exercise prescription for bone health should be based on. We also provide examples of exercise prescriptions for bone health in an illustrated, practical, easy-to-read manner. The prescriptions are written specifically for the growing child, the adult, and the older person. The focus is on exercises that can be done safely, in the convenience of the home, and without equipment. Since many people also have access to gym equipment, we outline specific activities using such equipment that can be incorporated into existing exercise programs. We include a chapter on exercise prescription for fall prevention, as this is being recognized as a most important factor in reducing fracture risk. The final chapter in this part tackles a tough problem—exercise prescription for people with osteoporosis. It might surprise many people that physical activity plays a crucial role in preventing further complications in this condition that has already reached epidemic proportions.

Although we are strong advocates of physical activity for health, excess activity can harm the skeleton. As the Romans said 'In medio stat virtute'—which can be loosely translated as 'the truth lies between the extremes.' This is merely a more interesting way of saying 'everything in moderation'! Part IV summarizes medical issues such as stress fractures and the effect of menstrual disturbance on bone. The latter is a topic of particular interest at present because of societal pressures for thinness, emphasis on appearance, and because of exciting new research in this area.

Part V reveals how you can be involved on a bone research team. If you are one of the large number of undergraduate and postgraduate students working in the field of physical activity and bone health, this part brings you up to speed without requiring that you plow through reams of literature (don't worry, there will be plenty of that later!). If you are a professor whose training is not in bone or exercise but you have been generous enough to advise a student with that interest, you should find this section particularly valuable.

We hope that you enjoy reading and using *Physical Activity and Bone Health* as much as we enjoy this field ourselves. After 130 years of working in this rapidly evolving, challenging territory (among us, not each), every day is still fun. So, welcome. There is plenty of room for visitors and for those who want to stay a while.

part I

Structure, Function, and Measurement of Bone

To understand the role of physical activity in bone health (and disease), we must first describe the target tissue—bone. Bone is a remarkable structure, stronger than oak, brick, or even concrete [1]. It resists bending as effectively as cast iron, weighs only one-third as much per unit of volume, and is far more flexible [2]. This flexibility helps bone, unlike brick or concrete, absorb sudden impacts without breaking. In part I we outline bone anatomy, physiology, biomechanics, and measurement. It may surprise you to learn that many advances have been made in each of these fields in recent years.

In chapter 1 (Anatomy) we describe bone as an organ and as a tissue. Our understanding of bone at the cellular level is much deeper now than it was just a few years ago. Understanding anatomy permits understanding of the site-specific effects of exercise that are discussed in later chapters. Studying anatomy also sheds light on why various parts of the skeleton respond differently to physical activities.

Chapter 2 (Physiology) reviews calcium homeostasis and the concept of mechanotransduction and bone modeling and remodeling. We summarize calcium homeostasis and explain why bone does not respond to increases in dietary calcium with directly proportional increases in bone strength. This section provides important background physiology for chapter 8, which discusses the role of nutrition and calcium in bone.

Mechanotransduction is the intriguing physiological mechanism whereby bone recognizes that loading is present and then responds to it. This feedback loop and communication system has only recently evolved from a theoretical concept to a well-described entity.

Chapter 3 (Biomechanics) moves us from the theoretical concept of the mechanostat to the real-life measurement of loading—biomechanics. In this chapter we simplify terminology (one of the major hurdles to understanding this subject), describe how material properties and structural arrangement contribute to bone strength, summarize the data regarding the ideal type of loading to stimulate bone in animal studies, and then consider how physical activity in humans can influence bone.

Chapter 4 (Measuring the Properties of Bone) describes the many advances in technology that allow us to better measure various properties of bone. DXA (dual energy X-ray absorptiometry), a state-of-the-art research tool only a decade ago, now seems to be approaching its middle years judging by the existence of numerous offspring such as QCT (quantitative computed tomography), pQCT (peripheral quantitative computed tomography), QUS (quantitative ultrasound), and QMR (quantitative magnetic resonance). Furthermore, we are no longer limited to measuring structure; bone biomarkers permit the measurement of bone metabolism, and mechanobiology shows us the effect of forces on bone.

We believe that time devoted to reviewing these fundamentals of bone biology is extremely worthwhile. In addition to helping you interpret the current literature more fully, it provides a platform for keeping abreast of new developments in bone science. So let us begin at the very beginning, by describing the anatomy of the fascinating tissue that is bone.

References

1. Koch JC. The laws of bone architecture. *Am J Anat* 1917;21:211-298.
2. Ascenzi A, Bell GH. *Bone as a mechanical engineering problem.* In: Bourne GH, ed. The biochemistry and physiology of bone. 2nd ed. New York: Academic Press, 1971: 311-346.

Chapter 1

Anatomy

Bone is a unique tissue with the principal responsibility of supporting loads that are imposed on it. This apparently simple task demands that bone have enormous strength and resilience while at the same time being lightweight and adaptable so that transportation is not a metabolic burden. This chapter briefly outlines the basic anatomy of bone as an organ (macroscopic anatomy) and as a tissue (microscopic anatomy). We aim to provide clear definitions of commonly used terms. If you are already familiar with these aspects of bone, you may want to move on to other chapters.

Bone's Organic Makeup

Bone consists of an organic component (20-25% by weight), an inorganic component (70% by weight), and a water component (5% by weight). The organic component is largely type I collagen but also includes bone cells and a small amount of noncollagenous protein, which is very important biologically. The inorganic component is mineral—almost all crystalline calcium hydroxyapatite [1].

Organic Matrix

The organic matrix of bone determines the structure and the mechanical and biochemical properties of bone. Ninety-eight percent of the organic matrix of bone is made up of type I collagen and noncollagenous proteins. The remaining 2% of the organic matrix of bone consists of cells—osteoblasts, osteocytes, and osteoclasts [1]. Cells and their role in bone homeostasis are discussed later in this chapter.

Collagen. Collagen, the major structural component of bone matrix, is a protein that consists of three polypeptide chains of about 1000 amino acids each. Its triple helix is formed from two identical alpha1(I) chains and a single alpha2(I) chain held together by cross-links of hydrogen bonding. Each molecule is aligned with the next in parallel order to form a collagen fibril. These fibrils are then grouped to form the collagen fiber. Small gaps within and between

TABLE 1.1

Noncollagenous proteins in bone and their roles

Noncollagenous protein	Notable aspects
Osteocalcin (the protein formerly known as Bone Gla Protein[1])	Makes up 10-20% of the noncollagenous protein in bone and is closely associated with the mineral phase. Thought to attract osteoclasts to sites of bone resorption. Serum osteocalcin serves as a biochemical marker of bone formation.
Osteonectin	Secreted by osteoblasts and platelets. Involved in the regulation of calcium concentration and enhances the attachment of bone resorbing cells to bone.
Integrins	A family of bone matrix proteins that includes osteopontin, bone sialoprotein, thrombospondin, and fibronectin. Integrins span the cellular membranes and provide a link between the extracellular matrix and the cytoskeleton of the cell. Integrins on osteoblasts, osteoclasts, and fibroblasts provide an anchor for these cells in the extracellular matrix.
Growth factors and cytokines	These include transforming growth factor-beta (TGF-beta), insulin-like growth factors (IGFs), and bone morphogenic proteins (BMP). Bind to bone mineral and matrix; released during osteoclastic bone resorption. Regulate bone cell differentiation, activation, growth, and turnover. Serve as coupling factors that link bone formation and bone remodeling. May play a role in control of matrix mineralization.
Others	These include phosphorylated sialoproteins, small proteoglycans, and glycoproteins.

[1]Bone γ-carboxyglutamic acid containing protein.

collagen molecules permit biological activity to take place (e.g., mineralization).

Noncollagenous Proteins. Noncollagenous proteins make up only a small proportion of bone by weight but have great biological significance. A detailed discussion of noncollagenous proteins is beyond the scope of this book, but the interested reader is referred elsewhere [1]. See table 1.1 for a list of the key noncollagenous proteins together with their proposed function so that you will recognize the place of these compounds in your wider reading.

Inorganic Phase: Hydroxyapatite

The inorganic component of bone consists mainly of platelike crystals of hydroxyapatite (20-80 nm long and 2-5 nm thick), which itself is composed of calcium and phosphate [1]. These crystals are found in and around collagen fibers. Small amounts of other substances such as carbonate, chloride, or fluoride may replace hydroxyapatite components and influence bone mechanical properties. For example, when there is excess fluoride, bone crystals increase in size and become more fragile. This is a reminder that there is an optimal crystal size distribution, as well as an optimal amount of mineral.

Macroscopic and Microscopic Appearance

On the macroscopic level, it is useful to think of the skeleton as consisting of two parts, the axial and the appendicular skeletons. The axial skeleton includes vertebrae, the pelvis, and other flat bones such as the skull and scapula. The appendicular skeleton includes all the long bones. Each long bone consists of two wider extremities (epiphyses), an essentially cylindrical shaft in the middle (diaphysis), and a zone between them (metaphysis) where remodeling of bone takes place during growth and development (figure 1.1). In this way bone can be described as an organ.

Bone can also be described as tissue at the microscopic level. Understanding the microscopic appearance of bone may require a great deal of effort, but the researcher who is aware of regional differences in type of bone (heterogeneity) is well placed to understand the varying effects of physical activity on bone.

There are two types of bone tissue, woven bone and lamellar bone. Woven bone is considered immature bone with collagen arranged randomly rather than in the uniform structure of lamellar bone. At birth, it makes up all the bone in the body; in later years it is found at sites of

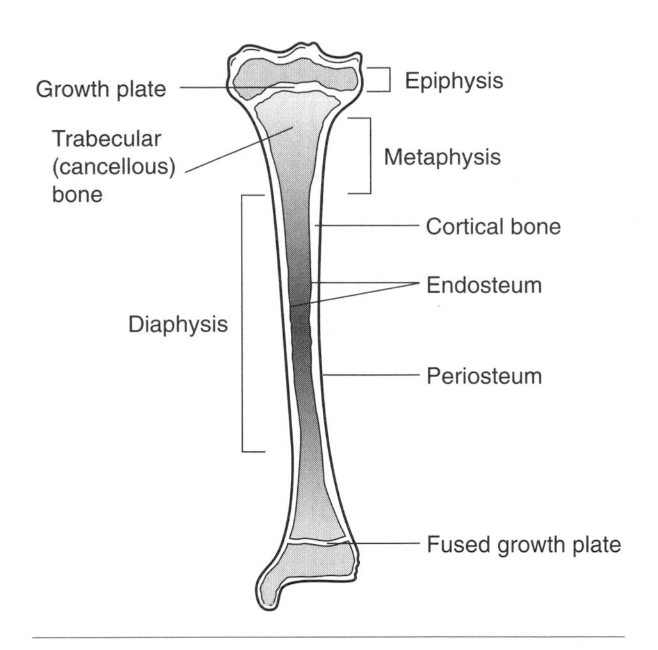

FIGURE 1.1 Schematic view of a growing long bone.

fracture healing or in response to extreme mechanical loads [2].

Lamellar bone is the name given to bone that eventually replaces woven bone. By the age of four years, most of the skeleton is lamellar bone. The collagen fibers arrange themselves along lines of force, which results in cortical bone having anisotropic properties, that is, it is able to withstand forces better in one orientation than in another.

Anatomically, both woven and lamellar bone can be organized into compartments as either cortical or trabecular bone. Cortical bone forms the external part of long bones and is made up of dense, calcified tissue (figure 1.2). The diaphysis of a long bone encloses the medullary cavity. Toward the metaphysis or epiphysis, cortical bone is thinner and the medullary cavity is replaced by cancellous or trabecular bone, which is characterized by an inner network of thin calcified trabeculae (figure 1.3).

An important difference between cortical and trabecular bone is in the way the bone matrix and cellular elements are arranged. These differences permit the two types of bone to function differently. Calcium takes up 80-90% of cortical bone volume but only 15-25% of trabecular bone volume. The trabecular arrangement permits bone marrow, blood vessels, and connective tissues to be in contact with bone—essentially the endosteum, which has an active metabolic role. The main function of cortical bone is for structure and protection [3]. Examples of the proportion of trabecular bone content at various skeletal sites are shown in table 1.2.

Cortical bone has two types of surfaces (figure 1.4). The inner surface that faces the bone marrow is the endosteum. The other, on the outer side, facing the soft tissue, is the periosteum. Cells lining the endosteum are metabolically active and very involved with bone formation and resorption. The periosteum has two layers; the inner, or cambium, layer contributes to appositional bone growth during bone development and is responsible for increasing the diameters of the long bones with aging.

Haversian bone is the most complex type of cortical bone. It consists of blood vessels that are

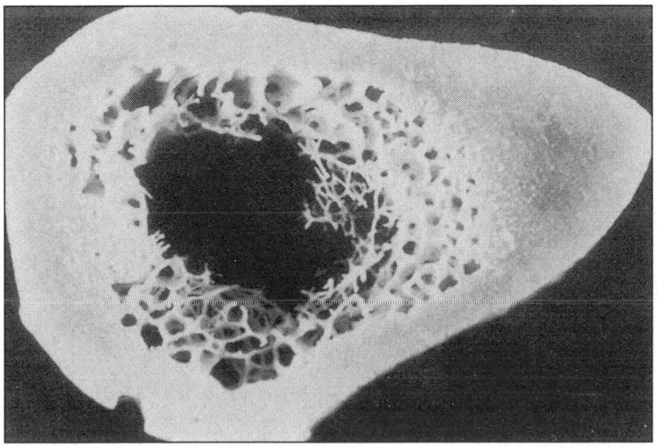

FIGURE 1.2 Cross-section of cortical bone—a solid structure arranged as ellipsoid cylinders.

Figure from "The Bone Organ System: Form and Function" by Thomas A. Einhorn in OSTEOPOROSIS, edited by Robert Marcus et al., copyright © 1996 by Academic Press, reproduced by permission of the publisher.

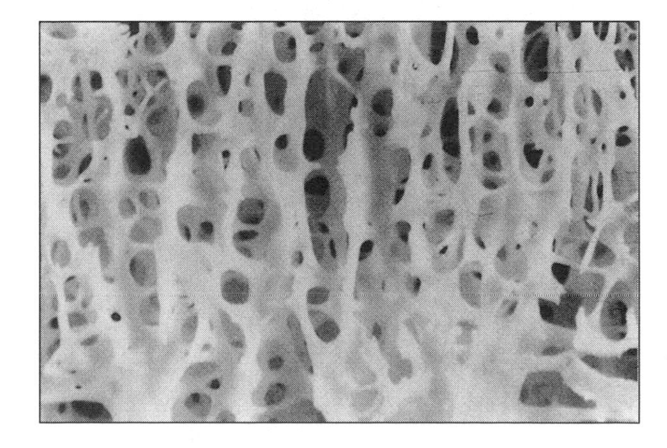

FIGURE 1.3 Trabecular bone consists of horizontal and vertical interconnecting plates called trabeculae.

Figure from "The Bone Organ System: Form and Function" by Thomas A. Einhorn in OSTEOPOROSIS, edited by Robert Marcus et al., copyright © 1996 by Academic Press, reproduced by permission of the publisher.

TABLE 1.2

Proportion of trabecular bone at various skeletal sites

Site	Trabecular bone (%)
Vertebrae	40
Intertrochanteric region of proximal femur	50
Femoral shaft	25
Distal radius	25
Midradius	1

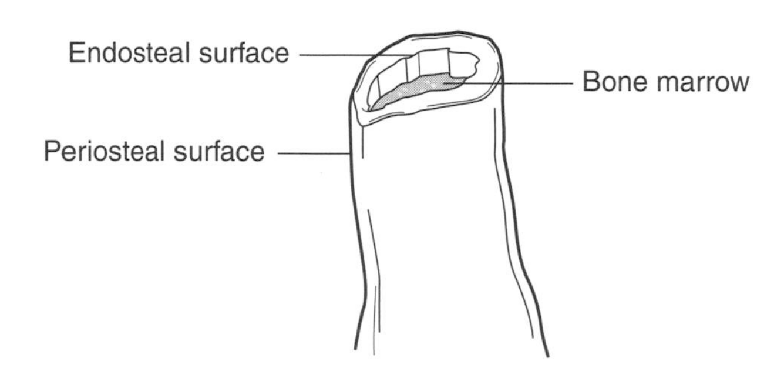

FIGURE 1.4 Schematic illustration of the surfaces of cortical bone.

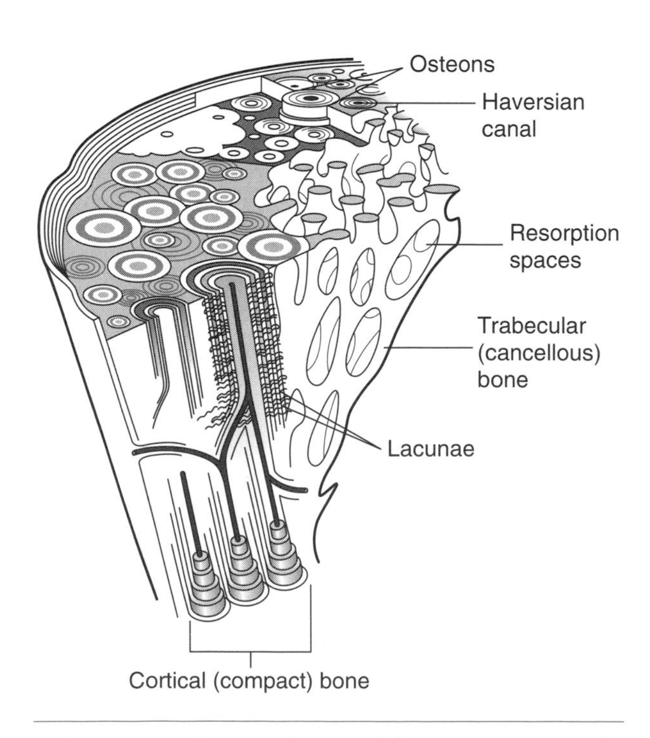

FIGURE 1.5 Diagram of some of the microstructure of cortical bone.

circumferentially surrounded by lamellae of bone (figure 1.5). This arrangement of cortical bone around a vessel is called an osteon. An osteon has also been defined as "an irregular, branching, and anastomosing cylinder composed of a neurovascular canal surrounded by cell-permeated layers of bone matrix" [1]. As illustrated in figure 1.5, osteons are usually aligned with the long axis of bone. They are the major structural units of cortical bone and are connected to one another by Volkmann's canals that run at right angles to the osteon.

Bone Cells: Osteoblasts, Osteocytes, Osteoclasts

Specialized bone cells regulate bone metabolism by responding to various environmental signals including chemical, mechanical, electrical, and magnetic stimuli [1]. The three cell types in bone are osteoblasts, osteocytes, and osteoclasts.

Osteoblasts and Bone Formation

The osteoblast is the bone cell that produces bone matrix—both collagen and ground substance. Osteoblasts are always found in clusters of about 100-400 at bone-forming sites (figure 1.6). They have a characteristic light microscope appearance with a round distinguishing nucleus at the base of the cell (opposite to the bone surface) and a prominent Golgi complex. This structure permits the osteoblast to synthesize various proteins.

Osteoblasts are found on the layer of bone matrix that they are themselves producing (called osteoid at that stage, before calcification). After about 10 days, osteoid becomes calcified. Alkaline phosphatase, a serum index of bone formation, is

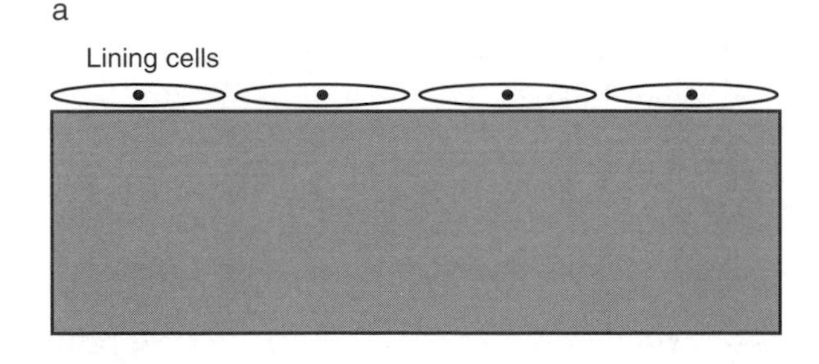

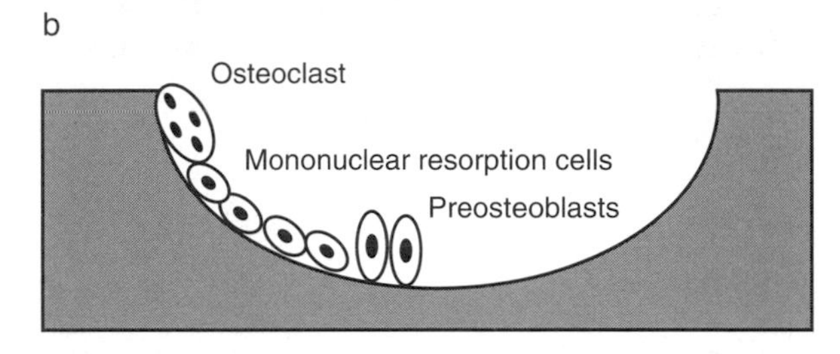

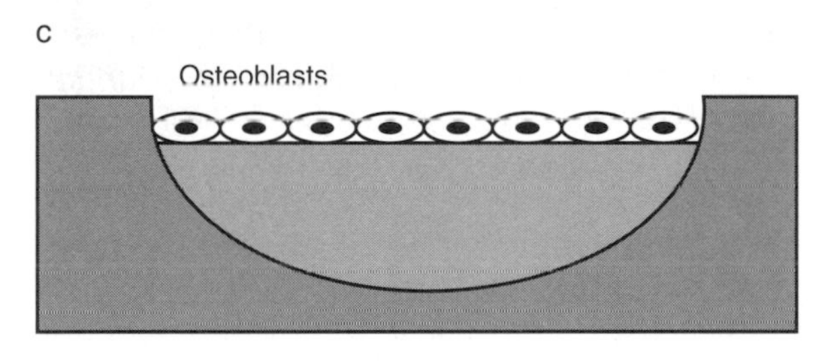

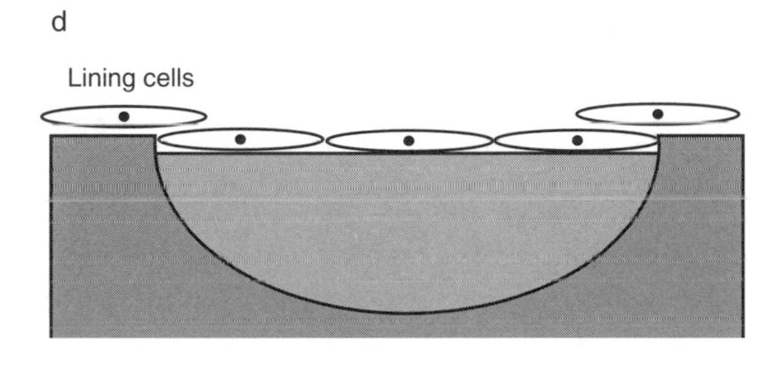

FIGURE 1.6 Remodeling cycle in trabecular bone: *(a)* inactive surface, *(b)* resorption by osteoclasts, *(c)* final resorption cavity, and *(d)* formation.

abundant on the plasma membrane of osteoblasts. This membrane also has receptors for parathyroid hormone. Osteoblasts express receptors for estrogen and 1,25-dihydroxy-vitamin D in their nuclei. With time, an osteoblast becomes either a flat lining cell or an osteocyte [3, 4].

Osteoblast function is controlled by endocrine, paracrine, and autocrine factors. Hormones such as parathyroid hormone (PTH), vitamin D_3, glucocorticoid hormones, growth hormone (GH), and gonadal steroids all act on the osteoblast [5]. PTH has been shown to mediate ion and amino acid transport, stimulate cyclic adenosine monophosphate (cAMP), regulate collagen synthesis, and bind a specific receptor [5]. Paracrine control of osteoblast activity occurs when nearby cells release locally acting factors such as PTH-related protein, members of the transforming growth factor beta (TGF-beta) family, fibroblast growth factors (FGFs), and insulin-like growth factors (IGFs). Autocrine control refers to factors that are produced by the osteoblast and eventually regulate their own activity. The key examples are IGF-I and IGF-II, which have proliferation- and differentiation-stimulating activity for osteoblasts [5]. Certain factors act both as paracrine and autocrine controls.

Osteocytes

Osteocytes are mature bone cells embedded deep within small bone cavities called osteocytic lacunae. These cells were originally osteoblasts, but they were buried in the forward march of their own bone matrix. The appearance of osteocytes is characterized by numerous long cell processes that are in contact with cell processes from other osteocytes at gap junctions, or with processes from bone lining cells or osteoblasts at the bone surface. These interconnecting processes permit osteocytes to form a complex network throughout the bone matrix, which makes them ideal candidates to provide the means of communication required in mechanotransduction (described in chapter 2 on page 13). This role has only recently been recognized; historically, the main role of osteocytes was thought to be the activation of bone turnover [3] and regulation of extracellular calcium.

Osteoclasts and Bone Resorption

The osteoclast is the bone cell responsible for bone resorption, which may be thought of as removal of old bone (see figure 1.6). This giant, multinucleated cell containing 4 to 20 nuclei looks very different from the osteoblast. It is usually found in contact with a calcified bone surface and within the resorption cavity it created, which is known as a Howship's lacuna [3]. Osteoclasts tend to operate singly or in pairs, but occasionally twice this number can be seen in the same lacuna.

The characteristic feature of the osteoclast on microscopy is the ruffled border surrounded by a ring of contractile protein. This border serves to attach the osteoclast to the bone surface

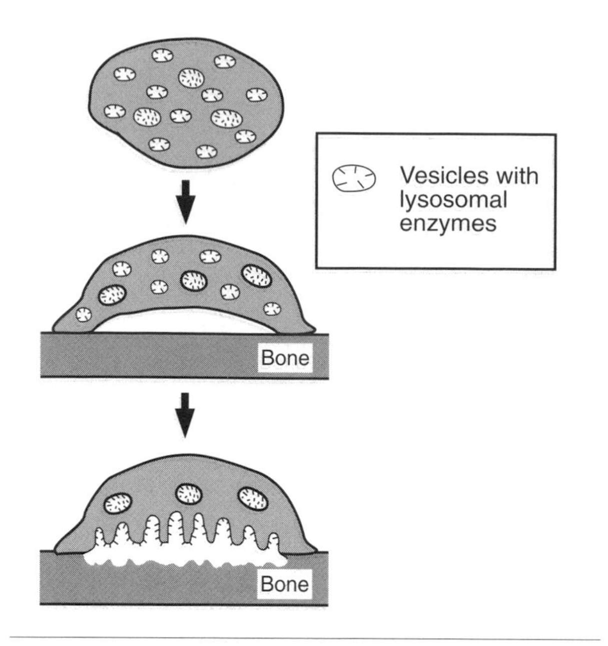

FIGURE 1.7 Osteoclast function. Once the osteoclast is attached to bone, lysosomal enzymes are secreted via the ruffled border into the extracellular bone-resorbing compartment.

and create what is known as the extracellular bone resorbing compartment or subosteoclastic bone resorbing compartment (figure 1.7). Lysosomal enzymes are actively synthesized in the osteoclast and then secreted, via the ruffled border, into the extracellular bone-resorbing compartment where a high concentration of enzymes develops to resorb bone [3]. The acid environment digests the noncollagenous link between hydroxyapatite crystals and collagen. The remaining collagen fibers are removed by collagenase or a factor called cathepsin.

As with the osteoblast, many hormones and factors influence the osteoclast. For example, osteoclast activity is stimulated by the proliferation of osteoclast progenitors (increasing the number of mature cells) and by the activation of mature cells to resorb bone. Similarly, osteoclast inhibition occurs by the converse of these simplified pathways. The systemic hormones (parathyroid hormone, 1,25-dihydroxyvitamin D_3, calcitonin) as well as the local factors (such as the interleukins, lymphotoxin, and tumor necrosis factor) appear to act at several of these steps [6].

Coupling of Osteoblast and Osteoclast Function

Histological studies have demonstrated [7] that osteoclastic bone resorption and osteoblastic bone formation are closely linked or "coupled" during bone remodeling (see figure 1.6). Studies using radio-labeled calcium or strontium to estimate the rates of bone formation and resorption have shown that when bone resorption increases, bone formation increases as well [8, 9]. The fact that osteoclastic bone resorption and osteoblastic bone formation follow each other is fundamental to the concept of the "basic bone multicellular unit" (BMU), which describes a packet of bone being resorbed or rebuilt [7]. As a result of coupling of osteoclast and osteoblast function, bone resorption initiates bone formation, which, under balanced conditions, restores lost bone (see figure 1.6). This underpins the physiological process of remodeling, which is discussed in chapter 2.

SUMMARY

- Bone consists largely of an organic component (20-25% by weight) and an inorganic component (70% by weight). The former consists mainly of collagen together with bone cells and a small amount of biologically important noncollagenous proteins. The latter is crystalline hydroxyapatite, which is formed by calcium and phosphate.
- Bone is arranged into an axial and an appendicular skeleton. The axial skeleton refers to the vertebrae, pelvis, and other flat bones. The appendicular skeleton includes all the long bones.
- At birth, bone is immature and woven, but by the age of four years it has developed into organized, lamellar bone.
- Bone is described as either cortical or trabecular. Cortical bone forms the external part of long bones and is made up of dense, calcified tissue. Trabecular

bone features an inner network of thin, calcified trabeculae. The two types of bone structure are designed to perform different functions. Strong cortical bone provides structure and protection. The more open weave of trabecular bone allows it to play a more active metabolic role.

- The three types of bone cells—osteoblasts, osteocytes, and osteoclasts—play different roles in the maintenance of bone homeostasis. Osteoblasts produce matrix, both collagen and ground substance. Osteocytes are mature bone cells that facilitate cellular communication and appear to transmit messages about bone loading. Osteoclasts are responsible for bone resorption. Bone resorption and formation are closely linked, or coupled, during bone remodeling.

References

1. Einhorn TA. *The bone organ system: Form and function*. In: Marcus R, Feldman D, Kelsey J, ed. Osteoporosis. San Diego, CA, USA: Academic Press, 1996: 3-22.
2. Forwood MR, Burr DB. Physical activity and bone mass: Exercises in futility? *Bone Miner* 1993;21:89-112.
3. Baron RE. *Anatomy and ultrastructure of bone*. In: Favus MJ, ed. Primer on the metabolic bone diseases and disorders of mineral metabolism. 3rd ed. Philadelphia: Lippincott-Raven, 1996: 3-10.
4. Lian JB, Stein GS. *Osteoblast biology*. In: Marcus R, Feldman D, Kelsey J, ed. Osteoporosis. San Diego, CA, USA: Academic Press, 1996: 23-60.
5. Puzas JE. *Osteoblast cell biology - lineage and functions*. In: Favus MJ, ed. Primer on the metabolic bone diseases and disorders of mineral metabolism. 3rd ed. Philadelphia: Lippincott-Raven, 1996: 11-16.
6. Mundy GR. *Bone-resorbing cells*. In: Favus MJ, ed. Primer on the metabolic bone diseases and disorders of mineral metabolism. 3rd ed. Philadelphia: Lippincott-Raven, 1996: 16-24.
7. Frost HM. *Dynamics of bone remodeling*. In: Bone biodynamics. Boston: Little, Brown, 1964: 286-292.
8. Harris WH, Heaney RP. Skeletal renewal and metabolic bone disease. *N Engl J Med* 1969;280:193-202.
9. Rodan GA. *Coupling of bone resorption and formation during bone remodeling*. In: Marcus R, Feldman D, Kelsey J, ed. Osteoporosis. San Diego, CA, USA: Academic Press, 1996: 289-299.

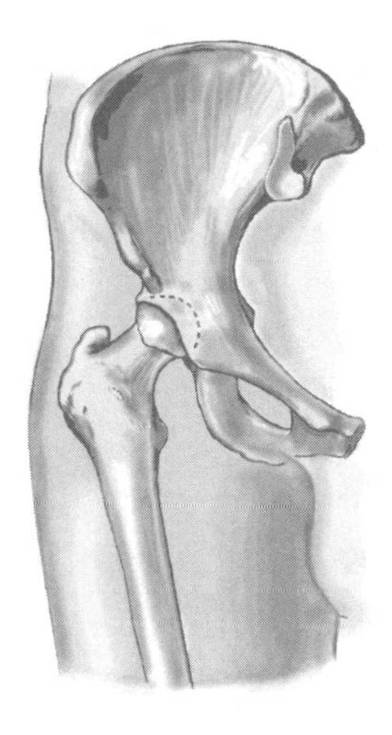

Chapter 2

Physiology

with Mark Forwood

Although Galileo observed bone's response to physical activity in the 17th century, only now are researchers grasping *how* bone responds to forces such as physical activity. Recent technological advances have permitted much closer investigations of bone tissue, which have triggered an explosion of knowledge in this relatively new, exciting area of exercise physiology. This chapter outlines three distinct areas of bone physiology: calcium homeostasis, mechanotransduction (the mechanism whereby physical activity influences bone cell function), and the process of modeling and remodeling.

Calcium homeostasis, although complex, cannot be ignored, as 98% of total body calcium exists in bone. We focus on the roles of hormones, particularly parathyroid hormone (PTH) and 1,25-dihydroxyvitamin D_3 (vitamin D). This chapter outlines how these key hormones maintain *serum* calcium levels. In chapter 7 we discuss in detail the role of a larger number of hormones that influence the level of bone mineralization.

Mechanotransduction is not yet a word on everyone's lips, but the topic of bone health was not a subject of broad discussion 10 years ago and yet it most definitely is today. Mechanotransduction is as fundamental to maintaining bone health as cardiac muscle cell contraction is to maintaining blood pressure. Let us explain why. Every physical activity, be it walking, running, or jumping, generates a force on the skeleton, on individual bones that make up the skeleton, and on the cells that constitute bone (see chapter 1). How these *bone cells* respond to applied forces (e.g., physical activity) is the process of mechanotransduction. How *whole bone* responds to applied forces (e.g., physical activity) is studied in the field of bone biomechanics, the subject of chapter 3.

Mechanotransduction influences groups of cells to act in concert, and the physiology of this coordinated bone cell action is labeled modeling and remodeling. This fundamental process and its implications for physical activity research is the third major topic in this chapter. Although bone physiology is still not completely understood, there is excitement in seeing current bone researchers unravel the mechanisms behind observations that date back almost 400 years!

Calcium Homeostasis

Bone consists largely of calcium, and 98% of total body calcium exists in bone. Here we introduce the key calcium metabolic hormones (parathyroid hormone and vitamin D) and explain their role in calcium homeostasis and bone mineral metabolism. This section provides some background for chapter 8, which discusses the relationship between dietary factors and bone. A comprehensive discussion of the endocrine control of calcium, and of other minerals such as phosphate and magnesium, is beyond the scope of this book.

Plasma calcium represents about 2% of total body calcium and exists in two forms—ionized and unionized. Ionized calcium makes up less than half of this and is regulated to between 1.8 and 3.0 mg/dl (1.1 to 1.3 mmol/L). Unionized calcium is bound to protein, mainly albumin and also complexed with anions. The remaining 98% of total body calcium is in the skeleton as crystalline molecules such as hydroxyapatite—$Ca_{10}(PO_4)_6(OH)_2$. Bone, therefore, while providing mechanical support, protection of internal organs, and aid of movement, also serves as a large reservoir for the central pool of calcium.

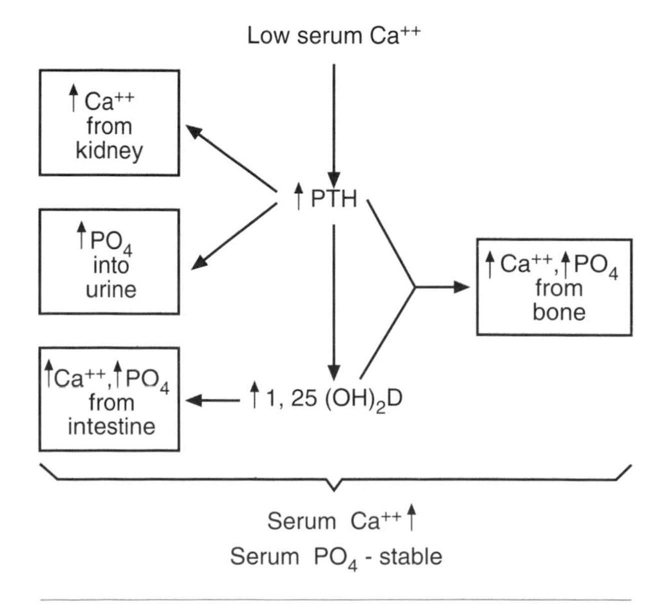

FIGURE 2.1 Diagram of the PTH and vitamin D response to lowered calcium.

The central pool of calcium is adjusted via a negative feedback mechanism (figure 2.1) that involves the alimentary tract, the kidneys, and bone. The so-called "calciotropic" hormones—parathyroid hormone, vitamin D, and to a lesser extent, calcitonin—maintain the equilibrium of the central pool. These are also known as the controlling hormones, as their secretion changes in response to plasma levels of ionized calcium concentrations. Endocrine disturbance found in association with physical activity is discussed in chapter 16. Specific endocrine disorders of bone that are generally not related to physical activity (e.g., primary hyperparathyroidism) are outside the scope of this book.

Parathyroid Hormone (PTH)

PTH, synthesized in the parathyroid gland as an 84-amino-acid protein, is stored in secretory granules in the chief cells of the parathyroid gland. One of the two major hormones of calcium homeostasis, it is secreted in response to falling serum calcium. It acts to increase ionized plasma calcium concentration in three ways [1, 2]:

- In the presence of sufficient 1,25-dihydroxyvitamin D, it stimulates bone resorption.
- It enhances intestinal calcium absorption by promoting renal biosynthesis of 1,25-dihydroxyvitamin D.
- It has a direct effect on the kidney to enhance fractional reabsorption of calcium in the distal tubule [3].

PTH also plays a role in a secondary, less precise method of controlling calcium. This process is called intestinal adaptation, and it refers to alimentary adaptation in response to the concentration of mineral passing through it. When calcium intake is low, PTH secretion increases and stimulates 1,25-dihydroxyvitamin D synthesis, which in turn enhances calcium absorption. This process is reversed during high intakes of calcium. This explains why the ingestion of high doses of supplemental calcium does not produce a large increase in the risk of kidney stones. The loss of intestinal adaptation may contribute to osteoporosis.

Vitamin D

Although vitamin D attempts to pass itself off as a vitamin, it is a hormone. Cholecalciferol, or

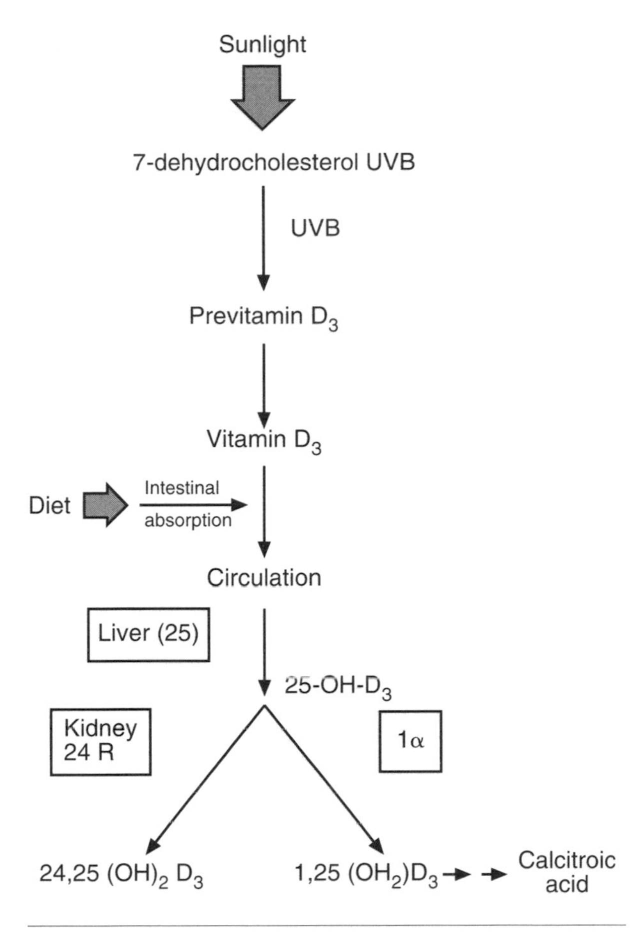

FIGURE 2.2 Metabolic pathway of vitamin D_3. UVB refers to the sun's ultraviolet radiation. Liver (25) refers to hepatic 25-hydroxylase enzyme. Kidney 24R and 1α refer to the respective hydroxylase enzymes in the kidney.

vitamin D_3, results from the conversion of 7-dehydrocholesterol in the skin during ultraviolet sun exposure. Vitamin D is transported in plasma bound to a vitamin D binding globulin. Vitamin D undergoes two key metabolic conversions, first in the liver, then in the kidney, to become the most biologically active metabolite, 1,25-dihydroxyvitamin D_3 (figure 2.2). The most important vitamin compounds are circulating 25(OH)D and 1,25(OH_2)D. A discussion of other vitamin D compounds can be found elsewhere [4].

1,25(OH_2)D circulates at concentrations 1000 times less than its precursor, 25(OH)D, and is relatively unstable. It is important to remember that 25(OH)D levels are considered the best clinical indicators of vitamin D nutritional status [5].

The two primary sources of vitamin D are endogenous production based on exposure to UV light and diet (see figure 2.2). Vitamin D_3 is produced in the skin and is available from animal sources, whereas vitamin D_2 is available only from dietary plant sources.

Levels of serum 25(OH)D, and subsequently calcium, decrease with age. This explains the need for vitamin D supplemented diets for the elderly, especially in countries where dermal production of vitamin D from sunlight exposure is limited. In North America, recommended oral intakes of vitamin D are approximately 200 IU up to age 50, 400 IU to age 70, and 700 IU over age 70 [5].

Production of 1,25-dihydroxyvitamin D_3 is regulated by serum calcium, phosphate, PTH, and the serum concentration of the hormone itself. PTH acts by inducing the renal conversion of the 25-hydroxylated vitamin D_3 derivative to the 1,25-dihydroxylated form.

The primary actions of 1,25$(OH)_2$D are on intestine, bones, and kidneys. Vitamin D influences calcium metabolism at a number of levels including calcium transport, renal calcium reabsorption, intestinal calcium absorption, and mobilization of calcium from bone. 1,25-dihydroxyvitamin D_3 plays a major role in calcium transport via a saturable transcellular absorption mechanism. In the intestine, vitamin D acts via the intracellular vitamin D receptors (VDRs), which exist primarily in the nucleus and bind Vitamin D. In chapter 8 we discuss the role of vitamin D as a nutrient and its role in bone mineralization and fracture risk reduction.

Calcitonin

Calcitonin is a small polypeptide secreted by the parafollicular cells of the thyroid gland. Its main biologic effect is to inhibit osteoclastic bone resorption, a property that has led to its use in the treatment of disorders characterized by increased resorption [6] including osteoporosis. At physiological levels of bone resorption, however, the effects of this hormone on bone appear to be minor. The secretion of calcitonin is regulated by blood calcium.

Mechanotransduction

Observations of bone's response to physical activity have a long history, dating back to Galileo in 1638 A.D. Since the 19th century these observations have been seen in the light of Wolff's Law [7], which described a relationship between the form of bone and its function. In recent times Frost, Lanyon, and others [8-10] have suggested that this response is controlled by a "mechanostat" that endeavors to keep bone strain at an optimal level by adjusting bone structure. More recently, Turner [11] proposed that bone cells react strongly to changes in their environment but eventually accommodate to steady-state signals. We use the word *mechanostat* to refer to the concept of bone regulation according to certain thresholds, and the word *mechanotransduction* to describe the physiological process that permits this to occur. To use an example from plant biology, growth is a concept as well as an action; photosynthesis is a process that allows growth to occur.

This section explains how cells within a microscopic piece of bone perceive mechanical loading and how they respond to loading. In chapter 3 (Biomechanics) we discuss how these signals are interpreted at the level of whole bone (e.g., how a femur responds to a jumping intervention).

Osteocytes and bone-lining cells form a vast, three-dimensional communicating network throughout bone that is vital for mechanotransduction (figure 2.3). These structures are often referred to as the osteocyte–bone-lining cell complex. This osteocyte–bone-lining cell network detects that bone is being loaded (i.e., bent or deformed in some way) and signals cells that will remove or add bone at appropriate sites [12-14]. This process by which the skeleton adjusts its structure to imposed demands is mechanotransduction. Mechanotransduction does not appear to involve any neural pathways.

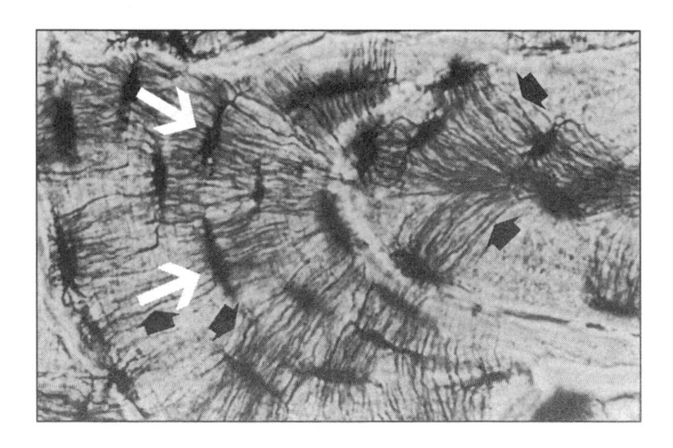

FIGURE 2.3 Photomicrograph of mature bone. Osteocytes (white arrows) form concentric layers around a central nutrient vessel. The canaliculi of the osteocytes (dark arrows) form a communication network among the cells.

Reprinted, by permission, from MJ Favus ed., 1999, *Primer on the metabolic bone diseases and disorders of mineral metabolism*, 4th ed. (Philadelphia: Lippincott Williams & Wilkins), 13.

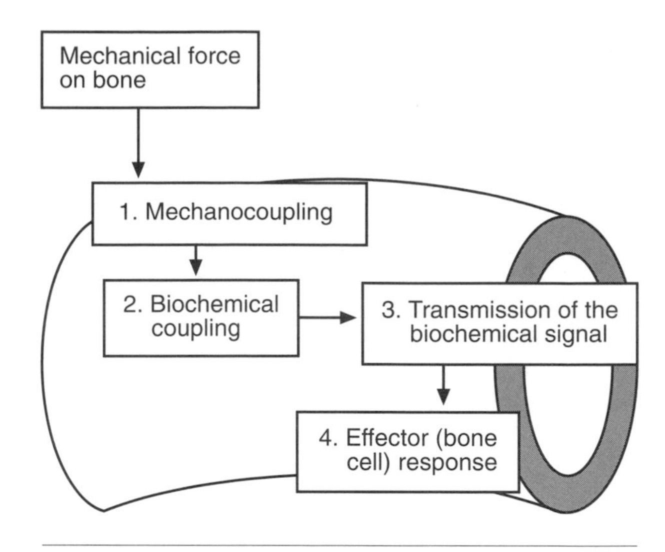

FIGURE 2.4 Schematic illustration of the four steps in mechanotransduction—the process whereby the skeleton responds to imposed demands.

Successful mechanotransduction requires four steps: (1) mechanocoupling, (2) biochemical coupling, (3) transmission of the biochemical signal, and (4) the effector response (figure 2.4).

Mechanocoupling

Mechanocoupling is the technical term for the process whereby load is signaled to cells. In bone, it refers to physical transduction of a local mechanical force (e.g., a bending force on a long bone) to a form that can be detected by cells [15]. Like fluid in a sponge, deformation of bone creates pressure gradients within bone canaliculae and interstitial spaces that cause tissue fluid to move. This fluid may flow past cell membranes of osteocytes and create fluid shear stress within the membrane [12, 13].

Fluid shear stresses generated by excessive deformation [16, 17-19] will activate a cellular response (described later). Because these tissues are viscoelastic, mechanically induced bone formation occurs once strain magnitude and strain rate reach a certain threshold [19, 20]. Osteocytes also derive their nutrients via displacement of fluid through the canaliculae. Decreased mechanical loading, or disuse, reduces the deformation of bone and therefore reduces the displacement of fluid through bone tissue. This may also activate a cellular response, but via different mechanisms than those activated by increased loading [21]. The concept of strain as it relates to the magnitude of the mechanical load, the number of loading cycles, and the strain rate is discussed in chapter 3.

Just as you may repeatedly bend a recalcitrant nail to break it, excessive deformation of bone can also cause fatigue damage [22-24] to cell processes and membranes. The signals generated by fatigue microcracks activate their own repair by remodeling (described later) [25, 26] through a mechanism involving programed osteocyte death (apoptosis) [26].

The two structures that detect that bone is being loaded (i.e., bent or deformed in some way) are the osteocyte and the bone-lining cell (figure 2.5, a and b) [12, 15, 27-29]. Precisely what constitutes the signal remains unknown, but it is most likely in the growth factor or cytokine family [30, 31].

Biochemical Coupling

Bone cells attach to the collagen matrix by binding to integrins, a form of glycoprotein, which span the membrane of cells. These integrins are also attached to the internal cytoskeleton that connects the extracellular matrix to the cytoplasm and the nucleus [32, 33]. This cytoskeleton maintains tension on the extracellular matrix. Thus, when fluid flows past a cell, it applies force via the integrins to the actin cytoskeleton within the cell and to the nucleus, which then alters gene expression in response [15, 34].

When osteoblastic cells are subjected to fluid shear stress, intracellular calcium (Ca2+) is released and this reorganizes the actin cytoskeleton [35]. The mechanism that triggers calcium release from the endoplasmic reticulum involves the protein phospholipase C (PLC). This was discovered when the inhibition of PLC using antibiotics completely suppressed the cytoskeletal reorganization induced by fluid flow in osteoblastic cells [35].

Transmission of the Biochemical Signal

When bone undergoes deformation because of physical activity, osteoblasts are stimulated in two ways. First, active osteoblasts and bone-lining cells can respond directly to mechanical strain [36-38] by changes in osteoblast gene expression and the reversion of bone-lining cells to osteoblasts [38].

If this were the only method of stimulating osteoblasts, bone would be very slow to respond to loading; active osteoblasts make up only 5% of the bone surface in adult humans. A second pathway whereby osteoblasts respond to loading involves intermediary biochemical compounds called second messengers [35, 39-41]. These substances are produced by activity-related bone deformation and stimulate osteoblast activity (see figure 2.5b).

While osteoblasts can communicate the message of loading via biochemical intermediaries,

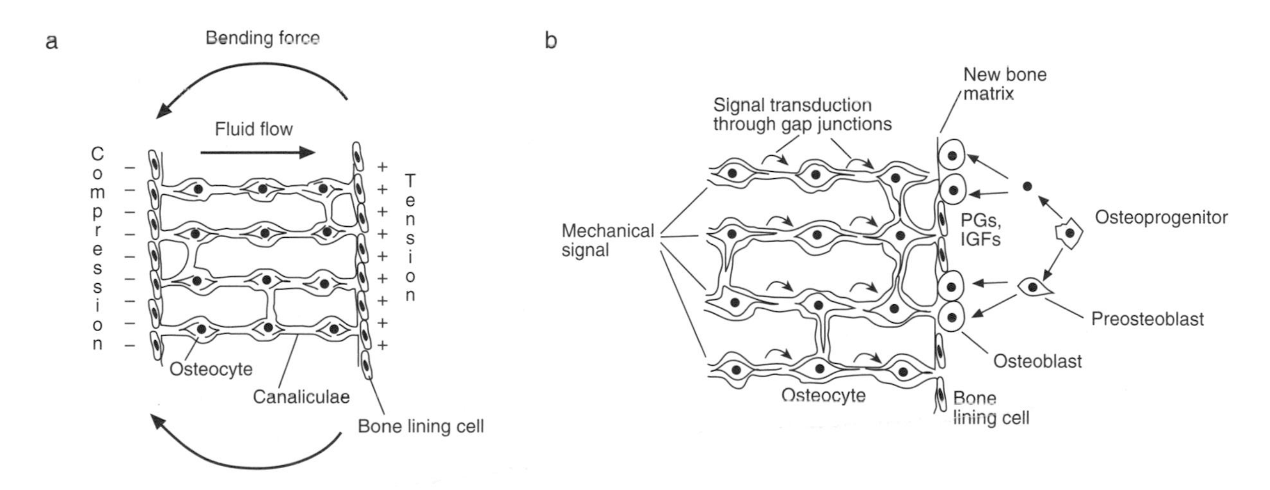

FIGURE 2.5 *(a)* Schematic representation of bone under bending loads. Bending causes a pressure gradient in the interstitial fluid that drives fluid from regions of compression to tension. This causes flow-related sheer stresses on cell membranes. *(b)* Schematic illustration suggesting how mechanical stimulus may be signaled to bone cells. The osteocyte network transmits messages via gap junctions to bone-lining cells. The cells then release paracrine factors that stimulate osteoprogenitor cells to divide and eventually produce osteoblasts that produce a new matrix.

Reprinted, by permission, from *Calcified Tissue International*, Mechanotransduction and the functional response of bone to mechanical strain, by RL Duncan and CH Turner, 47, 344-358, 1995, copyright of Springer-Verlag.

evidence indicates that they can also communicate directly with osteocytes [42]. Osteocytes can produce anabolic growth factors that are transported to the bone surface and recruit osteoprogenitor (bone precursor) cells.

Prostaglandins and particularly the cyclooxygenase enzyme (COX-2) play a crucial role in mechanotransduction. When rats had their prostaglandin pathways pharmacologically blocked prior to forces being applied to the vertebrae [43, 44] no bone formation occurred [44]. Control animals that underwent loading without pharmacological intervention responded to the loading with new bone formation. This indicates that the enzyme COX-2 and the prostaglandins it produces are essential to mechanotransduction [44].

The Effector

The final step in mechanotransduction requires an effector—a component of bone that produces new, or rearranged, bone. Some consider osteoblasts and osteoclasts the effector cells [15]. Others have proposed that the effector for bone strength changes are really the bone modeling and remodeling systems that provide organized, site-specific formation and resorption that isolated cell units cannot achieve [45, 46].

Let us first review what is known about single-cell effectors. After a mechanical load is placed on bone (to mimic physical activity), osteocytes and bone-lining cells release prostacyclin (see "Transmission of the Biochemical Signal" on page 15), which is followed five minutes later by an increase in glucose-6-phosphate dehydrogenase (G6PD) [47] and 6 to 24 hours later by increases in RNA synthesis and IGF-I messages in osteocytes [42]. Increased collagen and mineral apposition on the bone surface begins three to five days after the bout of loading [29]. From days 5 through 12, the bone formation rate is increased largely due to increases in bone-forming surface [48]. Thus, loading activates discrete packets of osteoprogenitor cells to differentiate and synthesize osteoid—a process that takes about four days.

Three studies have reported the appearance of osteoblasts a few days after a single session of loading [38, 49-51]. Osteoblast-like cell surface activity (measured by recording the percentage of surface covered by osteoblasts) can occur on day two and peak on day three after rat tibial loading using a four-point bending device that applies load (and can be considered to replicate physical activity). The whole process of creating a mature osteoblast from an osteoprogenitor cell takes from 60 to 72 hours [52].

In life, physical activity causes many cells to behave in a coordinated fashion. When bone is growing, this is called modeling. When linear growth has finished, bone cells are "coupled," or linked, in the process of remodeling. These processes report bone cell behavior on a larger scale, which is the final subject of this chapter.

Modeling and Remodeling

Bone modeling is an organized bone cell activity that allows bone growth and adjusts bone strength through the strategically placed, non-adjacent activity of osteoblasts and osteoclasts [53]. Modeling improves bone strength not only by adding mass, but also by expanding the periosteal and endocortical diameters of bone. Examples of modeling include change in bone shape with growth and the slow, continuous periosteal and endocortical expansion of bone throughout adult life [54].

When bone mass is added regionally via the process termed macromodeling, geometric properties are improved. For example, increased endosteal diameter increases the moment of inertia of bone and, therefore, of strength in bending and torsion (see chapter 3). When alignment of trabeculae occurs within cancellous bone in response to regular loading, this is called minimodeling. Minimodeling is a change in orientation of trabeculae only, not a change in frank mineralization or bone size [53, 55]. The archlike appearance of trabeculae at the femoral neck is an example of minimodeling (figure 2.6) [56].

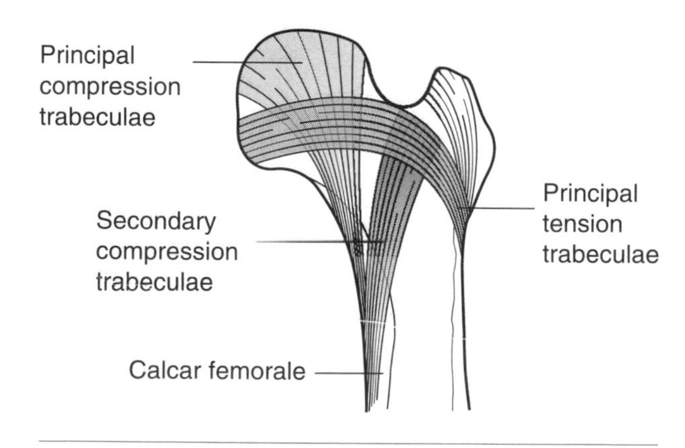

FIGURE 2.6 Schematic drawing of trabecular systems of the proximal femur. Trabeculae align in response to regular loading.

Both types of modeling can act in concert to strengthen bone.

Remodeling is a stereotyped, quantal, organized bone cell activity based on the basic multicellular unit (BMU). The activity is a locally coordinated, sequential activity of osteoclasts, osteoblasts, and their precursors [45, 57]. Remodeling provides a mechanism whereby bone can both prevent and repair fatigue damage (mechanical aspect). For example, when antiresorptive drugs suppress remodeling, microdamage accumulates in bone tissue and reduces its ability to absorb energy [58]. Remodeling is also vital for calcium homeostasis (metabolic aspect). A key feature of remodeling is that it replaces damaged tissue with an equal amount of new bone tissue in the healthy adaptive skeleton [57]. In the aging and osteoporotic skeleton, however, the balance between the amount of bone resorbed and formed is shifted in favor of resorption, so that insufficient bone is formed to refill the resorption cavity. A net loss of bone results, and eventually bone strength and integrity are compromised.

Comparison of Modeling and Remodeling

When the sensor (osteocyte–bone-lining cell complex discussed earlier) detects substantial deformation that requires increased bone strength, the response is very specific. As most bending deformation takes place far from the neutral axis of a long bone, new bone is added to this site—the periosteal surface (figure 2.7a). Modeling places new bone to maximize stiffness [54] and minimize deformation (maximizes the effect on moment of inertia). At this time, remodeling increases in order to replace fatigue-damaged bone more rapidly [59]. During major increases in loading, both geometric properties and bone mass increase as woven bone is added to the periosteal surface in response to excessive strain [20, 60]. (Geometric properties and bone strength are discussed further in chapter 3.) Woven bone formation enables a rapid increase in strength because it can be deposited faster than lamellar bone, and because it is located on the periosteal surface of bone [59]. When the sensor detects decreased deformation, it decreases bone mass by removing the endocortical surface that has the least deformation first, thus producing a minimum negative effect on moment of inertia (figure 2.7b) [59].

The balance between modeling and remodeling differs between the growing and the nongrowing skeleton. In the former, modeling is the dominant mode; in the latter, remodeling is dominant. The mechanisms that control the

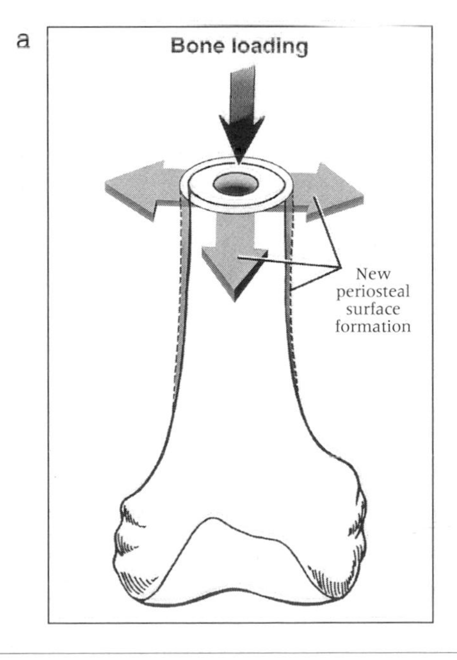

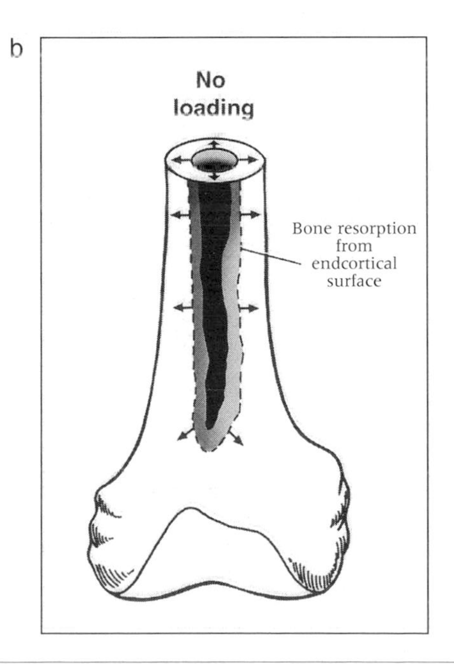

FIGURE 2.7 Diagram of cortical bone undergoing *(a)* substantial mechanical loading and deformation and *(b)* insufficient loading to maintain bone mineral. When bone is added, it strengthens maximally on the periosteal surface (represented by thick arrows). When bone is resorbing, it diminishes bone strength least when removed from the endocortical surface (represented by thin arrows).

patterns of cell behavior that these terms refer to are not understood, but the processes, and their response to mechanical loading, can be clearly described [61]. In modeling, osteoblasts continue to function at one site for a number of years. In the remodeling phase, they only operate at one location for a discrete, relatively short time [61]. This mechanism permits optimum bone gain during growth.

The Bone Remodeling Transient

The process referred to as the bone remodeling transient is very relevant to those interested in physical activity and bone health. Unfortunately, it is often overlooked by those unfamiliar with bone physiology. The bone-remodeling transient must be considered in studies that involve exercise, calcium, or medication [62].

During the remodeling process a proportion of bone remodeling units are in the resorption phase, while others are in either the formation or mineralization phase. Sites where remodeling is occurring contain regions where a temporary loss of bone occurs during the resorption period. This is termed the remodeling space. In most adults this represents about 2% of the skeleton. Undermineralized bone can be found in these "midprocess" remodeling spaces, and this bone is not measured by DXA. By definition, then, measuring the amount of bone mineral will always underestimate the amount of bone tissue to some extent, and the greater the proportion of sites being remodeled, the greater the underestimation (figure 2.8) [62].

Interventions (drugs, nutrients, or physical activity) that alter bone activation rate can thus result in short-term changes in the amount of measurable bone mineral without necessarily leading to sustained changes in the amount of bone tissue. For example, because calcium is known to inhibit bone resorption, the number of bone sites beginning the remodeling process during a given time period decreases when calcium intake increases [63]. The short-term result of an increased calcium intake, therefore, would be that previously activated sites would mineralize normally, fewer new sites would be activated, and the size of the remodeling space would decrease. This causes an apparent increase in the amount of bone mineral measured, when there is very little change in the amount of bone tissue present (only a small amount of the resorption space is filled with bone). When the remodeling cycle in the previously activated sites was completed, mineral content would level off at a new plateau. This higher plateau would be maintained as long as the high calcium intake persisted.

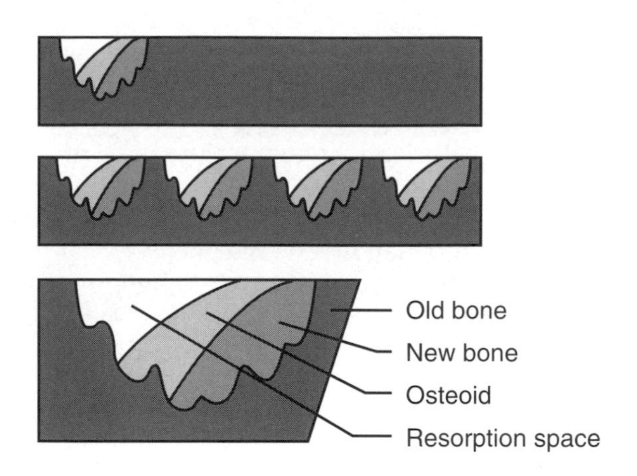

FIGURE 2.8 Schematic illustration of how bone remodeling influences the amount of mineralized bone present. In the upper panel, the undermineralized osteoid and the remodeling space make up a far smaller proportion of bone than in the lower panel, where there are four times as many bone remodeling units.

Reprinted, by permission, from JA Kanis, 1994, *Osteoporosis*. (Oxford, England: Blackwell Science), 30.

This "transient" apparent increase in bone mineral disappears when the intervention is removed. For example, if calcium intake returned to baseline levels, the remodeling activation rate would increase, and for a period there would be more sites entering the remodeling cycle (i.e., being resorbed) than there would be previously activated sites being mineralized. This would result in an increase in the remodeling space and consequently a transient decrease in the measured amount of bone mineral. As the number of sites undergoing resorption and formation again came into balance, bone mineral would be measured at a new plateau lower than that reached when calcium intake was high.

To reduce the effect of the remodeling transient on intervention study outcomes, it is necessary to study changes that occur during the first remodeling cycle of an intervention study separately from those that occur subsequently, when any effects of the transient have already been accounted for. This period is typically about 10 months long [64], though in postmenopausal women it is approximately 6 months long. This time period depends on the rate of bone turnover, which is influenced by such factors as age and estrogen status.

SUMMARY

- This chapter focused on three key aspects of bone physiology: calcium homeostasis, mechanotransduction, and bone remodeling.
- Ninety-eight percent of total body calcium exists in bone. Plasma calcium, the remaining 2% of body calcium, exists in an ionized (metabolically active) and an unionized form
- Plasma-ionized calcium is adjusted via a negative feedback mechanism that involves the intestine, bone, and the kidneys. Parathyroid hormone, in the presence of sufficient vitamin D, stimulates bone resorption, enhances renal synthesis of vitamin D, and enhances renal reabsorption of calcium. Vitamin D, synthesized in the skin during ultraviolet light exposure, acts on intestine, bone, and kidney.
- Mechanotransduction is the process whereby forces acting on bone cells (such as physical activity) influence bone cell behavior. The four components of mechanotransduction—mechanocoupling, biochemical coupling, signal transmission, and the effector response—are currently being unraveled.
- Bone remodeling is a stereotyped organized bone cell activity that takes place in the basic multicellular unit (BMU). Bone remodeling has a mechanical and a metabolic role.
- The coupling of bone resorption and formation in remodeling underpins bone physiology. The "uncoupling" of those processes, or changes in the balance between resorption and formation, can lead to disease states.

References

1. Geusens P, ed. *Osteoporosis in clinical practice: a practical guide for diagnosis and treatment*. London: Springer-Verlag, 1998:188.
2. Martin TJ, Findlay DM, Moseley JM. *Peptide hormones acting on bone*. In: Marcus R, Feldman D, Kelsey J, ed. Osteoporosis. San Diego, CA, USA: Academic Press, 1996: 185-204.
3. Aurbach GD, Marx SJ, Spiegel AM. *Parathyroid hormone, calcitonin and the calciferols*. In: Wilson JD, Foster DW, ed. Williams textbook of endocrinology. Philadelphia: W B Saunders, 1992: 1397-1476.
4. Feldman D, Malloy PJ, Gross C. *Vitamin D: metabolism and action*. In: Marcus R, Feldman D, Kelsey J, ed. Osteoporosis. San Diego, CA, USA: Academic Press, 1996: 205-227.
5. Heaney RJ. *Nutrition and osteoporosis*. In: Favus MJ, ed. Primer on the metabolic bone diseases and disorders of mineral metabolism. 3rd ed. Philadelphia: Lippincott Williams & Wilkins, 1999: 270-272.
6. Deftos LJ, Roos BA, Oates EL. *Calcitonin*. In: Favus MJ, ed. Primer on the metabolic bone diseases and disorders of mineral metabolism. 4th ed. Philadelphia: Lippincott Williams & Wilkins, 1999: 99-103.
7. Wolff J. *The law of bone transformation*. Berlin: Hirschwald, 1892.
8. Lanyon LE, Goodship AE, Pye CJ, et al. Mechanically adaptive bone remodeling. *J Biomechanics* 1982;15:141-154.
9. Frost HM. Bone "Mass" and the "Mechanostat": A proposal. *Anat Rec* 1987;219:1-9.
10. Currey JD. *The mechanical adaptations of bones*. Princeton University Press, 1984.
11. Turner CH. Toward a mathematical description of bone biology: the principle of cellular accommodation. *Calcif Tissue Int* 1999;65:466-471.
12. Aarden EM, Burger EH, Nijweide PJ. Function of osteocytes in bone. *J Cell Biochem* 1994;55:287-299.
13. Weinbaum S, Cowin SC, Zeng Y. A model for the excitation of osteocytes by mechanical loading-induced bone fluid shear stresses. *J Biomech* 1994;27:339-360.
14. Frost HM. Obesity, and bone strength and "Mass": A tutorial based on insights from a new paradigm. *Bone* 1997;21:211-214.
15. Duncan RL, Turner CH. Mechanotransduction and the functional response of bone to mechanical strain. *Calcif Tissue Int* 1995;57:344-358.
16. Lanyon LE. Functional strain in bone tissue as an objective, and controlling stimulus for adaptive bone remodeling. *J Biomech* 1987;20:1083-1093.

17. Frost HM. A determinant of bone architecture: The minimum effective strain. *Clin Orthop Rel Res* 1983;175:286-292.
18. Rubin CT. Skeletal strain and the functional significance of bone architecture. *Calcif Tissue Int* 1984;36:S11-S18.
19. Turner CH, Forwood MR, Otter MW. Mechanotransduction in bone: do bone cells act as sensors of fluid flow? *FASEB J* 1994;8:875-878.
20. Turner CH, Forwood MR, Rho J-Y, et al. Mechanical strain thresholds for lamellar and woven bone formation. *J Bone Miner Res* 1994;9:87-97.
21. Knothe Tate ML, Knothe U, Niederer P. Experimental elucidation of mechanical load-induced fluid flow and its potential role in bone metabolism and functional adaptation. *Am J Med Sci* 1998;316:189-95.
22. Carter DR, Hayes WC. Fatigue life of compact bone - I Effects of microstructure and density. *J Biomech* 1976;9:211-218.
23. Frost HM. Some ABC's of skeletal pathophysiology. 5. Microdamage physiology. *Calcif Tissue Int* 1991;49:229-231.
24. Burr D, Forwood MR, Fyhrie DP, et al. Bone microdamage and skeletal fragility in osteoporotic and stress fractures. *J Bone Miner Res* 1997;12:6-15.
25. Mori S, Burr DB. Increased intracortical remodeling following fatigue damage. *Bone* 1993;14:103-109.
26. Verbogt O, Gibson GJ, Schaffler MB. Loss of osteocyte integrity in association with microdamage and bone remodeling after fatigue in vivo. *J Bone Miner Res* 2000;15:60-67.
27. Skerry TM, Bitensky L, Chayen JL, et al. Early strain-related changes in enzyme activity in osteocytes following bone loading in vivo. *J Bone Miner Res* 1989;4:783-788.
28. Cowin SC. Candidates for the mechanosensory system in bone. *J Biomech Eng* 1991;113:191-197.
29. Dodds RA, Ali N, Pead MJ, et al. Early loading-related changes in the activity of glucose 6-phosphate dehydrogenase and alkaline phosphatase in osteocytes and periosteal osteoblasts in rat fibulae in vivo. *J Bone Miner Res* 1993;8:261-267.
30. Hauschka PV, Mavrokos AE, Iafrati MD, et al. Growth factors in bone matrix. *J Biol Chem* 1986;261:12665-12674.
31. Kawata A, Mikuni-Takagaki Y. Mechanotransduction in stretched osteocytes—temporal expression of immediate early and other genes. *Biochem Biophys Res Commun* 1998;246:404-8.
32. Pavalko FM, Otey CA, Simon KO, et al. a-Actinin: a direct link between actin and integrins. *Biochem Soc Trans* 1991;19:1065-1069.
33. Bockholt SM, Burridge K. Cell spreading on extracellular matrix proteins induces tyrosine phosphorylation of tensin. *J Biol Chem* 1993;268:14565-14567.
34. Schwartz MA, Ingber DE. Integrating with integrins. *Mol Biol Cell* 1994;5:389-393.
35. Chen NX, Ryder KD, Pavalko FM, et al. Ca2+ regulates fluid shear-induced cytoskeletal reorganisation and gene expression in osteoblasts. *Am J Physiol* 2000;Cell Physiol 278:C989-C997.
36. Burger EH, Veldhuijzen JP. *Influence of mechanical factors on bone formation, resorption and growth in vitro.* In: Hall BK, ed. Bone Vol 7: Bone Growth-B. Boca Raton, FL: CRC Press, 1993: 37-56.
37. Harter LV, Hruska KA, Duncan RL. Human osteoblast-like cells respond to mechanical strain with increased bone matrix production independent of hormonal regulation. *Endocrinology* 1995;136:528-535.
38. Forwood MR, Owan I, Takano Y, et al. Increased bone formation in rat tibiae following a single short period of dynamic loading in vivo. *Am J Physiol* 1996;270 (Endocrinology and Metabolism 33):E419-E423.
39. Sandy JR, Farndale RW. Second messengers: regulators of mechanically induced tissue remodelling. *Eur J Orthod* 1991;13:271-278.
40. VandenBergh MFQ, DeMan SA, Witteman JC, et al. Physical activity, calcium intake and bone mineral content in children in The Netherlands. *Journal of Epidemiology and Community Health* 1995;49:299-304.
41. Watson PA. Function follows form: generation of intracellular signals by cell deformation. *FASEB J* 1991;5:2013-2019.
42. Lean JM, Jagger CJ, Chambers TJ, et al. Increased insulin-like growth factor I mRNA expression in rat osteocytes in response to mechanical stimulation. *Am J Physiol* 1995;268:E318-E327.
43. Chow JW, Chambers TJ. Indomethacin has distinct early and late actions on bone formation induced by mechanical stimulation. *Am J Physiol* 1994;267:E287-92.
44. Forwood MR. Inducible cyclo-oxygenase (COX-2) mediates the induction of bone formation by mechanical loading in vivo. *J Bone Miner Res* 1996;11:1688-93.
45. Parfitt AM. *Skeletal heterogeneity and the purposes of bone remodeling.* In: Marcus R, Feldman D, Kelsey J, ed. Osteoporosis. San Diego, CA, USA: Academic Press, 1996: 315-329.
46. Frost HM. Why do marathon runners have less bone than weight-lifters? A vital-biomechanical view and explanation. *Bone* 1997;20:183-189.
47. El Haj AJ, Minter SL, Rawlinson SCF, et al. Cellular responses to mechanical loading in vitro. *J Bone Miner Res* 1990;5:923-932.
48. Forwood MR, Turner CH. The response of the rat tibiae to incremental bouts of mechanical loading:

A quantum concept for bone formation. *Bone* 1994;15:603-609.

49. Pead MJ, Skerry TM, Lanyon LE. Direct transformation from quiescence to bone formation in the adult periosteum following a single brief period of bone loading. *J Bone Miner Res* 1988; 3:647-656.
50. Chow JWM, Jagger CJ, Chambers TJ. Characterization of osteogenic response to mechanical stimulation in cancellous bone of rat caudal vertebrae. *Am J Physiol* 1993; 265 (Endocrinol Metab 28):E340-E347.
51. Boppart MD, Cullen DM, Yee JA, et al. Time course for osteoblast appearance after in vivo mechanical loading. *J Bone Miner Res* 1996; 11 (supplement:? - abstract #340.
52. Lian JB, Stein GS. *Osteoblast biology*. In: Marcus R, Feldman D, Kelsey J, ed. Osteoporosis. San Diego, CA, USA: Academic Press, 1996:23-60.
53. Frost HM. Skeletal structural adaptations to mechanical usage (SATMU).1.Redefining Wolff's law: The bone modeling problem. *Anat Rec* 1990;226:403-413.
54. Cordey J, Schneider M, Belendez C, et al. Effect of bone size, not density, on the stiffness of the proximal part of normal and osteoporotic human femora. *J Bone Miner Res* 1992;2:S437-S444.
55. Cowin SC, Sadegh AM, Luo GM. An evolutionary Wolff's law for trabecular structure. *J Biomech Eng* 1992;114:129-136.
56. Singh M, Nagrath AR, Maini PS. Changes in trabecular pattern of the upper end of the femur as an index of osteoporosis. *J Bone Joint Surg* 1980;52-A:457-467.
57. Frost HM. Some effects of the basic multicellular unit-based remodeling on photon absorptiometry of trabecular bone. *Bone Miner* 1989;7:47-65.
58. Mashiba T, Hirano T, Turner CH, et al. Suppressed bone turnover by bisphosphonates increases microdamage accumulation and reduces some biomechanical properties in dog rib. *J Bone Miner Res* 2000;15:613-20.
59. Kimmel DB. A paradigm for skeletal strength homeostasis. *J Bone Miner Res* 1993;8:S515-S522.
60. Rubin CT, Gross TS, McLeod KJ, et al. Morphologic stages in lamellar bone formation stimulated by a potent mechanical stimulus. *J Bone Miner Res* 1995;10:488-495.
61. Parfitt AM. The two faces of growth: benefits and risks to bone integrity. *Osteoporos Int* 1994;4:382-398.
62. Heaney RP. The bone-remodeling transient: Implications for the interpretation of clinical studies of bone mass change. *J Bone Miner Res* 1994;9:1515-1523.
63. Barr SI, McKay HA. Nutrition, exercise, and bone status in youth. *Int J Sport Nutr* 1998;8:124-142.
64. Heaney RP. *Design considerations for osteoporosis trials*. In: Marcus R, Feldman D, Kelsey J, ed. Osteoporosis. San Diego, CA, USA: Academic Press, 1996: 1125-1143.
65. Kanis JA. *Osteoporosis*. Oxford, England: Blackwell Science, 1994:1–254.

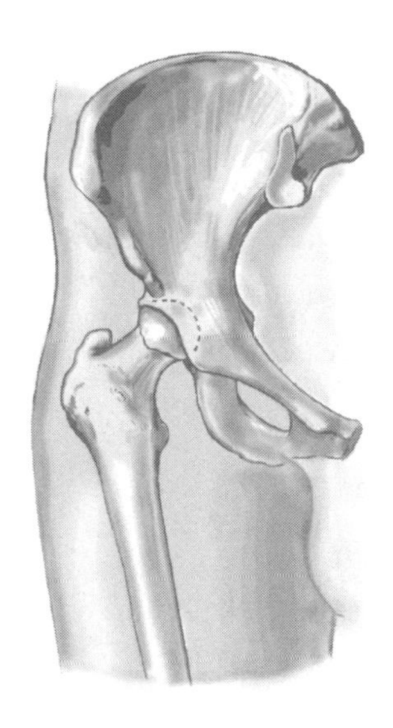

Chapter 3

Biomechanics

with Ari Heinonen

To design a strong, resilient building, engineers must consider a number of factors, including the materials available for construction (material properties) and the size, shape, and geometry of the structure (structural properties). They must then consider the maximum loads that the building will be exposed to and optimize the use of materials and structural design so as to prevent the building's collapse.

Mother Nature has engineered a unique material (bone) into a very functional and attractive structure (the skeleton) that is strong enough to withstand the demands of intense physical activity and externally affecting forces, adaptive enough to respond to changes in these demands, and lightweight enough to allow effective, energy-saving locomotion. The mechanical competence of bone is a function of both its intrinsic material properties (mass, density, stiffness, and strength) and its gross geometric characteristics (size, shape, cortical thickness, cross-sectional area, and trabecular architecture) [1].

For example, the tubular shape of long bones responds ideally to the torsional and bending loading imposed on the diaphyses. The widened, metaphyseal ends of long bones and short bones are designed for support and to dissipate contact forces. The complex trabecular bone architecture in these end regions illustrates how well the skeleton is designed to meet its mechanical needs.

Human bone building commences as early as five to seven weeks in utero, when the involuntary contraction of newly formed muscle fibers begins to shape the cartilaginous endoskeleton. From that time on, bone continues to respond, either positively or negatively, to its milieu so that at any age the integrity of the human skeleton is a product of genetics and its "lifestyle" history. Factors that affect skeletal strength and design include genetics, physical activity, hormones, and dietary factors. Of these, the mechanical loading activity of bone is vitally important, as illustrated in figure 3.1.

The overall aim of this chapter is to explain the association between mechanical loading and bone strength. The chapter has four parts. The first two describe the material and structural properties that influence bone strength. The third

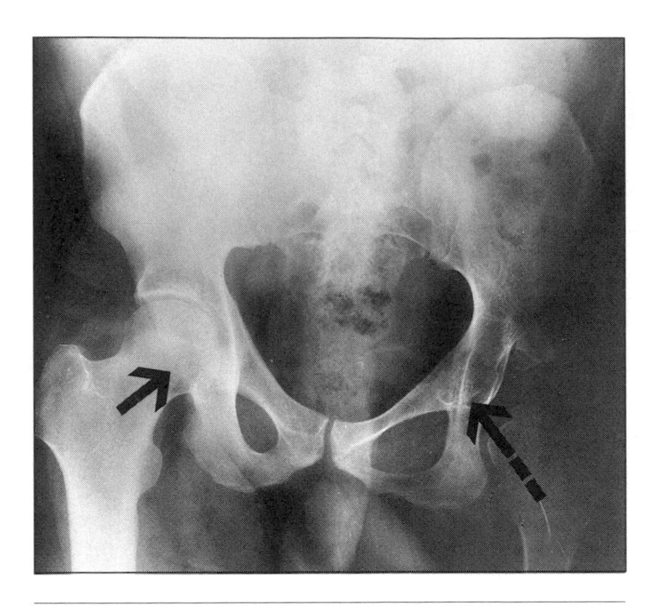

FIGURE 3.1 The importance of mechanical loading on bone structure—a radiograph of the pelvis of a 22-year-old man who was born without an upper femur. Note that the acetabulum and the pelvis that were not subjected to mechanical loading (dotted arrow) because of the absent upper femur are smaller and less dense than the corresponding anatomy on the normal side (solid arrow).
Photo courtesy of Dr. S. Houston.

section discusses bone's response to local mechanical loading. This section summarizes the many animal experiments that examined the type of loading that increases bone strength. The final section describes how physical activity in humans generates loads on bone.

Material Versus Structural Properties of Bone

It is important to understand the difference between the way the skeleton behaves at the whole bone or structural level and the properties of bone as a material at the tissue level. During loading activities such as walking, running, or jumping, whole bone as a structure will experience forces or loads that result in bone deformation. The structural behavior of bone in response to a given load or set of forces is a function of both its geometric properties and the material properties of the bone tissue itself. When the forces applied to whole bone are greater than the structure can withstand, structural failure, or fracture, may occur.

On the other hand, the material behavior of bone is not related to the geometry of the bone specimen. If increasing loads are applied to a specimen of bone tissue, it will eventually undergo permanent deformation or yield. The material will ultimately fail when it can no longer withstand the increased stress imposed upon it. Bone is a heterogeneous material, and a discussion of bone as a material is quite complex. Studies of material behavior might refer to the mechanical properties of trabecular or cortical bone sections, single trabeculae, or the calcified bone matrix.

To illustrate the difference between material and structural properties, consider two methods of bridging a 65-ft (20-m) space between the rooftops of adjacent skyscrapers. One bridge consists of a single row of bricks cemented end to end. The other is 500 stacked layers of cardboard 165 ft (50 m) wide. The cemented bricks have far greater material properties than the cardboard (per unit volume), while the cardboard provides greater support for the person crossing due to its thickness and width. With tubular structures, such as pipes in engineering and long bones in the skeleton, structural properties (diameters, thickness of the walls) play a major part in the ability to resist loads.

The organic and inorganic components of bone (see chapter 1) determine its material properties. The organic component, primarily type I collagen, provides tensile strength. Abnormal collagen matrix leaves bone brittle, irrespective of the amount of bone mineral present. The inorganic component, mineral, resists compressive forces. When bone mineral is insufficient, bone bends easily. Children suffering from rickets (vitamin D deficiency) have insufficient bone mineralization resulting in bowed tibiae.

Functional tests under controlled conditions using a defined volume of bone can be used to measure the material behavior of bone as a tissue. When a load, or "stress," is applied, bone is deformed. This deformation is referred to as "strain."

Material Properties of Bone

In this section we explain the concepts of stress, strain, and stiffness, which are fundamental to understanding the material properties of any building material. Remember that these properties are all independent of size.

Stress and Strain. The material behavior of a bone specimen is determined in laboratory experiments. A load is applied under controlled conditions and the relationship between stress and strain is determined. From this relationship, other material properties of bone can be assessed.

Stress, the force applied per unit area, can be classified as tensile, compressive, or shear. Stress is measured in units of Newtons per square meter (N/m^2) or Pascals (Pa). *Strain* describes the deformation of a material and refers to the relative change in the bone dimension under study (i.e., length, width, or angulation). Strain is nondimensional and is calculated by dividing change in bone dimension by original bone dimension. It is sometimes expressed as a fraction or percentage (strain of 1 = 100%, strain of 0.001 = 0.1%, or 1000 microstrain [$\mu\epsilon$]). Strain is greatest at the point of highest loading and dissipates along the length of the long bone. For example, during walking or running, the highest measurable strain would occur at the calcaneus and distal tibia. In addition, during locomotion the greatest strain is generated at the cortex under compression (figure 3.2).

The intrinsic material properties of bone include the concepts of stiffness and strength. The amount of force required to deform a structure is termed its *stiffness* and is represented by the slope of the stress-strain curve (figure 3.3). The strength of the structure can be defined as the load at the yield or failure points, or as the ultimate load, depending on the circumstances. Material strength is an intrinsic property of bone and is independent of its size. Therefore, in the life sciences bone strength is often reported in units of force and in engineering studies in terms of stress or as intrinsic strength.

A technical term for the slope of the stress-strain curve (in the elastic region) is Young's modulus, or the elastic modulus. Conceptually, it represents inherent resistance to loading (i.e., stiffness). The relationship between axial stress and strain and Young's modulus can be represented mathematically as

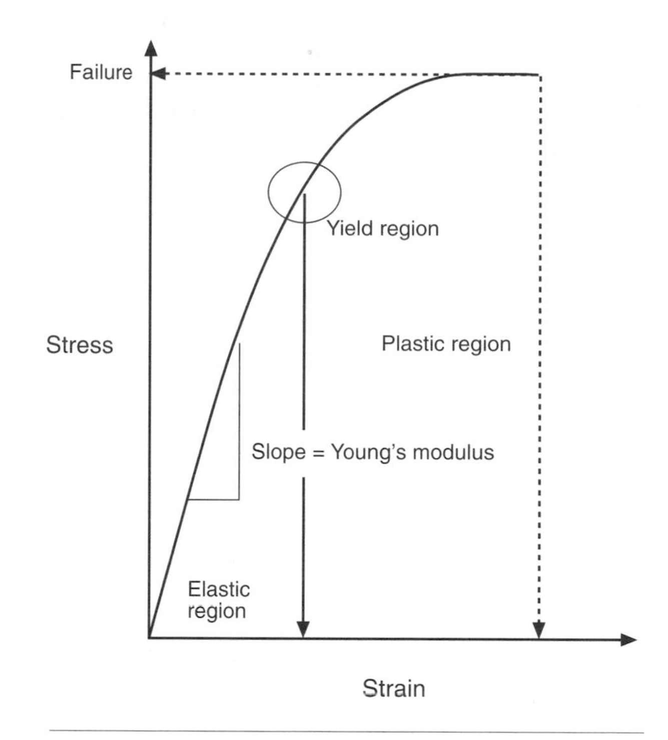

FIGURE 3.3 Stress-strain curve for a specimen of bone—testing the material properties.

$$\epsilon = S/E$$

in which ϵ = strain, S = stress, and E = Young's modulus.

The stress-strain curve allows us to compare different materials by examining the slope of the curve and the other strength parameters. Thus, the modulus of steel is ten times that of cortical bone. This translates into a tensile strength of steel that is about five times that of cortical bone [52].

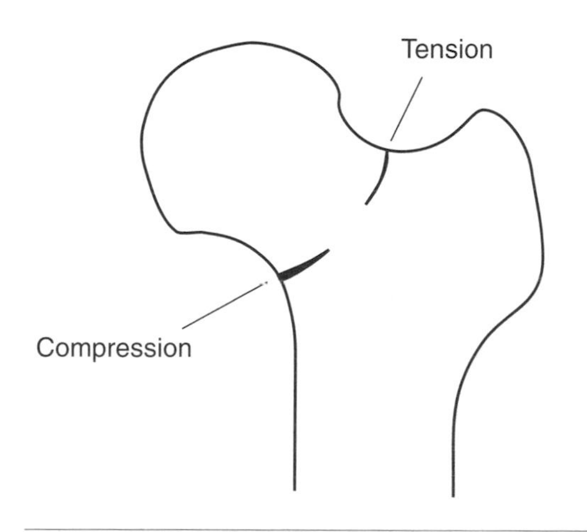

FIGURE 3.2 Illustration of the compression side and tension side of a bone—the femoral neck.

Bone Mineral Mass. *Bone mass* is a determinant of bone material properties; the distribution of bone mass and bone geometry are connected to bone strength and stiffness. Further, the strength and stiffness of bone is a function of density. Stiffness is proportional to density cubed and to strength squared. Although bone mass is only one component of overall bone strength [3], it does explain more than 80% of that variable [4, 5]. Because bone mass is highly correlated with dual energy X-ray absorptiometry (DXA) results, this technique is used to measure bone mass and give an estimate of fracture risk in humans. DXA technology and its clinical implications are discussed fully in chapter 4.

Structural Properties of Bone

Structural (or geometric) characteristics of whole bone include size, shape, cortical thickness, cross-sectional area, and trabecular architecture. Appendicular bone adapts to mechanical loads by endosteal resorption and periosteal apposition of bone tissue. This increases bone diameter and thus provides greater resistance to loading. This adaptive process allows bone to be strong enough to resist compression, tension, and shear stresses, yet lightweight enough for efficient and economic locomotion. In addition, the skeleton has its shock absorbers. For example, long bones are slightly curved so as to bend under an impact load, and vertebrae are cushioned by intervertebral discs.

Cross-Sectional Moment of Inertia. A hollow cylinder provides the least mass and the greatest strength during bending and torsional loading. In bending, both the cross-sectional area and the distribution of bone mass around a neutral axis (usually the bone center) affect the bone's mechanical behavior (figure 3.4). These concepts are important for understanding cross-sectional moment of inertia (CSMI).

If you were given a set amount of material with which to build a cylindrical structure of maximal strength (e.g., a beam), your best bet would be to make it hollow with as large a diameter as possible. In the skeleton, minimal weight and maximal CSMI is achieved when cross-sectional bone area is as far from the neutral axis as possible (as in figure 3.4). A larger CSMI requires less bone area and bone mass to maintain the mechanical competence of bone. In other words, a larger CSMI results in a stronger and stiffer bone because much of the bone mass is distributed at a distance from the neutral axis—a rather complex label for a relatively simple concept. CSMI for a circular cross section is represented mathematically as

$$\text{CSMI} = (\pi/4) \times (r_1^4 - r_2^4)$$

in which CSMI = cross-sectional moment of inertia, r_1 = outer radius of the cylinder, and r_2 = inner radius of the cylinder. Note that a small increase in new bone at the periosteal surface increases CSMI considerably as the CSMI is proportional to the fourth power of the radius.

The mechanical competence of *whole* bone can be characterized by the deformation it undergoes during loading. Deformation is characteristically measured and plotted as a load-deformation curve (figure 3.5). Generally, a linear relationship exists between the imposed load and the amount of bone deformation until the bone reaches its *yield* point. Prior to reaching

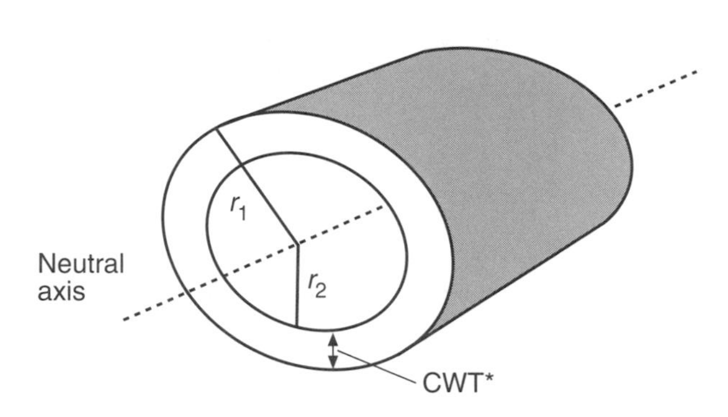

FIGURE 3.4 Depiction of cross-sectional moment of inertia (CSMI) for a circular hollow tube, where r_1 is the outer radius, r_2 is the inner radius, and CWT is the cortical wall thickness.

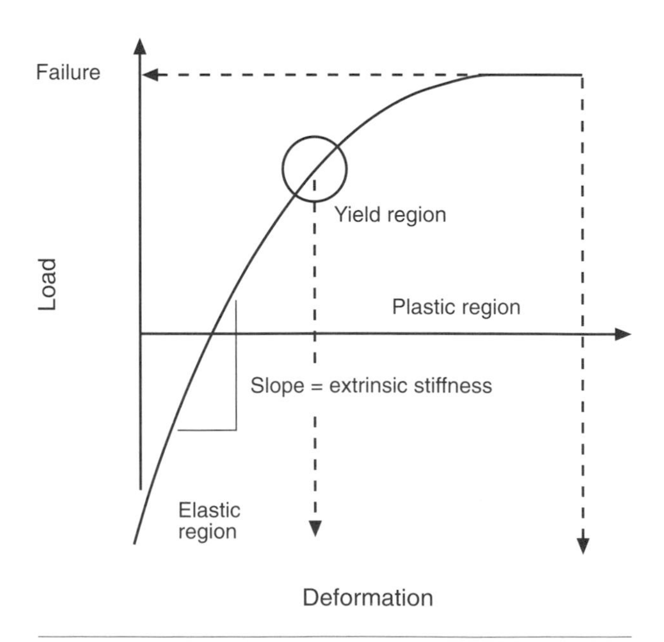

FIGURE 3.5 Load-deformation curve. This is similar to a stress-strain curve, but it is used to assess whole bone. The extrinsic stiffness of whole bone depends on both its material properties and its structural geometry.

this point, the bone is in its *elastic region* (if unloaded, it would return to its original shape). That is, applied forces in this region cause only temporary deformation. Beyond the yield point, the slope of the curve plateaus; the area beneath this part of the curve is called the *plastic* region. It marks the point at which permanent deformation occurs and in extreme cases the point at which local fractures or other damage may result. If the load continues to increase, *failure load* may be reached, and the structure may fail completely. A perfect clinical example of this is a greenstick fracture in which the incomplete fracture is notable on the convex surface (figure 3.6).

Whole Bone Strength. The mechanical competence of whole bone may improve with exercise training by improving bone geometry without any measurable changes in bone mass [7, 8]. This seems reasonable since cross-sectional moments of inertia are more important for resisting bending loads than are bone mass or density under loading conditions and during bending [9]. Therefore, it is important to include or estimate other bone geometric properties, including CSMI, whenever possible when determining whole bone strength. Thus, an increase in bone strength results from (1) increased bone mass; (2) a positively adaptive change in bone size, shape, or microarchitecture; or (3) a combination of these factors (figure 3.7).

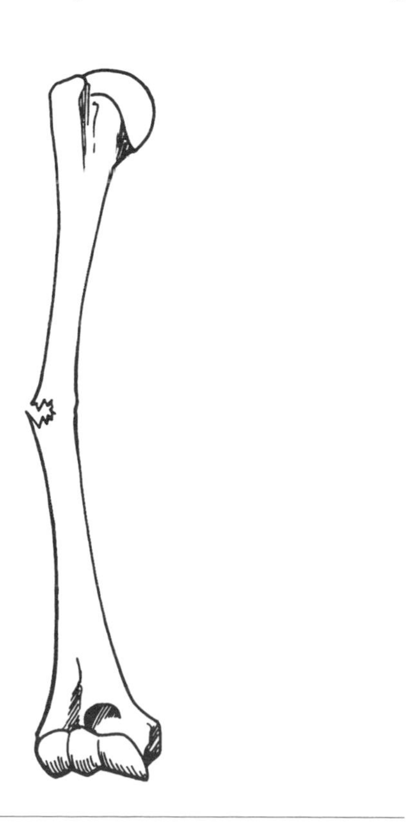

FIGURE 3.6 Schematic illustration of a greenstick fracture of a long bone.

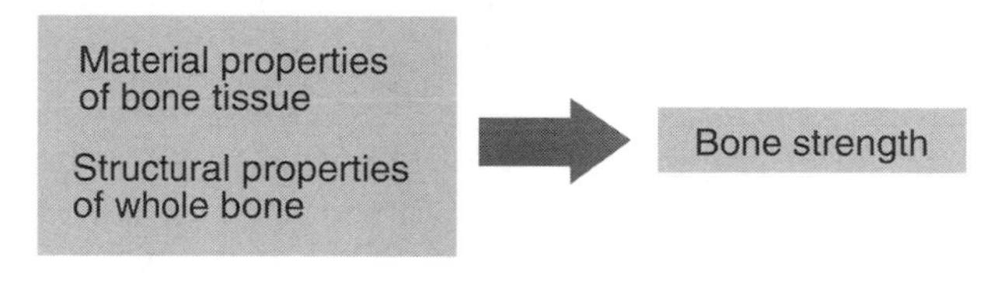

FIGURE 3.7 Components of bone strength.

An increase in bone strength results in reduced strain on the skeleton as greater mass or altered architecture will bend less in response to a given load. Conversely, when bone does not experience high loads (in a sedentary lifestyle, for example), bone mass is decreased to reduce the metabolic cost of maintaining and transporting mineral [10].

Bone's Response to Local Mechanical Loading

Bone's ability to adapt to physical demands was recognized more than a hundred years ago by a German anatomist, Julius Wolff. Although commonly given sole credit for his theories on bone's adaptation to mechanical loading, Wolff was strongly influenced by the work of Wilhelm Roux (1885). Wolff developed a theory relating physical forces and bone structure, referred to, not surprisingly, as Wolff's Law [11]. Wolff's Law states that bone will optimize structure so as to withstand functional loading and ensure the metabolic efficiency of locomotion. More specifically, Wolff's Law states that the function of those cells responsible for mechanically adaptive modeling and remodeling is to ensure that bone mass, geometry, and material properties are appropriate to the applied load.

The skeleton's response to a load depends on the strain magnitude, rate, distribution, and cycles in the target bone [1,12]. Mechanically adaptive bone modeling and remodeling can be regarded as part of a homeostatic mechanism that regulates functional strains throughout the skeleton. Based in part on variations in the longitudinal curvature and cross-sectional nature of different bones, the strains encountered during functional loading vary in magnitude and distribution.

Strain Magnitude

Strain magnitude can be defined as the amount of relative change in bone length under mechanical loading. Much of what we know about the influence of functional loads on bone comes from research with animal models. Early work using the turkey ulna clearly showed that bone formation increased with larger strain magnitudes [13]. Several researchers suggested that this response is controlled by a "mechanostat" [14, 15] that endeavors to keep bone strain within an optimal level and adjusts the structure of bone in order to do so [14].

The mechanostat theory describes a control system in which a minimum effective strain (MES) is necessary for bone maintenance [16]. Figure 3.8 illustrates the variable response of bone depending on the extent of loading. In the trivial loading zone (< 50 - 200 $\mu\epsilon$ = microstrain) virtually no mechanical stimulus to bone and remodeling occurs. Studies conducted during bed rest [10] and space flight [17] reveal that this results in a net loss of bone over time. Strains in the physiologic loading zone (200-2000 $\mu\epsilon$) maintain remodeling at a steady state that, in turn, maintains bone strength [18, 19]. If loading-induced local strain exceeds the MES (1500-2000 $\mu\epsilon$), bone enters a state of overuse. In the overload zone (2000-3000 $\mu\epsilon$) modeling is stimulated and new bone is added in response to the mechanical demand. This results in increased bone strength. Finally, in the pathological overload zone (> 4000 $\mu\epsilon$), bone suffers microdamage and woven (unorganized) bone is added as part of the repair process.

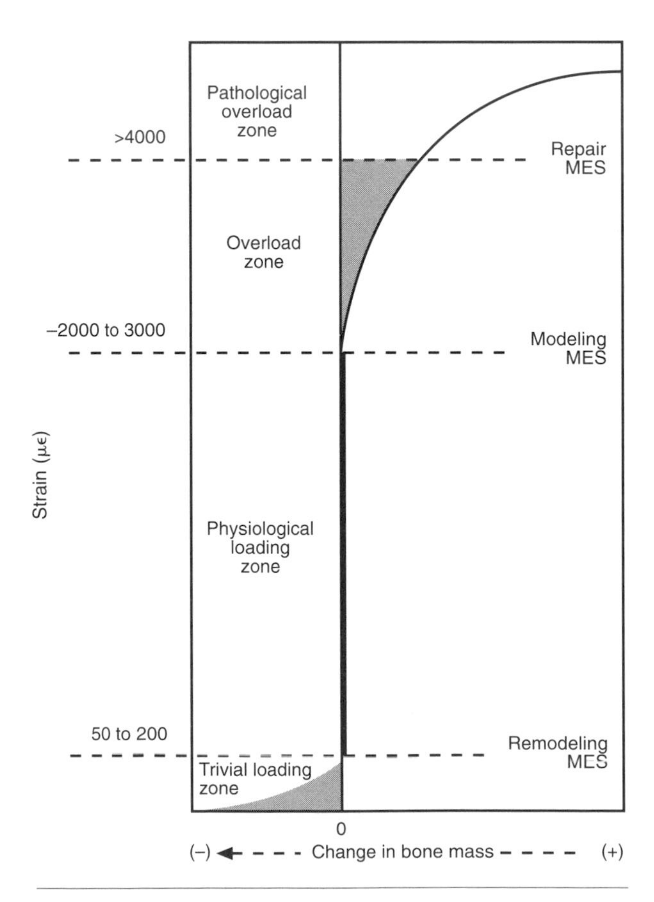

FIGURE 3.8 The mechanostat theory relating strain to bone mass.

Reprinted from *Bone*, 17, MR Forwood and CH Turner, Skeletal adaptations to mechanical usage: results from tibial loading studies in rats, 1975-2055, copyright 1995, with permission from Elsevier Science.

According to MES theory, bone responds at set-points. The set-point appears to be genetically controlled and modified by past site-specific loading and several biochemical factors. A change in the set-point thresholds would elicit a similar adaptive response [20]. Because the response to mechanical strain is site specific, one region of the skeleton may experience a net loss of bone while at the same time another region experiences a net gain.

Although the fracture strain of cortical bone is about 25,000 $\mu\epsilon$ in tension or compression, individuals cannot produce peak longitudinal compression and tension in cortical bone that exceeds 1500-3000 $\mu\epsilon$ even with the most vigorous voluntary activity [21]. This limiting strain range persists with little relative change between birth and skeletal maturity. During this time period peak loads on many bones may increase more than 20 times [22].

A few studies have measured strain in vivo by attaching strain gauges to human bone. In other studies, tibial strains were shown to be higher during vigorous activities such as sprinting and hill running, compared to walking [23, 24]. Activities that elicit high peak forces (or high strain magnitude) may have a greater effect on bone mass than activities associated with a large number of loading cycles [25].

Strain Rate

Strain rate, the rate at which strain develops and releases, determines bone's adaptive response. As strain rate is proportional to dynamic load magnitude, peak strain magnitude may be equally as osteogenic as strain rate. Research has shown a strong correlation among remodeling, strain rate, and strain magnitude [26]. However, a high magnitude but static load (continuous strain rate) with an unusual load distribution was no more effective than disuse in protecting bone from atrophy [27]. Martin and Burr [28] have suggested that high magnitude strain together with high strain rate is most effective for a maximal adaptive bone response. Umemura and colleagues [29] compared jump training with running training in rats and found that jumping was associated with a higher strain rate and magnitude and was more effective than running for eliciting a positive bone response.

Strain Distribution

Strain distribution refers to the way strain is distributed across a section of bone. A prolific group of researchers well known for their animal work on the adaptive response of bone to mechanical loading introduced the error strain distribution hypothesis [30]. This theory states that bone cells maintain the skeleton's structural competence by making architectural adjustments to eliminate or reduce perceived deviations from normal dynamic strain distributions [13]. That is, unusual strains of uneven distribution are more important for osteogenesis than the strain repetitions or strain magnitudes that result from everyday activity. The more unusual the strain distribution at a particular bone site, the greater its potential to increase bone mass at that site [30]. Further, a number of studies have shown that bone will respond at a lower strain magnitude if loading differs from its usual pattern [31]. As in animals, strain rate and activities associated with an unusual strain distribution (for example, volleyball, squash, and gymnastics), rather than strain magnitude may provide the most important osteogenic stimulus for humans [32, 33].

Strain Cycles

Strain cycles denote the number of load repetitions that change bone dimensions at a given magnitude. Although a minimum number of loading cycles is required for a positive bone response, the number of strain cycles appears to be less important than strain magnitude or strain rate [18]. Rubin and Lanyon [27] showed that after only 36 consecutive loading cycles (a total of 72 seconds per day), further loading cycles did not stimulate further osteogenesis, suggesting a threshold above which additional loading cycles produce no additional bone formation. In a rat jumping study, 5 jumps per day generated the same osteogenic response as did 20, 50, and 100 jumps [34].

Loading in Various Activities

To demonstrate how theory translates into practice, consider the daily loading in three individuals: a nonathlete, a recreational athlete, and an elite gymnast. A nonathlete who walks about the office or climbs a few stairs achieves the threshold of adaptation after only a small number of loading cycles. A recreational athlete may benefit from additional strains of higher magnitude associated with jogging or running. The elite gymnast, who experiences extremely high strain magnitudes and strain rates in an unusual distribution, creates the ideal environment for mechanical adaptation of bone. Thus, generally speaking, the gymnast will have the strongest bones, and both the gymnast and the recreational athlete will have stronger bones than the nonathlete.

How Physical Activity Generates Loads on Bone

The magnitude of the physical force and the rate at which this force is applied are determined primarily by the movement conditions (velocity of the segments in motion, number of repetitions, muscular activities) and the boundary conditions (anthropometry, fitness level, performance surface, climate/weather, and type of shoe) [35, 36, 37].

Bone tissue is a viscoelastic and anisotropic material. This means that its elastic properties depend on the orientation of its various structural materials. It is therefore sensitive to loading direction, and its behavior varies depending on the direction of the applied load. In contrast, the elastic properties of isotropic materials such as rubber are the same regardless of the applied load.

The skeleton is subjected to forces produced by gravity (weight bearing), by muscles, and by other external factors. As a result of the gentle curvature of bones, they are subjected to a combination of axial compressive forces and bending forces. The long slender structures of the appendicular skeleton (the tibia for example) are designed to resist bending loads that can be applied in a number of ways. A good analogy is bending a narrow tree branch that is grasped at both ends. A convex surface is produced on one side and a concave surface on the other (figure 3.9). The material on the convex side of the branch experiences tensile (or stretching) strains while the material on the concave side experiences

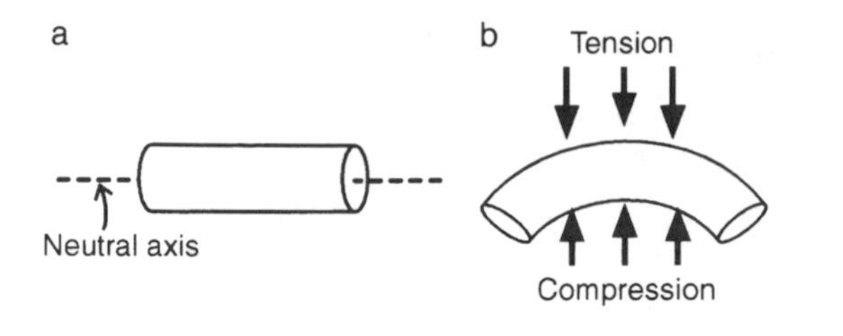

FIGURE 3.9 Longitudinal stresses on a tree branch with *(a)* no external force and *(b)* external bending force.

compressive strains. In addition to these bending forces, bone also experiences compressive forces from axial loading. Thus, even higher forces are created on the concave surface. The combined stresses on bone can be described as a sum of the stresses from axial compression and bending stresses.

In general, bone tissue can handle the greatest loads in a longitudinal direction (i.e., compressive forces), the direction of habitual loading, and lesser loads when they are applied across the bone surface (i.e., bending or shear forces). None of this is straightforward, as the degree of anisotropy varies with the anatomical region (cortical or trabecular bone) and with the magnitude and direction of the mechanical load [38].

Ground Reaction Forces

To study mechanical loading associated with weight-bearing physical activity in humans, it is necessary to measure ground reaction forces. Force, measured in Newtons (N), has traditionally been described as a push or a pull exerted by one body on another. It is characterized by point of application, magnitude, and direction. Any time a body exerts a force on a second body, the body exerting the force experiences a reaction force. For example, in walking, the foot exerts a force on the ground and the ground exerts a counterforce on the foot that is equal to the applied force and opposite in direction. These ground reaction forces (GRFs) can be measured on a force platform. The higher the impact of the exercise, the greater the ground reaction forces. For example, the GRFs associated with jumping are much higher than those associated with walking.

Ground reaction force data can be displayed as a force-time curve (figure 3.10), which illustrates the change in force over time. The integral of this curve (that is, the area under the curve), is impulse (Ns) or a change in momentum. For example, the larger the impulse generated against the floor from a vertical jump, the greater the change in momentum, and the higher the jump. Jumping can be divided into three phases for GRF analysis: takeoff, flight, and landing. For single jumps, the takeoff phase is the time before the flight phase; landing is the time after the flight phase. The flight phase consists of the aerial portion, while the landing phase begins at the point of contact to the point of lowest displacement.

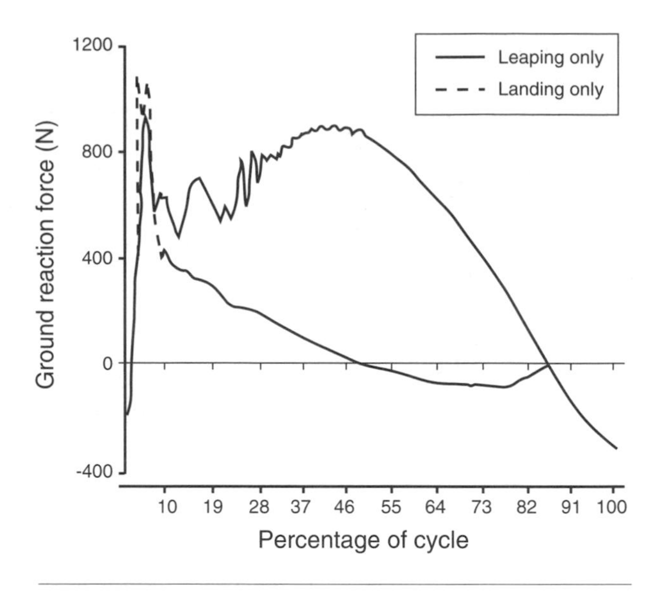

FIGURE 3.10 Typical ground reaction force curve for a countermovement jump.

The curve begins with the value of the vertical component equal to body mass. This stationary value, until about 0.1 sec, is followed by a period of unweighting, until about 0.4 sec, after which the force rises to its peak. The force then falls to zero as the jumper leaves the ground. There is then a large reaction force at the point of landing. A rigid landing causes a relatively large ground reaction force lasting for a relatively short time. On the other hand, by flexing the hip, knee, and ankle during landing, the time interval over which the landing force is absorbed increases and the magnitude of the force is reduced.

Ground reaction forces from typical daily activities help to model developing bone [14] and activities that elicit high ground reaction forces are known to be osteotrophic for adults [39, 40]. Ground reaction forces measured for two-foot plyometric jumps in children exceeded six times body weight [41]. In adults, jumping, running, and aerobics induced ground reaction forces three to six times body weight [42].

Without muscle activity GRFs are transmitted from the foot and along the lower limbs to the hips through a series of action-reaction forces (a system of three rigid links). However, forces applied to bone are primarily the result of muscular contraction [28]. Therefore, with the additional influence of muscles and lever arms, GRFs may reach 10 times body weight at certain skeletal sites [43].

To illustrate, implant forces at the hip from gravitational force and muscle tension were 2.5 to 3.0 times the ground reaction forces during takeoff from the ground in a jump [44]. Further, in another study in which a strain gauge was imbedded in hip prostheses, walking produced forces at the hip that were about twice the ground reaction forces [45]. Increases in strain are linearly associated with increased ground reaction forces [44]. This relationship indicates that GRFs can serve as a surrogate for strain and provides some indication of the kinds of activities that will stimulate a positive bone response.

Tensile, Compressive, and Shear Stress on the Lower Body

Physical forces produce a variety of stresses that can be described by their various components. When two forces are directed away from each other along a straight line they produce *tensile stress*. An excellent example of this is the patella, a non-weight-bearing bone that receives its mechanical stimuli entirely from the quadriceps activity pulling against the fixed tibial attachment of the patellar tendon [46]. *Compression* is produced when two forces are directed toward each other along a straight line. This causes bone to shorten and widen and to absorb maximal stress in a plane perpendicular to the compressive load. For example, the hip joint must absorb a compressive stress of approximately three to seven times body weight during walking [47, 48]. *Shear* stress is produced when two forces are directed parallel to each other but not along the same line, for example, upon braking from fast running when the foot is contacting the floor while body mass is still moving forward over the foot. This produces shear stresses in the tibia.

Bending occurs when the convex surface of bone is in tension and the concave surface is in compression. For example, bending occurs in the forearm during a biceps curl. *Torsion* refers to shear stresses occurring along the entire length of bone. For example, if during braking from running the upper body is turned, the tibia experiences torsional loading. It is not uncommon for one or more of these stresses to be acting on bone to produce complex loading patterns.

Forces at the Proximal Femur

To illustrate the complex nature of the forces acting on any bone, we examine forces at the proximal femur, a site of osteoporotic fracture. The pelvis rests on both femora like a supported beam during normal standing with body weight distributed equally over both hip joints. The proximal end of each femur acts as an eccentrically loaded cantilever undergoing vertical compression at the femoral head. This produces a bending moment in the femoral neck around the anatomical axis of the femoral shaft and a shear stress along the head and the neck of the femur [49] as shown in figure 3.11. A resisting moment within the neck counteracts the bending moment by setting up tension and compression stresses. Just as in the small branch, compressive forces act on the inferior portion of the femoral neck while tension stresses occur in the superior portion of the neck (figure 3.11).

Contraction of the powerful hip abductors (gluteus medius) counteracts body weight during normal stance. This muscle also produces a

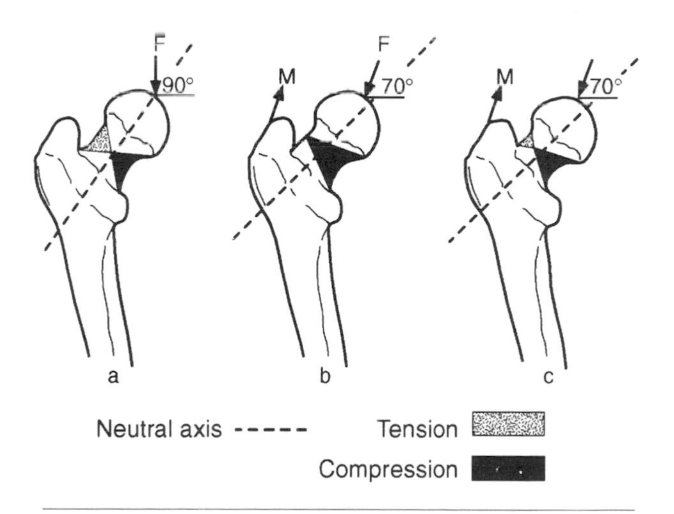

FIGURE 3.11 Tension and compression at the femoral neck. *(a)* Standing on both limbs, tension and compression are equally distributed around the neutral axis. *(b)* Standing on one limb, the tension component is eliminated by strong muscular force (M). *(c)* Standing on one limb, when muscular force is weak, relative tension is greater than in *b* but less than in *a*, where no muscle forces are applied.

compressive load on the superior aspect of the femoral neck to reduce tensile forces. As tensile forces are more likely to cause fractures, this muscular reduction in load also reduces the potential for injury in the femoral neck.

In general, muscles provide the necessary moment equilibrium in all joints. As importantly, they also compensate for, and reduce, the bending stress in bones while increasing the axial compressive load.

SUMMARY

- The skeleton is composed of a material (bone tissue) arranged in a structure (whole bones) that can withstand the demands of intense physical activity and adapt to changes in the loading environment while still being lightweight enough to permit locomotion at reasonable metabolic efficiency. Engineering principles can be applied to the skeleton.
- The *material* properties of bone depend on organic and inorganic compounds, which, between them, provide both resilience to bending and resistance to compression. When testing material properties of bone, stress refers to the force per unit area and strain refers to the change in material length per unit of length.
- *Structural* properties of bone refer to bone size, shape, cortical thickness, cross-sectional area, and trabecular architecture. Because bone is structured to optimize its function, different parts of the skeleton have very different structures. Cross-sectional moment of inertia refers to the ratio of the column diameter of the cylinder or beam to wall thickness. A higher ratio represents a stronger structure.
- We emphasize the differences between material and structural properties because both can be altered to change whole bone strength. In research, because DXA scan results are a common outcome measure, there is a tendency to focus on the bone mass component of bone strength as if they were synonymous. Bone strength, however, can be altered by adjusting bone mass, bone architecture and geometry, and bone size. This is an important consideration, as physical activity may act to strengthen bone mainly by the latter mechanisms, rather than merely by bone mass.
- Bone responds to loading. Wolff's Law states that whole bone will optimize structure to withstand functional loading and ensure the metabolic efficiency of locomotion. Specifically, Wolff's Law states that the function of those cells responsible for mechanically adaptive modeling and remodeling is to ensure that bone mass, geometry, and material properties are appropriate to the applied load.
- The mechanostat theory describes a control system in which a minimum effective strain (MES) is necessary for bone maintenance. In addition to strain magnitude, it appears that strain rate and unusual strain distribution play a vital role in stimulating osteogenesis. These factors appear to be more important than the number of strain cycles.
- To measure forces on bone in vivo requires measurement of ground reaction forces. Ground reaction forces from typical daily activities help to model developing bone [43]. In children and adults, running and jumping induce ground reaction forces three to six times body weight. With the additional influence of muscles and lever arms, these forces may reach 20 times body weight at certain skeletal sites [50].

References

1. Carter DR, Hayes WC. Bone compressive strength: The influence of density and strain rate. *Science* 1976;194:1174-1175.
2. Turner CH, Burr DB. Basic biomechanical measurements of bone: a tutorial. *Bone* 1993;14:595-608.
3. Mosekilde L. Vertebral structure and strength in vivo and in vitro. *Calcif Tissue Int* 1993;53:S121-S126.
4. Lotz JC, Hayes WC. The use of quantitative computed tomography to estimate risk of fracture of the hip from falls. *J Bone Joint Surg* 1990;72-A:689-700.
5. Johnston J, Slemenda C. Determinants of peak bone mass. *Osteoporosis Int* 1993;Suppl. 1:S54-55.
6. Heinonen A. *Exercise as an Osteogenic Stimulus*. Jyvaskyla: University of Jyvaskyla, 1997.
7. Adami S, Gatti D, Braga V, et al. Site-specific effects of strength training on structure and geometry of ultradistal radius in postmenopausal women. *J Bone Miner Res* 1999;14:120-124.
8. Järvinen TLN, Kannus P, Sievänen H. Have the DXA-based exercise studies seriously underestimated the effects of mechanical loading on bone? (Letter). *J Bone Miner Res* 1999;14:1634-1635.
9. Kimmel DB. A paradigm for skeletal strength homeostasis. *J Bone Miner Res* 1993;8:S515-S522.
10. Bloomfield S. Changes in musculoskeletal structure and function with prolonged bed rest. *Med Sci Sports Exerc* 1997,29:197-206.
11. Wolff J. *The law of bone transformation*. Berlin: Hirschwald, 1892.
12. Lanyon L. Using functional loading to influence bone mass and architecture. *Bone* 1996;18:37S-43S.
13. Rubin CT, Lanyon LE. Regulation of bone mass by mechanical strain magnitude. *Calcif Tissue Int* 1985;37:411-417.
14. Frost HM. Bone "Mass" and the "Mechanostat": A proposal. *Anat Rec* 1987;219:1-9.
15. Lanyon LE, Goodship AE, Pye CJ, et al. Mechanically adaptive bone remodeling. *J Biomech* 1982;15: 141-154.
16. Frost HM. The mechanostat: A proposed pathogenetic mechanism of osteoporoses and bone mass effects of mechanical and nonmechanical agents. *Bone Miner* 1987;2:73-85.
17. Oganov V, Grigor'ev A, Voronin L. Bone mineral density in cosmonauts after 4.5-6 month long flights aboard orbital station MIR. *Avik Ekolog Med* 1992;26:20-24.
18. Lanyon LE. Functional strain in bone tissue as an objective, and controlling stimulus for adaptive bone remodeling. *J Biomech* 1987;20:1083-1093.
19. Turner C. Homeostatic control of bone structure: an application of feedback theory. *Bone* 1991;12:203-217.
20. Frost H. Perspectives on a "paradigm shift" developing in skeletal science. *Calcif Tissue Int* 1995;56:1-4.
21. Reilly D, Burnstein A. The mechanical properties of cortical bone. *J Bone Joint Surg* 1974;56a:1001-1022.
22. Frost HM. Vital biomechanics: Proposed general concepts for skeletal adaptations for mechanical usage. *Calcif Tissue Int* 1988;42:145-156.
23. Milgrom C, Burr D, Fyhrie D. A comparison of the effect of shoes on human tibial axial strains recorded during dynamic loading. *Foot Ankle* 1998;19:85-90.
24. Burr D, Milgrom C, Fyhrie D. In vivo measurement of human tibial strains during vigorous activity. *Bone* 1996;18:405-410.
25. Whalen R, Carter D, Steele C. Influence of physical activity in the regulation of bone density. *J Biomech* 1988;21:825-837.
26. O'Connor PJ, Lanyon LE, Macfie H. The influence of strain rate on adaptive bone remodeling. *J Biomech* 1982;15:767-781.
27. Rubin CT, Lanyon LE. Regulation of bone formation by applied dynamic loads. *J Bone Joint Surg* 1984;66-A:397-402.
28. Martin RB, Burr DB. *Structure, function and adaptation of compact bone*. New York: Raven, 1989.
29. Umemura Y, Ishiko T, Tsujimoto H, et al. The effects of jump training on bone hypertrophy in young and old rats. *Int J Sports Med* 1995;16:364-367.
30. Lanyon LE. Relationship with estrogen of the mechanical adaptation process in bone. *Bone* 1996;18:375-435.
31. Lanyon LE. Functional strain as a determinant for bone remodeling. *Calcif Tissue Int* 1984;36:S56-S61.
32. Fehling PC, Alekel L, Clasey J, et al. A comparison of bone mineral densities among female athletes in impact loading and active loading sports. *Bone* 1995;17:205-210.
33. Heinonen A, Oja P, Kannus P, et al. Bone mineral density in female athletes representing sports with different loading characteristics of the skeleton. *Bone* 1995;17:197-203.
34. Umemura Y, Ishhiko T, Yamauchi T, et al. Five jumps per day increase bone mass and breaking force in rats. *J Bone Miner Res* 1997;12:1480-1485.
35. Lees A. Methods of impact force absorption when landing from a jump. *Eng Med* 1980;40:653-663.
36. Nigg B. Biomechanics, load analysis and sports injuries in the lower extremities. *Sports Med* 1985;2:367-379.
37. Ricard M, Veatch S. Effect of running speed and aerobic dance jump height on vertical ground reaction forces. *J Appl Biomech* 1994;10:14-27.

38. Brown T, Ferguson A. Mechanical property distributions in the cancellous bone of the proximal femur. *Acta Orthopedica Scandinavica* 1980;51:429-437.

39. Heinonen A, Kannus P, Sievänen H, et al. Randomised control trial of effect of high-impact exercise on selected risk factors of osteoporotic fractures. *Lancet* 1996;348:1343-1347.

40. Bassey EJ, Ramsdale SJ. Increase in femoral bone density in young women following high-impact exercise. *Osteoporos Int* 1994;4:72-75.

41. McKay H, Tsang G, Heinonen A, et al. Ground reaction forces in children jumping. *(submitted)* 2001.

42. Heinonen A, Sievänen H, Kannus P, et al. Effects of unilateral strength training and detraining on bone mineral mass and estimated mechanical characteristics of the upper limb bones in young women. *J Bone Miner Res* 1996;11:490-501.

43. Burdett R. Forces predicted at the ankle during running. *Med Sci Sport Exer* 1992;14:308-318.

44. Bassey EJ, Littlewood JJ, Taylor SJ. Relations between compressive axial forces in an instrumented massive femoral implant, ground reaction forces, and integrated electromyographs from vastus lateralis during various 'osteogenic' exercises. *J Biomech* 1997;30:213-223.

45. Bergmann G, Graichen F, Rohlmann A. Hip joint loading during walking and running, measured in two patients. *J Biomech* 1993;26:969-90.

46. Sievanen H, Heinonen A, Kannus P. Adaptation of bone to altered loading environment: a biomechanical approach using X-ray absorptiometric data from the patella of a young woman. *Bone* 1996;19:55-9.

47. Riegger C. *Mechanical properties of bone.* In: Gould J, Davies G, ed. Orthopedics and Sports Physiotherapy. St. Louis: C.V. Mosby Co., 1985: 3-49.

48. Nordin M, Frankel VH. *Basic biomechanics of the musculoskeletal system.* (12th ed.) Philadelphia: Lea and Febiger, 1989.

49. Singleton M, Leveau B. The hip joint: Structure, stability and stress: A review. *Phys Ther* 1975;55:957-973.

50. Heinonen A, Sievänen H, Kyröläinen H, et al. Mineral mass, bone size and estimated mechanical strength of lower limb in triple jumpers. *Bone* 2001; 12 (in press).

51. Forwood MR, Turner CH. Skeletal adaptations to mechanical usage: results from tibial loading studies in rats. *Bone* 1995; 17(4 Suppl):1975-2055.

52. Hayes WC, Bouxsein ML. *Biomechanics of cortical and trabecular bone: Implications for assessment of fracture risk.* In Mow VC, Hayes WC, ed. Basic orthopedic biomechanics 2nd edition. Philadelphia: Lippincott-Raven publishers, 1997:69-105.

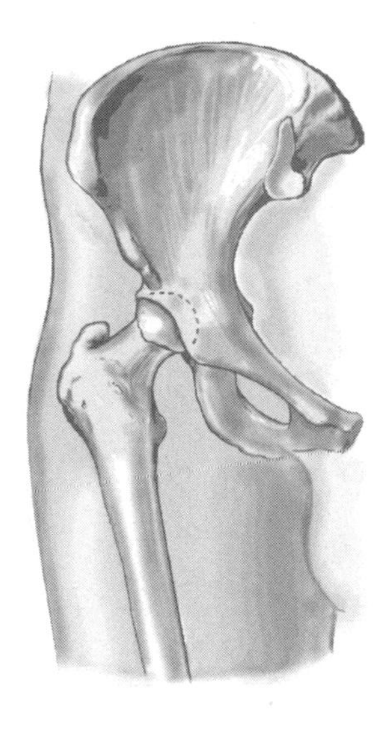

Chapter 4

Measuring the Properties of Bone

All branches of biological science and medicine depend on accurate and precise measurement of biological variables. Galileo said, "Measure what can be measured and make measurable what cannot." Bone has many interesting attributes that need to be measured for scientific and clinical reasons.

Areas of interest for scientists include bone's size and shape (anatomy), strength (biomechanics), metabolic activity (biochemistry and physiology), and the development and adaptation capabilities of bone. Some of these aspects can be examined in the laboratory using bones excised from animals or in live animal studies. The same properties of bone can also be measured, albeit with more limitations, in humans. Anatomy can be studied using various forms of imaging, mechanical strength can be estimated by applying the laws of physics to dimensions obtained using imaging techniques, and bone metabolism can be estimated from biochemical markers found in blood and urine.

Clinicians see bone differently. Because they see bone as tissue that provides structure and permits the human body to move, clinicians focus on preserving or reestablishing those functions. Imaging can play an important clinical role, for example, in predicting fracture risk in older adults (see chapter 15).

Clearly, imaging provides important information about bone. This chapter focuses on the three most commonly used methods of imaging bone for research and clinical purposes: dual energy X-ray absorptiometry (DXA), quantitative ultrasound (QUS), and quantitative computerized tomography (QCT). However, because bone is much more than a conglomerate of minerals, this chapter also describes measures of bone metabolism (bone biomarkers) and bone strength. The field of bone biomarkers is advancing rapidly, and this technology has provided a great deal of insight already into the mechanisms underpinning bone behavior. See table 4.1 for key definitions in the measurement of bone mineral.

TABLE 4.1

Key definitions in the measurement of bone mineral

Term	Definition
Accuracy	Generally defined as the DXA value expressed as a percentage of the specified BMC[a].
Precision	A measure of the degree of variability found in multiple measures of the same item. Usually expressed as a coefficient of variation.
Stability	A similar concept to precision, but generally used to imply a longer time frame than precision. Thus, may reflect daily measurements for 12 months. Usually expressed as a coefficient of variation.
Reliability	A nonspecific term for minimal measurement error.
Coefficient of variation	Standard deviation of repeated measures of a single case is divided by the mean and expressed as a percentage.

[a]Accuracy can also be defined against the ash content of cadaver specimens, but this is rarely done except in a research setting.

Dual Energy X-Ray Absorptiometry

The use of X-ray in the diagnosis of bone disease has a long history. Measurement of bone began with radiogrammetry, which was used to assess cortical thickness from a standard AP radiograph using a caliper or reticle. A relatively fast and precise quantitative evaluation of bone mineral only became possible with the development of absorptiometric techniques. In the last half century increasingly sophisticated systems evolved from single photon absorptiometry (SPA) to dual photon absorptiometry (DPA) to dual energy X-ray absorptiometry (DXA). In this section we focus on DXA, as it is currently the most widely used method for bone mineral assessment.

In the following sections we will outline fundamental concepts of DXA and define terms such as *bone mineral content* and *bone mineral density,* report the strengths and limitations of DXA as a tool in research and clinical medicine, review the literature that demonstrates the association between BMD and the risk of osteoporotic fracture as measured by DXA, and help the reader interpret the results of DXA scans.

Basic Concepts and Definitions

Dual energy X-ray absorptiometry (figure 4.1) came about as the result of the discovery that the radiation spectrum of gadolinium-153 required for bone absorptiometry could be simulated using an X-ray tube in specific ways. The use of photons produced by X-ray tubes permitted greater precision and shorter scanning times compared with the use of an isotope source [1]. DXA considers the body in two compartments, bone and nonbone, and uses X-ray beams of two distinct energy levels to distinguish the relative composition of each compartment.

DXA measurements are based on the decrease in photon energy of the photon beam as it passes through bone and nonmineralized soft tissue. The degree to which the X-ray beam is attenuated depends on the energy of the incident photons, the length of the path of the beam, and the material properties of the tissue being studied. The attenuation of the X-ray beam is measured at two

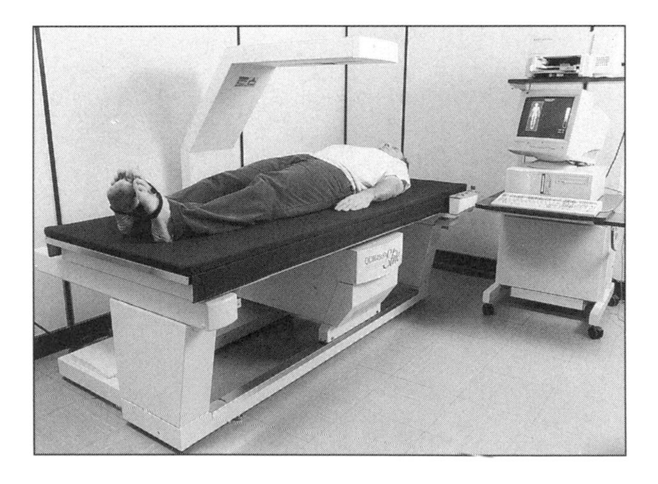

FIGURE 4.1 Dual energy X-ray absorptiometer, commonly known as a DXA scanner. This particular instrument is a Hologic 4500 scanner.

different locations [2]. One site is the region of interest (ROI), which is comprised of bone and soft tissue; this is where BMD will be determined. The other location is in close proximity to the bone region (termed the "baseline region") but contains only soft tissue. To determine BMD, the system assumes that the X-ray absorbed by the soft tissue component is the same in the bone and nonbone regions [3]. This includes the assumption that the ratio of lean mass to fat mass is constant [2]. Testing the accuracy of these assumptions proved that the inhomogeneity in fat distribution may result in measurement error—especially in the elderly [3].

DXA measures all bone within a given area but does not assess spatial orientation or alignment within this area. This is an extremely important concept. Although DXA provides a precise evaluation of bone mass, it does not assess the architecture of the region or material properties of bone. Bone mass, density, architecture, and material properties are all important components of bone strength.

Results of DXA measurement, bone mineral content (BMC, g), and areal bone mineral density (BMD, g/cm^2) vary at different skeletal sites. Most DXA manufacturers' systems assess clinically relevant regions including the posteroanterior (PA) lumbar spine (L1-L4 or L2-L4) (figure 4.2a), total proximal femur and its regions (femoral neck, trochanteric region, Ward's area, and, with Hologic systems, the intertrochanteric region (figure 4.2b and c), and the radius. Some densitometers are capable of measuring total body bone and soft tissue.

BMC (g)—Bone mineral content refers to total grams of bone mineral as hydroxyapatite within a measured bone region.

BMD (g/cm^2)—Bone mineral density refers to grams of bone mineral per unit of bone area scanned. BMD is an areal density, and volume determinations cannot be made with this technology.

The difference between BMC and BMD is often overlooked, but it is important, particularly when comparing bone of different size. For example, when comparing bone mineral in two athletes, the larger person will have greater BMC, even if her BMD is lower. If this is not considered, the researcher or clinician may draw incorrect conclusions about bone mineral. Also, children increase bone strength through changes in size as well as mineral content, so measuring only BMD in pediatric studies would fail to capture the full extent of changes that improve bone strength.

Advantages of DXA Measurement

The advantages of DXA include the low level of radiation exposure, its accuracy and precision,

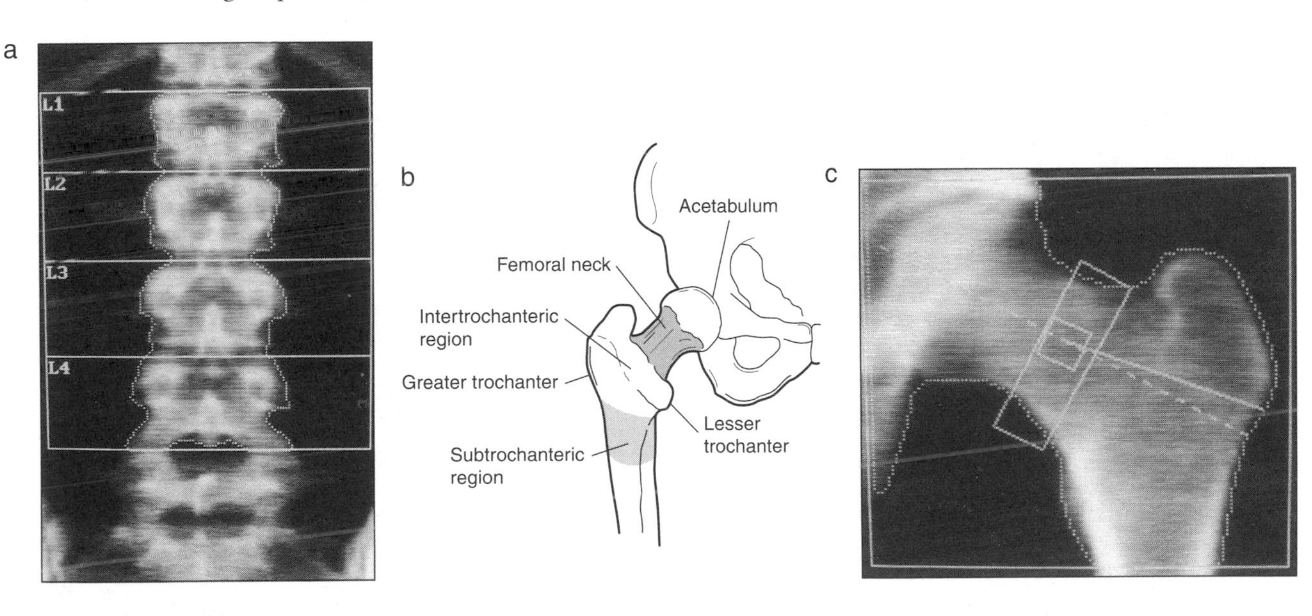

FIGURE 4.2 *(a)* DXA scan of the posteroanterior (PA) lumbar spine (L1–L4). *(b)* Gross anatomy of the proximal femur. *(c)* DXA scan of the proximal femur showing the location of subregions.

its ability to provide data regarding soft tissue composition as well as bone mineral, and its acceptability to patients. Also, it has the versatility to measure bone mineral at a range of sites in both the axial and appendicular skeleton.

Radiation exposure is an important issue in the use of imaging modalities. DXA scanning provides a safe means with which to assess both adults and children for single and repeat measurements. The radiation surface dose due to DXA is in the order of 10-30 μSv (1 μSv = 0.1 mrem). When this surface dose is corrected according to protocols of the International Commission on Radiological Protection, the effective dose equivalent for a PA spine scan or a hip scan is approximately 1 μSv. Natural radiation dose in western communities in the United States is about 3000 μSv per year. Thus, a single DXA examination is equivalent to about 0.03% of the natural annual dose of radiation. This is substantially lower than the radiation exposure resulting from a dental bite-wing X-ray (2% of annual dose), chest X-ray (4%), and mammogram (25%) [1].

DXA provides excellent accuracy, short-term precision, and long-term reliability (stability) when the operator maintains high-quality control standards. That said, any instrument is only as good as the quality assurance procedures that are implemented in the laboratory or clinic, and the care and experience of the user. Although DXA scans are fairly robust, patients must be positioned in a reproducible way to permit reproducible serial scan measurements and to permit comparison of patient data against company normative data. Similarly, technicians must standardize the analysis procedures in every instance using the same region of interest.

Quality control (QC) procedures include system calibration at installation, daily quality control procedures, and ongoing maintenance and adjustments following service and repair. A systems technician routinely calibrates the system at installation to minimize variability among same-manufacturer instruments [4]. In addition, the user should perform quality control procedures. This normally entails scanning an anthropomorphic phantom daily or before each use. QC results (figure 4.3) should be studied prior to subject or patient data acquisition to certify the stability of the instrument [4, 5].

The accuracy of BMC measurement is generally defined as the DXA value expressed as a percentage of the specified BMC [6]. Accuracy can also be defined against the ash content of cadaver specimens [7]. BMC as measured by DXA correlated well with, but systematically underestimated, the weight of cadaveric specimens by about 5-9% [7-10].

Although clinical and research applications of DXA, such as fracture risk estimation, are based on the DXA measurements (i.e., accuracy against the unique phantom) rather than the biological mineral content of bone itself (accuracy against ash weighing), the main clinical problem is the *variable* accuracy among instrument types. This means that until now, results obtained on the common brands of densitometers could not be compared readily. However, manufacturers are collaborating to minimize this problem by providing "standard BMD" results (sBMD) in addition to their own readings (see the following section, "Limitations of DXA Measurement").

Short-term precision and long-term stability are generally expressed as the coefficient of variation (CV) of a number of measurements (short-term or long-term, respectively). Although these vary among manufacturers, the in vivo precision of the examination of the PA lumbar spine by DXA is 0.5-1.5% [1, 8-10]. The precision of BMD at the proximal femur is between 1% and 2% [11]. Longitudinal precision of instruments from one manufacturer at multiple centers was 1.1% for measurements of the femoral neck, in vivo [10].

In addition to measuring bone mineral, DXA total body scans measure lean mass and fat mass [11, 12]. To determine the percentage of fat in the soft tissue region, a standard curve with known proportions of fat mass and fat-free mass is plotted as a function of R_{st}. R_{st} is the averaged ratio of beam attenuation at low and high energy levels [13]. This ratio is determined by measuring the attenuation pixel by pixel, for each photon energy in the soft tissue region [3] and comparing it to known calibration standards [15]. While total body soft tissue can be measured with excellent precision, error in fat mass estimates increased when the anterior-posterior depth of the subject was greater than 12 centimeters [14]. Large amounts of body fat have also been shown to decrease the accuracy of bone mineral measurements [13,14]. Finally, and as mentioned previously, fat and lean mass cannot be assessed for pixels containing bone [15]. Total mass can be calculated for bone "points"; however, the contributions of fat and lean must be estimated from tissue measured near the bone

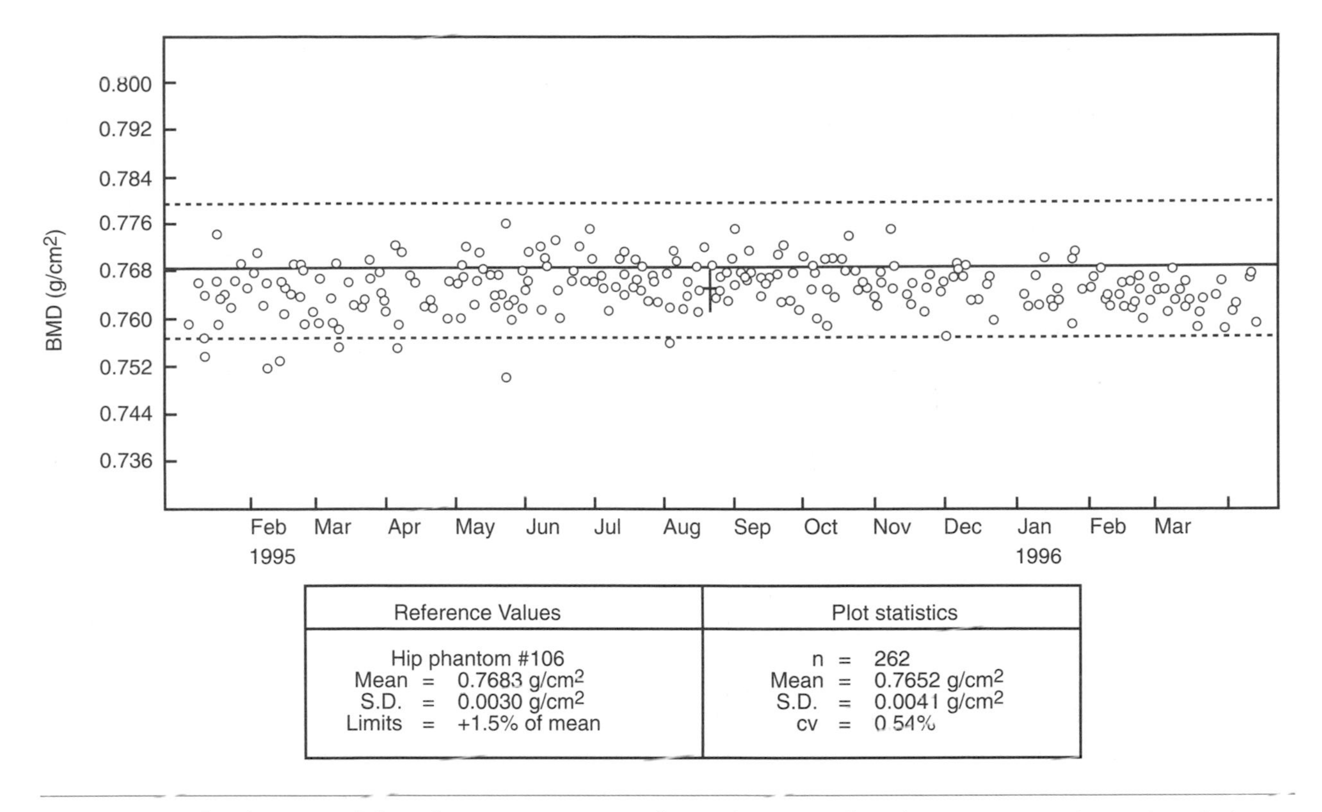

Reference Values	Plot statistics
Hip phantom #106 Mean = 0.7683 g/cm2 S.D. = 0.0030 g/cm2 Limits = +1.5% of mean	n = 262 Mean = 0.7652 g/cm2 S.D. = 0.0041 g/cm2 cv = 0.54%

FIGURE 4.3 Plot showing stability of DXA measurement of an anthropomorphic phantom over a 12-month period.

[15]. These values are then interpolated linearly across those regions containing bone.

Soft tissue estimation is a useful research capability, as both lean mass and fat mass are determinants of bone mineral [16, 17] (see chapter 6) as well as being of interest in their own right. Soft tissue composition as measured by DXA correlates well with results obtained by other methods such as underwater weighing [12, 18].

Patient acceptability of DXA is high as it requires no special preparation and is relatively cheap, painless, and rapid. Scan acquisition time varies among instruments and manufacturers. Older DXA instruments that use single-beam or "pencil-beam" technology, such as the Hologic QDR-1000W, perform a total body scan in about 20 minutes. Newer instruments, however, use multiarray or "fan-beam" technology and perform a total body examination in approximately five minutes and lumbar spine and proximal femur scans in less than one minute.

Clinicians and researchers like DXA because the instruments are widely and relatively readily available. DXA results are numerical and do not require subjective evaluation. Nevertheless, the numerical data require careful interpretation [5] (see "Interpreting DXA Scan Data" on page 41).

Limitations of DXA Measurement

The limitations of DXA include those that are inherent to every densitometer. DXA provides no measure of bone architecture. It assesses both trabecular and cortical bone but does not differentiate between them. In addition, DXA measures calcification unrelated to bone strength and provides a two-dimensional estimate of bone's three-dimensional structure.

DXA provides no measure of bone microarchitecture (trabecular number, connectivity, and orientation) [19], yet osteoporosis involves a change in bone architecture as well as BMD [20]. For example, if two discontinuous ends of a bony trabeculum gained mineral thickness as a result of therapy, DXA would interpret this as an improvement in bone mineral, despite there being no increase in bone strength (figure 4.4).

DXA scans cannot differentiate between cortical and trabecular bone, and most measure both types even though, generally speaking, the more metabolically active trabecular bone is of particular interest. Because QCT (described later) can distinguish between trabecular and cortical bone, it has greater sensitivity for detecting bone loss than does DXA [21, 22].

Another disadvantage of DXA is that it interprets calcified tissue, such as that lining the aorta in elderly people or osteophytes on vertebral bodies, as bone mineral and thus overestimates bone mineral, particularly when measuring spinal BMD in elderly patients. Software and hardware have been developed to overcome this and permit a lateral DXA scan, but this strategy was not wholly successful and is generally not used in clinical practice. A problem with both AP and lateral DXA scans is that vertebral fractures (arising from extremely low BMD) cause spurious elevation of the BMD result.

Another important limitation of DXA is that it measures both BMC and the area of bone scanned. BMC itself is highly dependent on bone size and therefore is not the most useful measure of bone mineral. The quotient of BMC and area is a more useful measure because it partially corrects for the effect of size [23], although it fails to assess the skeleton in three dimensions. If two bones are made of precisely the same material, the larger one will be measured to have greater BMD by DXA (figure 4.5).

Density is defined in physics and in common usage as mass per unit volume, but DXA measures mass per unit *area*. Thus, the term *bone mineral density* for the quotient of BMC and area is literally not correct. The quotient of BMC and area is more appropriately described as *areal* BMD. The terms *areal BMD* and *areal density* are increasingly used in technical papers to acknowledge the deficiency of this less-than-ideal, but entrenched, "bone density" term.

To overcome the problem of areal BMD measures and their inherent limitations, Dennis Carter and colleagues [24] derived equations that attempted to better correct for size differences provided by BMD. Termed "bone mineral apparent density" (BMAD) [24], this measure has not achieved popular use as it has neither the precision of BMD [25] nor has it been validated as a predictor of fracture risk.

DXA as Predictor of Bone Strength and Fracture Risk

As discussed in chapter 3, bone mass, bone architecture, and bone material properties all contribute to bone strength as do bone geometry and the ratio of cortical to trabecular bone at any site. Despite these various inputs into bone strength, bone mass as measured by DXA explains up to 90% of the variance in breaking strength of excised bone [32]. To provide a ballpark sense of how well DXA predicts fracture, we share John Kanis' comparison, which shows that DXA predicts fracture better than blood cholesterol levels predict heart attacks [33]. However, DXA measurements do not reflect structural changes due to aging, growth, or mechanical loading.

To improve prediction even further, engineering principles have been brought into play in an attempt to account for bone geometry and architecture in overall bone strength (table 4.2). Although DXA does not assess bone strength directly, it is possible to use it to estimate CSMI [34]. Cross-sectional moments of inertia and modulus of elasticity can be estimated using BMC and bone width from DXA scans, as discussed in chapter 3. A limitation of these calculations is that assumptions must be made regarding bone shape and structure. Computer software has been devel-

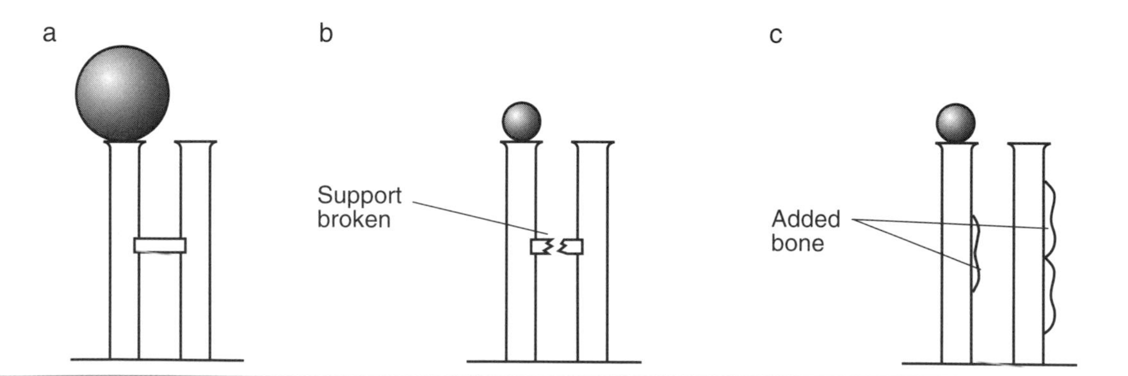

FIGURE 4.4 Schematic representation of bone loading and structural support. *(a)* A single connected strut permits a trabeculum to bear four times more weight than it could if the strut were not connected. *(b)* However, an intervention that added mineral to the accessory strut *(c)*, but did not bridge the gap, would provide no additional strength, despite an increase in bone mineral measured by DXA.

	Block A (1 cm)	Block B (2 cm)
Total material density (g/cm³)	1 g/cm³	1 g/cm³
Total material (bone mineral content) (g)	1 gram	8 grams
Bone area (cm²)	1 cm²	4 cm²
Areal bone mineral density (g/cm²)	1 g/cm²	2 g/cm²

FIGURE 4.5 Areal BMD by DXA is greatly size dependent. Even though the two blocks are made of the same material, the large block has twice the areal BMD as measured by DXA.

oped to derive geometric properties of the hip from the digital DXA data using a single-plane bending stress analysis to provide a hip structural analysis (HSA). Although HSA requires exact positioning of the hip, it predicted breaking strength at the mid-femoral neck better than BMD in excised femur [92]. While much of this research is preliminary, it is extremely important because it acknowledges that bone strength cannot be measured accurately by DXA alone.

Cross-sectional and longitudinal studies have shown that DXA measurement of BMD predicts fracture risk at the group level [35-38]. Depending on the skeletal site, fracture risk has been shown to increase approximately 1.5 to 2.0 times for every 1.0 standard deviation that an individual's BMD by DXA is below the mean [35-41]. Specific examples of BMD and relative risk of fracture (in age-matched people) are shown in table 4.3.

TABLE 4.2

Bone measurements related to fracture risk
• Bone mineral
• Cortical thickness
• Trabecular structure
• External geometry
• Strength estimates (e.g., cross-sectional moment of inertia)

Interpreting DXA Scan Data

At first glance, DXA scan reports (figure 4.6) can seem difficult to interpret, but that need not be the case. Three methods are commonly used to report BMD. The most direct method provides the unadjusted score, in g/cm², as assessed by DXA. As age is a major predictor of BMD, this unadjusted score is useful for comparing subjects of the same age and sex, or comparing the results obtained to age and sex norms. In research, unadjusted BMD score can be used to compare groups that are matched by age and sex. As clinicians would need to remember the normal values for each age and sex group to use the unadjusted score, it is not the preferred form for clinical interpretation of DXA scans.

A BMD z (or standard) score is defined as the deviation of the BMD raw score (in this case, the BMD in g/cm²) from the mean BMD result for the age- and sex-matched controls divided by the standard deviation of the age-matched controls [19]. Mathematically this would be written

$$\text{z score} = \frac{(\text{BMD}) - (\text{Age-specific mean BMD})}{\text{Standard deviation of BMD of age-matched controls}}$$

TABLE 4.3

Relative risk of fracture at spine, hip, and forearm with various BMD scores			
BMD t score	**Spine**	**Hip**	**Forearm**
t score = 0 (normal)	1 (i.e., no increased risk)	1 (i.e., no increased risk)	1 (i.e., no increased risk)
t score = –1	1.9 times greater risk	2.4 times greater risk	2.7 times greater risk
t score = –2	3.8 times greater risk	4.8 times greater risk	5.4 times greater risk

The z score principle is applied to a range of biological and statistical methods.

When interpreting a DXA scan result (figure 4.6), you will come across a t score (or T score) as well as a z score. This score standardizes the patient's raw-score BMD according to the World Health Organization (WHO) definitions of osteoporosis (t score < –2.5) and osteopenia (–2.5 < t score < -1.0) [42]. T scores use the same standardizing principle as z scores, but the raw BMD score is compared with mean peak bone density of 25- to 35-year-old healthy, sex-matched individuals, not age-matched individuals. If osteoporosis were defined only on the basis of age-matched data, the implication would be that the incidence of osteoporosis did not increase with age, which is not the case [42, 43]. In clinical practice, t scores of less than –2.5 are widely used as an indication to start pharmacotherapy for osteoporosis.

You may wonder how the WHO study group decided on a t score of –2.5 to diagnose osteoporosis. As BMD is normally distributed, a t score of –2.5 means the patient's BMD is 2.5 standard deviations below the mean BMD of 25- to 35-year-olds. Probability theory indicates that this represents the lowest 2% of the population at that age. With aging, the proportion of the population that has a t score of less than –2.5 increases, but it still represents a relatively small proportion of the population. Prospective studies have shown that those with such low t scores are at greater fracture risk. Note that other techniques of measurement (i.e., ultrasound) lack these norms, which limits their usefulness in predicting fracture. We believe that there is currently no ideal way of expressing the data numerically. For this reason people who interpret the DXA results to patients must understand the concepts of relative risk and the limitations of the technology.

Quantitative Ultrasound

Recent years have seen an increasing interest in the use of quantitative ultrasound for the assessment of skeletal status (figure 4.7). This appears to be due to the increasing interest in bone research overall and the fact that ultrasound has the potential to provide information about bone architecture, which may be a determinant of fracture risk. In addition, QUS is cheaper than DXA, instruments are available for field testing (and thus attractive to researchers), and, significantly, the test is without radiation.

The fact that QUS eliminates exposure to ionizing radiation makes it especially attractive to researchers conducting pediatric studies that assess bone. Although commercial manufacturers of QUS have not yet developed a system specifically for children, some systems (such as the Lunar Achilles) now have variable-sized "shims" available to accommodate feet of different sizes and shapes.

It is important to note that devices currently available vary considerably with respect to the site measured and the physical principles inherent in their operation [44]. Although the calcaneus, patella, radius, and finger have all been used for QUS measurement, the calcaneus is most commonly assessed [45].

Commercially available calcaneal ultrasound systems use immersion/wet (Lunar Achilles) or contact/dry (Hologic Sahara, CUBA, and EBSE Scientific QUS-2) technology. The basic mode of operation is somewhat similar among systems.

Two ultrasound transducers (a transmitter and receiver) are positioned on each side of the tissue to be measured (e.g., the calcaneus). The ultrasound transducers measure primarily one type of bone, depending on the site. At the patella and calcaneum, trabecular bone is predominantly measured, whereas cortical bone predominates at the tibia. Outcome measures include

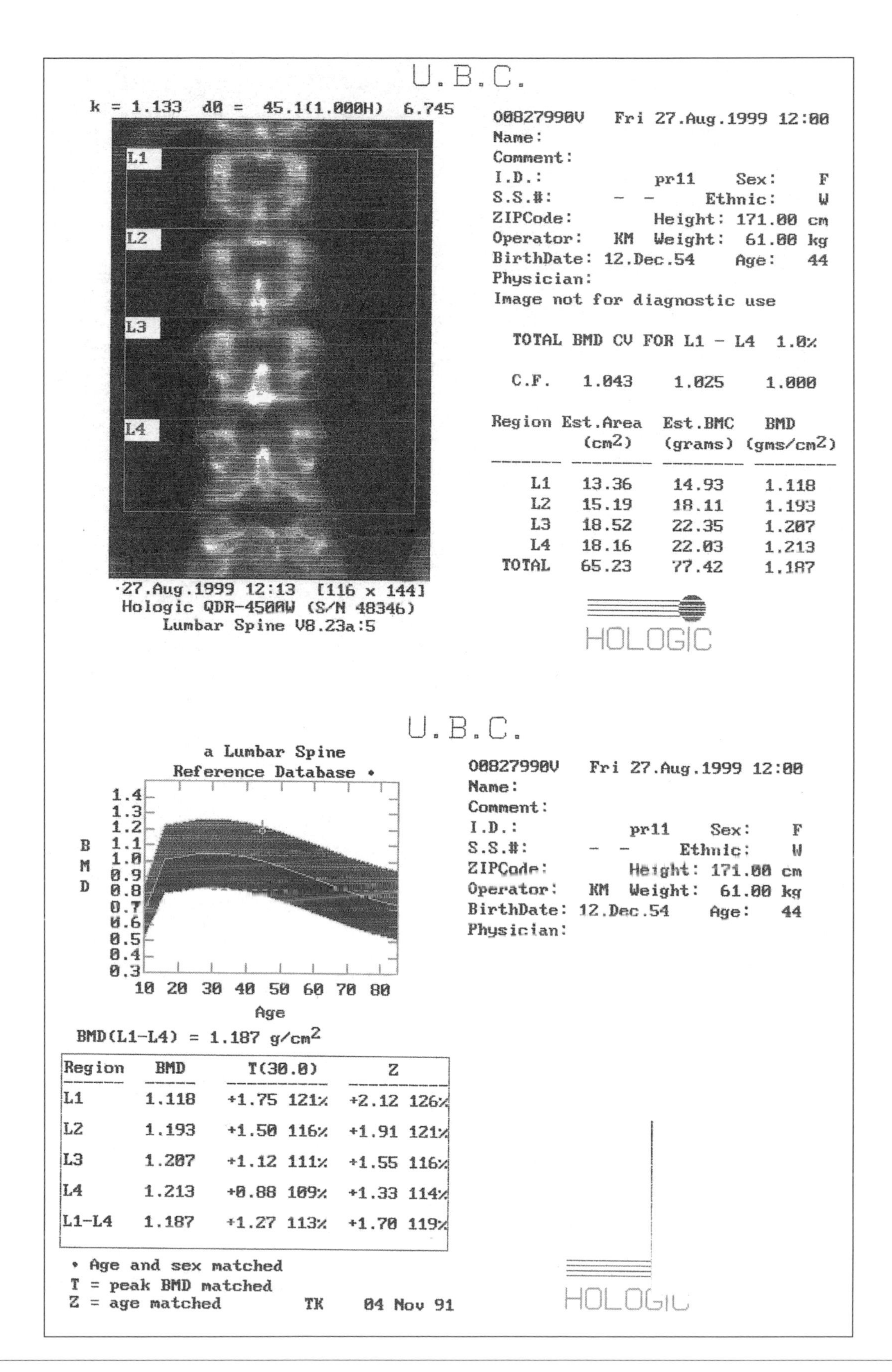

U.B.C.

k = 1.133 d0 = 45.1(1.000H) 6.745

L1
L2
L3
L4

·27.Aug.1999 12:13 [116 x 144]
Hologic QDR-4500W (S/N 48346)
Lumbar Spine V8.23a:5

00827990V Fri 27.Aug.1999 12:00
Name:
Comment:
I.D.: pr11 Sex: F
S.S.#: - - Ethnic: W
ZIPCode: Height: 171.00 cm
Operator: KM Weight: 61.00 kg
BirthDate: 12.Dec.54 Age: 44
Physician:
Image not for diagnostic use

TOTAL BMD CV FOR L1 - L4 1.0%

C.F. 1.043 1.025 1.000

Region	Est.Area (cm2)	Est.BMC (grams)	BMD (gms/cm2)
L1	13.36	14.93	1.118
L2	15.19	18.11	1.193
L3	18.52	22.35	1.207
L4	18.16	22.03	1.213
TOTAL	65.23	77.42	1.187

HOLOGIC

U.B.C.

a Lumbar Spine
Reference Database •

B M D: 1.4, 1.3, 1.2, 1.1, 1.0, 0.9, 0.8, 0.7, 0.6, 0.5, 0.4, 0.3
Age: 10 20 30 40 50 60 70 80

00827990V Fri 27.Aug.1999 12:00
Name:
Comment:
I.D.: pr11 Sex: F
S.S.#: - - Ethnic: W
ZIPCode: Height: 171.00 cm
Operator: KM Weight: 61.00 kg
BirthDate: 12.Dec.54 Age: 44
Physician:

BMD(L1-L4) = 1.187 g/cm2

Region	BMD	T(30.0)		Z	
L1	1.118	+1.75	121%	+2.12	126%
L2	1.193	+1.50	116%	+1.91	121%
L3	1.207	+1.12	111%	+1.55	116%
L4	1.213	+0.88	109%	+1.33	114%
L1-L4	1.187	+1.27	113%	+1.70	119%

• Age and sex matched
T = peak BMD matched
Z = age matched TK 04 Nov 91

HOLOGIC

FIGURE 4.6 DXA scan printout consisting of an image of the lumbar spine as measured; raw scores of BMC, area, and BMD; and a normal range that the result can be compared against.

Finer Point: Limitations of Using More Than One DXA Scanner

Limitations inherent to using more than one DXA densitometer are relevant to both clinical practice (e.g., when comparing sequential results in a patient) and research study (e.g., when a study spans several research centers).

Until recently, BMD results for the same person measured on instruments from different manufacturers may have differed by up to 20% [1]. In an attempt to allow comparisons among manufacturers, the three most common instrument manufacturers (Hologic, Lunar, and Norland) established cross-calibration factors via the International Standardization Committee [26]. The standardized BMD (sBMD) score permits comparison of results obtained on different manufacturers' instruments. The sBMD is expressed in mg/cm^2 to avoid confusion with manufacturer-specific BMD expressed in g/cm^2 [26-28]. Essentially, the BMD result obtained on each manufacturer's densitometer can be converted to an sBMD score. Although the method will undoubtedly need some fine-tuning [29] and may not be suitable for some types of research, it should improve the comparability of BMD data collected on different instruments. Although conversion formulas [26] that reduce the range of error among systems have been published, it is recommended that, if possible, the same software version and the same instrument be used to assess an individual over time.

Similarly, different models of densitometers by the same manufacturer may yield different results [30]. The older Hologic QDR-1000W densitometer used single-beam X-ray mode, whereas newer Hologic densitometers have both single-beam and the more rapid fan-beam (QDR-2000) or fan-beam-only capabilities (QDR-4500). Results obtained from the QDR-2000 using single-beam mode correlate better with the QDR-1000W than do results obtained using fan-beam mode [26, 30].

The measurement of soft tissue composition by DXA is very much instrument, and software, dependent. Results may vary substantially among different models of densitometer from the same manufacturer [30, 31] and even among different versions of software [13].

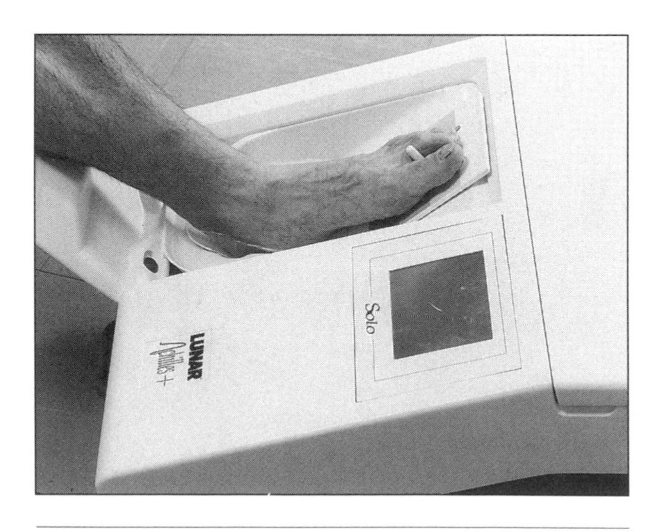

FIGURE 4.7 Heel ultrasound instrument.

ultrasound transmission velocity (UTV) and the frequency dependency of the attenuation of ultrasound signal, called broadband ultrasound attenuation (BUA). Ultrasound transmission velocity is usually expressed as the quotient of the time taken to pass through the bone in question and the dimension that it passed through. Thus, its units are meters per second (m/s). If the dimension includes soft tissue surrounding bone, the measure is called speed of sound (SOS); if it includes bone only (i.e., in laboratory setting), it is called ultrasound velocity through bone (UVB). The precision error of UTV measurements is about 0.3-1.5% [19].

Exactly what ultrasound measures presents a puzzle for bone researchers. It is believed that the parameters measured by ultrasound, BUA, and UTV are influenced by bone density and by bone microarchitecture, which includes trabecular number, connectivity, and orientation. Ultrasound velocity was correlated with bone specific gravity and porosity [46]. Velocity is largely influenced by trabecular separation [47]. BUA is dependent on trabecular orientation as it is as much as 50% higher along the axis parallel to the principal orientation of the trabeculae [19]. If QUS can detect architectural change, the modality may provide information about fracture risk in addition to that provided by DXA.

Prospective studies have now clearly demonstrated that calcaneal QUS measurements are predictive of fracture risk in postmenopausal women [48, 49]. In a prospective study of 710 older adults (mean age ± SD: 83 ± 6 yr), both BUA and SOS of the calcaneus predicted risk for hip fracture as well as any fracture [50]. Two papers also provide evidence that QUS (obtained at various sites) can identify the patient with prevalent vertebral fractures with the same effectiveness as conventional bone mass measurements at the spine, hip, or forearm [51, 52]. Interestingly, when researchers have used models combining both DXA and QUS measures, both SOS and BUA maintain their ability to identify patients with vertebral fractures, independently of the effect of BMD. This suggests that QUS measures characteristics of bone strength that are potentially independent of density [19].

Laboratory studies with human cadaveric specimens have shown that BUA and SOS of the calcaneus correlate strongly with proximal femur strength tested to failure [53]. The potential role of calcaneal as well as tibial SOS for the evaluation of skeletal strength of the proximal femur was also tested in human cadaveric specimens and compared with femoral bone mineral density [44]. Tibial and calcaneal SOS and calcaneal BUA were strongly correlated with tibial BMD ($r^2 = 0.55$-0.69, $p < 0.0001$). While tibial and calcaneal BMD, calcaneal BUA, and SOS were strongly correlated with femoral strength ($r^2 = 0.60$-0.78), tibial SOS was only weakly associated with femoral BMD and strength [44].

Although recent studies have begun to clarify the role of QUS for the assessment of the architectural mechanical quality of trabecular bone [54], these imaging systems display a number of inconsistencies [55]. Some researchers suggest that based on the lack of precision [55] and the high incidence of false positive (for osteoporosis) results, the role for QUS in clinical practice should be to identify individuals that require more extensive assessment [56]. Further evaluation of QUS is needed for the full utility of this assessment modality to be clearly understood.

Quantitative Computed Tomography

Quantitative computed tomography (QCT) (figure 4.8) determines the volumetric density (mg/cm^3) of trabecular or cortical bone [19]. However, even with QCT, the cross-sectional information refers to anatomical bone dimensions and includes marrow spaces. Calculation of true density requires bone volume with the marrow excluded. The term *apparent density* is therefore often used for QCT measurements.

true density—The density of a section of cortical bone excluding any marrow cavity (figure 4.9).

The primary use of QCT has been to measure trabecular bone mineral density in the vertebral bodies. Because vertebral fractures are common, this site is of particular interest. The advantages of QCT for measuring the spine are its precise three-dimensional anatomic localization, its ability to distinguish cancellous from trabecular bone, and its ability to exclude bony artifacts such as

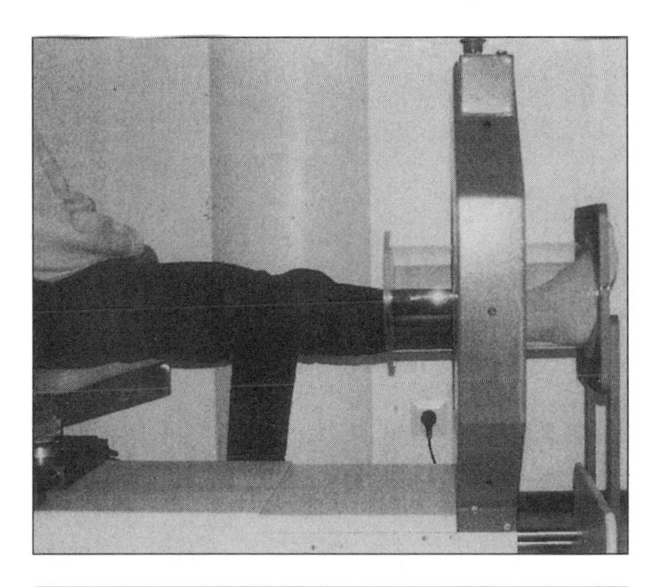

FIGURE 4.8 Quantitative computed tomography (QCT) scanner.

Reproduced from *J Bone Miner Res* 1998; 13:871-882 with permission of the American Society for Bone and Mineral Research.

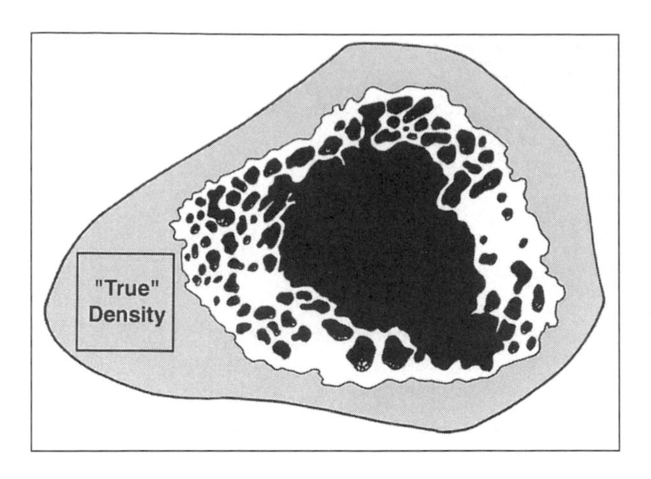

FIGURE 4.9 True density refers to cortical bone excluding any marrow cavity.

aortic calcification from measurement [57]. Trabecular bone has a greater metabolic rate than cortical bone and responds more quickly to changes in the environment (e.g., to menopausal hormone changes). Because of this, QCT has been used to assess vertebral fracture risk, to monitor age-related bone loss, and to follow up patients with clinical conditions such as osteoporosis and other metabolic bone diseases.

In the only study prospectively evaluating spinal QCT and fracture risk, a QCT result two standard deviations below the normative value predicted future vertebral fracture 40% better than did spinal DPA [41]. Decrements between normal subjects and patients with vertebral fractures are significantly higher when measured by QCT than when measured by DXA. Also, QCT is generally superior for vertebral fracture discrimination. Several studies have shown that postmenopausal spinal bone loss is almost 2% per year as measured by QCT, whereas it appears to be about 1% as measured by lateral DXA, and 0.5% as measured by standard PA DXA [19, 21].

QCT has some limitations when compared with DXA. Because of the complex architecture of the femoral neck and the dramatic three-dimensional variation in its density, QCT is not yet commonly used to measure this site. QCT has relatively poor precision, in vivo, with errors of 2-4% and accuracy errors of 5-15% [19]. Finally, for assessment of the spine, a single-energy low-dose QCT setting results in an organ dose of 200 mrem [57] as compared to an effective dose equivalent of 1.6 mrem for a PA lumbar spine DXA scan. Given that studies of apparently healthy populations of children often require repeat measurements, this method may be untenable for these populations.

When investigating physical activity and bone health, there may be a role for peripheral quantitative computed tomography or pQCT. This technique employs small purpose-built scanners to measure BMC and BMD of the peripheral skeleton using an X-ray source. As with vertebral QCT, pQCT measures volumetric tissue density of appendicular bone without the superimposition of any other tissues [19] (figure 4.10). Peripheral QCT is easy to use; it can assess cortical

a
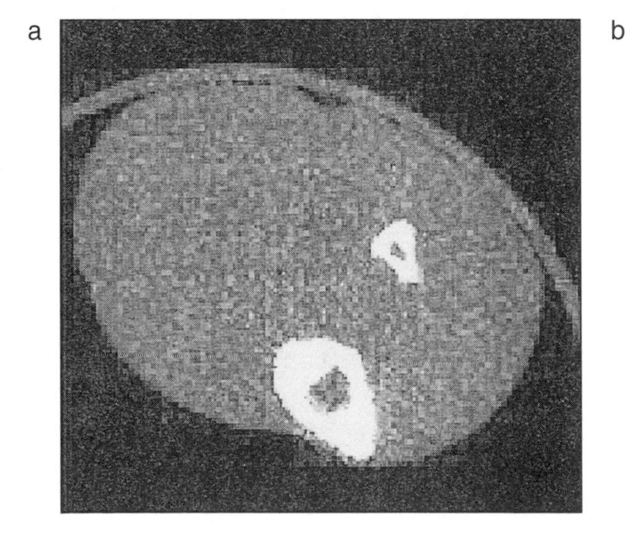

b
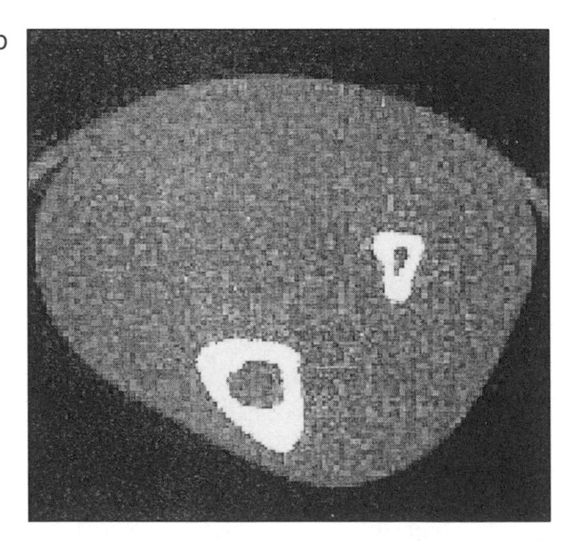

FIGURE 4.10 Image of the distal tibia obtained using pQCT in *(a)* an elite triple jumper and *(b)* a control athlete. Note the greater cortical bone dimensions in the triple jumper.

and trabecular bone separately; and it measures BMD, BMC, and the axial cross-sectional area.

When using pQCT, precision errors (coefficient of variation – CV) in short-term studies, in vitro, range from 0.5% for trabecular density to 2.2% for cortical density [93]. pQCT accuracy, in vitro, against ashing was about 2%. An in vivo performance evaluation of one pQCT system (Norland/Statec XCT 3000) showed that this instrument was precise, linear, and stable [93]. In one study, only the radial cortical area and BMC significantly distinguished between women with non-traumatic vertebral fractures and healthy postmenopausal women; these two parameters also had the highest age-adjusted odds ratio for fracture risk. These data suggest that pQCT of cortical rather than trabecular bone at the radius may have the greatest diagnostic sensitivity. However, in another report the radius displayed the poorest precision, ranging from 7.6% for trabecular area to 6.5% for cortical density to 2.2% for trabecular density [93].

Exercise researchers are showing a surge of interest in pQCT at present as investigators acknowledge that bone strength is influenced by the type of bone present and its size and shape, as well as bone mass. This was clearly demonstrated in a controlled exercise intervention study of postmenopausal women in which DXA could not detect any difference between subject and control groups, but pQCT revealed that exercising subjects had increased cross-sectional area of cortical bone and cortical BMC [58]. Järvinen and colleagues from the UKK Institute in Finland also concluded that pQCT may be more sensitive to bone change than DXA as a result of an exercise study in rats [59]. They recently made a strong case for inclusion of pQCT in all bone studies in which structural alterations may be relevant [60].

Measuring Bone Metabolism: Biochemical Markers

Throughout the adult skeleton teams of osteoclasts and osteoblasts repeat the process of bone remodeling in perpetuity. This process of remodeling is accelerated in certain conditions, such as in women after menopause, resulting in more rapid bone loss.

Numerous biochemical bone markers (biomarkers) provide information about bone turnover. Biochemical markers, considered "global images" of remodeling, primarily represent the activity of highly metabolic cancellous bone [61]. They provide only a snapshot of the overall status of skeletal remodeling. They do not provide an estimate of bone mass or distinguish between cortical or trabecular bone remodeling or identify the geographic locations where remodeling is taking place. Biomarkers have been used widely to investigate metabolic bone disease and less widely to examine the relationship between bone and exercise. Eventually, however, they may provide crucial information about the mechanisms related to the maintenance of bone's physical properties.

Lindsay [61] and Garnero [62] provide useful updates on the status of biomarkers. Although biomarkers are playing an increasingly important role in understanding bone physiology, some are concerned about their clinical utility. This concern centers around the considerable variability in the assessments, which are influenced by technical factors and, of greater concern, biological factors. The magnitude of the error can range from 5-9% for markers of formation to 15-47% for markers of bone resorption. The normal range of biologic variability within individuals (due to diurnal variation), and between individuals may be as much as 30% on either side of the mean [61]. An in-depth discussion of all of the biochemical markers currently available is well beyond the scope of this book. However, we will introduce a few commonly used markers of formation and resorption. A list of other commercially available biomarkers is provided in table 4.4.

Bone Formation Markers

Serum osteocalcin and skeletal alkaline phosphatase are the most familiar bone formation markers. Osteocalcin (previously called bone-Gla protein, BGP) is a noncollagenous protein synthesized and released by osteoblasts. Approximately 1% of the organic matrix of bone is osteocalcin. The fraction of newly synthesized osteocalcin that is released into the circulation is considered to be a sensitive indicator of osteoblast synthetic activity.

The common form of alkaline phosphatase (ALP) is as an enzyme related to the cell membranes of liver, bone, kidney, and placenta cells. Levels of total alkaline phosphatase in serum remain the most common index of bone formation in clinical use [63]. However, immunoassays to

TABLE 4.4

List of commercially available biochemical markers of bone formation and resorption

Biochemical markers of formation	Biochemical markers of resorption
Serum osteocalcin (OC)	Urinary hydroxyproline (HYP)
Serum bone-specific alkaline phosphatase (bAP)	Urinary total pyridinoline (Pyr)
Serum total alkaline phosphatase (ALP)	Urinary total deoxypyridinoline (dPyr)
Serum procollagen type I carboxyterminal propeptide (PICP)	Urinary-free pyridinoline (f-Pyr)
Serum procollagen type I N-terminal propeptide (PINP)	Urinary-free deoxypyridinoline (f-dPyr)
	Urinary collagen type I cross-linked N-telopeptide (NTx)
	Urinary collagen type I cross-linked C-telopeptide (CTx)
	Serum carboxyterminal telopeptide of type I collagen (ICTP)

assess both osteocalcin and bone alkaline phosphatase (bAP) in serum have become the preferred tools in clinical research [63]. Propeptides (N and C) of collagen types I, II, and III can be measured by immunoassay in serum as by-products of collagen synthesis. Turnover of collagen type I, as a major component of bone, provides an index of bone formation. This is a relatively new assay and is not yet widely used.

The reader may intuit that higher levels of bone formation markers might be associated with increased bone formation and increased bone mineral. The relationship between bone formation markers and bone mineral, however, is not so straightforward. Because bone formation is coupled with bone resorption in the process of remodeling (see chapter 2), a rise in bone formation is, in most cases, associated with an increase in bone remodeling. Bone mineral may decrease due to increasing bone turnover at a time when bone biochemistry reveals an increase in markers of formation.

Bone Resorption Markers

Most biochemical tests to assess bone resorption involve quantifying collagen degradation products that appear in urine. The urinary cross-links of pyridinium are thought to be the most sensitive markers of bone resorption [64]. Pyridinoline (Pyr) and deoxypyridinoline (Dpd) are excreted unmetabolized into the urine where they can be measured by high pressure liquid chromatography (HPLC) or by enzyme immunoassay [65]. Dpd is derived predominantly from type I collagen of bone, whereas Pyr is present in type I collagen of bone and type II collagen of cartilage, as well as in other tissues in smaller amounts.

Although bone turnover markers identified menopausal women who were at risk of higher bone loss [66], several researchers found that a single measurement of bone turnover markers did not predict bone loss [67, 68]. A few studies have identified biochemical markers of resorption as significant predictors of fracture, independent of BMD [62, 69, 70] and of bone loss in older women [71]. The likelihood of fracture increases as the marker result increases [61].

Effect of Exercise on Biochemical Markers

To date, biomarkers have not proven themselves to be clinically useful for such things as prediction of stress fractures (figure 4.11) [72]. Nevertheless, the biochemistry of these compounds continues to be investigated. Recently, Zanker used biomarkers to show that nutrition, not estrogen deficiency, was a key factor in bone loss of exercise-associated amenorrhea [73].

Levels of biomarkers are influenced by exercise of various intensity and duration [74-80]. When physical activity is performed consistently, no correlation exists between physical activity level and biomarkers [77]. This suggests that if biomarkers are to play a role in physical activity and bone research, it may be in the role of detecting change, rather than predicting absolute levels of bone turnover.

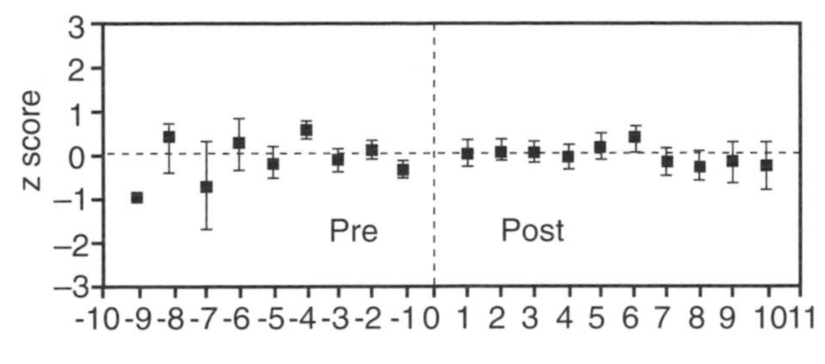

FIGURE 4.11 There was no clear association between biomarkers and onset of stress fracture [72].

In a study of the acute effects of 30 minutes of treadmill walking on bone turnover in young men, no change in the levels of osteocalcin or bone alkaline phosphatase occurred after 32 hours [81]. Both urinary cross-links increased by about 40% after exercise.

Studies of the long-term effects of an exercise intervention on bone markers have generally found no change [82-84] or an increase [85, 86] in levels of osteocalcin over 9 to 12 months in older men and women whose BMD increased during the period under investigation. On the other hand, Dalsky and colleagues [87] found that while osteocalcin decreased after nine months of high-intensity exercise intervention, lumbar spine BMC increased.

In the first longitudinal studies of resistance training, serum levels of the bone markers osteocalcin and bone alkaline phosphatase increased [74, 88-90] as soon as four weeks after beginning the resistance program. How-ever, resistance training had no effect on plasma PICP in this longitudinal study or in a cross-sectional report [91]. Bone markers seem to suggest that resistance training increases bone formation but does not necessarily alter bone turnover.

SUMMARY

- To study bone, we need to measure its various properties. Today, the researcher can do this with modalities such as DXA, QUS, and QCT.
- Dual energy X-ray absorptiometry (DXA) is the most common measure of bone mineral content (BMC, g) and bone mineral density (BMD, g/cm^2). DXA is accurate and precise, provides little radiation exposure, and can measure soft tissue as well as bone. The instrument, however, cannot differentiate between trabecular and cortical bone, and it measures a two-dimensional area rather than the three dimensions of bone. Thus, DXA will measure a larger bone of identical material as having greater BMD. Because different manufacturers' densitometers provide different measures of BMD, patients need to be studied on the same instrument, and researchers undertaking multisite collaborations must take great care with instrument calibration and quality control.
- Quantitative ultrasound (QUS) is becoming more widely used in both clinical and research circles. The modality is cheaper than DXA, lightweight (and therefore portable), and employs no radiation. Although it is not clear which bone parameter QUS measures, some have postulated that the modality reflects bone microarchitecture, such as trabecular number, connectivity, and orientation. A small number of studies have shown that QUS predicts fracture risk, and that this ability is maintained after adjusting for subjects' BMD. QUS may measure a component of fracture risk that is independent of BMD.
- Quantitative computed tomography (QCT) measures apparent volumetric density of trabecular or cortical bone and has been largely used to measure BMD in vertebral bodies. Although this modality is valid for vertebral fracture prediction, the relatively high radiation dose and its limited accuracy and precision mitigate against its widespread use in this role. Furthermore, it cannot readily provide hip data.
- Peripheral QCT (pQCT) is a recent technology that permits QCT evaluation of the appendicular skeleton that may prove to be of increasing use in exercise-related research.

• Imaging modalities are not the only ways of understanding bone behavior. Bone biomarkers provide information about bone turnover, bone resorption, and bone formation. This field is only in its infancy, and its present clinical usefulness is a subject of debate. As more accurate, less variable bone biomarkers are being developed, this field seems to have an exciting future.

References

1. Genant HK, Faulkner KG, Gluer C-C. Measurement of bone mineral density: current status. *Am J Med* 1991;91:49S-53S.
2. Engelke K, Glüer CC, Genant HK. Factors influencing dual X-ray bone absorptiometry (DXA) of spine and femur. *Calcif Tissue Int* 1995;56:19-25.
3. Formica C, Loro ML, Gilsanz V, et al. Inhomogeneity in body fat distribution may result in inaccuracy in the measurement of vertebral bone mass. *J Bone Miner Res* 1995;10:1504-11.
4. Khan KM, Henzell SL, Prince RL, et al. Instrument performance in bone density testing at five Australian centres. *Aust NZ J Med* 1997;27:526-530.
5. White C, Pocock N. Bone density and osteoporosis: crunching more than numbers (editorial). *Aust NZ J Med* 1997;27:519-520.
6. Gluer CC, Blake G, Blunt BA, et al. Accurate assessment of precision errors: How to measure the reproducibility of bone densitometry techniques. *Osteoporos Int* 1995;5:262-270.
7. Ho CP, Kim RW, Schaffler MB, et al. Accuracy of dual-energy radiographic absorptiometry of the lumbar spine: Cadaver study. *Radiology* 1990; 176:171-173.
8. Lilley J, Walters BG, Heath DA, et al. In vivo and in vitro precision of bone densitometry measured by dual-energy x-ray absorption. *Osteoporos Int* 1991;1:141-146.
9. Mazess R, Chesnut III CH, McClung M, et al. Enhanced precision with dual-energy x-ray absorptiometry. *Calcif Tissue Int* 1992;51:14-17.
10. Orwoll ES, Oviatt SK, Nafarelin Bone Study Group. Longitudinal precision of dual-energy x-ray absorptiometry in a multicenter study. *J Bone Miner Res* 1991;6:191-197.
11. Mazess RB, Barden HS, Bisek JP, et al. Dual-energy x-ray absorptiometry for total-body and regional bone-mineral and soft-tissue composition. *Am J Clin Nutr* 1990;51:1106-1112.
12. Pritchard J, Nowson C, Strauss B, et al. Evaluation of dual energy X-ray absorptiometry as a method of measurement of body fat. *Eur J Clin Nutr* 1993;47:216-228.
13. Van Loan MD, Keim NL, Berg K, et al. Evaluation of body composition by dual energy x-ray absorptiometry and two different software packages. *Med Sci Sports Exerc* 1995;27:587-591.
14. Laskey MA, Lyttle KD, Flaxman ME, et al. The influence of tissue depth and composition on the performance of the Lunar dual-energy X-ray absorptiometer whole-body scanning mode. *Eur J Clin Nutr* 1992;46:39-45.
15. Kelly TL, Berger N, Richardson TL. DXA body composition: theory and practice. *Appl Radiat Isot* 1998;49:511-3.
16. Salamone LM, Glynn N, Black D, et al. Body composition and bone mineral density in premenopausal and early perimenopausal women. *J Bone Miner Res* 1995;10:1762-1768.
17. Douchi T, Oki T, Nakamura S, et al. The effects of body composition on bone density in pre- and postmenopausal women. *Maturitas* 1997;27:55-60.
18. Van Marken Lichtenbelt WD, Fogelholm M, Ottenheijm R, et al. Physical activity, body composition, and bone density in ballet dancers. *Brit J Nutr* 1995;74:439-451.
19. Genant HK, Engelke K, Fuerst T, et al. Review: Noninvasive assessment of bone mineral and structure: state of the art. *J Bone Miner Res* 1996;11:707-730.
20. Parfitt AM, Matthews C, Villaneuva A. Relationships between surface, volume, and thickness of iliac trabecular bone in aging and osteoporosis. *J Clin Invest* 1983;72:1396-1409.
21. Guglielmi G, Grimston SK, Fisher KC, et al. Osteoporosis: Diagnosis with lateral and posteroanterior dual X-ray absorptiometry compared with quantitative CT. *Radiology* 1994;192:845-850.
22. Cann CE, Genant HK, Kolb FO, et al. Quantitative computed tomography for prediction of vertebral fracture risk. *Bone* 1985;6:1-7.
23. Mazess RB, Barden HS. Interrelationships among bone densitometry sites in normal young women. *Bone Miner* 1990;11:347-356.
24. Carter DR, Bouxsein ML, Marcus R. New approaches for interpreting projected bone densitometry data. *J Bone Miner Res* 1992;7:137-145.
25. Mazess RB, Barden H, Mautalen C, et al. Normalization of spine densitometry. *J Bone Miner Res* 1994;9:541-548.
26. Genant HK, Grampp S, Gluer CC, et al. Universal standardization for dual x-ray absorptiometry: Patient and phantom cross-calibration results. *J Bone Miner Res* 1994;9:1503-1514.

27. Hanson J. Letter to the editor: Standardization of femur BMD. *J Bone Miner Res* 1997;12:1316-1317.

28. Steiger P. Letter to the editor: standardization of spine BMD measurments. *J Bone Miner Res* 1995;10:1602-1603.

29. Hui SL, Gao S, Zhou X-H, et al. Universal standardization of bone density measurements: a method with optimal properties for calibration among several instruments. *J Bone Miner Res* 1997;12:1463-1470.

30. Abrahamsen B, Gram J, Hansen TB, et al. Cross calibration of QDR-2000 and QDR-1000 dual-energy x-ray densitometers for bone mineral and soft tissue measurements. *Bone* 1995;16:385-390.

31. Drinkwater DT, Bailey DA, Faulkner RA, et al. Comparison of QDR-1000W and QDR-2000 scanners. *J Bone Miner Res* 1993;8:S352 [Abstr].

32. Martin RB. Determinants of the mechanical properties of bones. *J Biomech* 1991;24:79-88.

33. Kanis JA. *Osteoporosis*. Oxford: Blackwell Science, 1994:1-254.

34. Heinonen A, Kannus P, Sievanen H, et al. Randomised controlled trial of effect of high-impact exercise on selected risk factors for osteoporotic fractures. *Lancet* 1996;348:1343-1347.

35. Hui SL, Slemenda CW, Johnston Jr CC. Age and bone mass as predictors of fracture in a prospective study. *J Clin Invest* 1988;81:1804-1809.

36. Cummings SR, Black DM, Nevitt MC, et al. Bone density at various sites for prediction of hip fractures. *Lancet* 1993;341:72-75.

37. Ross PD, Davis JW, Epstein RS, et al. Pre-existing fractures and bone mass predict vertebral fracture incidence in women. *Ann Int Med* 1991;114:919-923.

38. Melton LJ, Atkinson EJ, O'Fallon WM, et al. Long-term fracture prediction by bone mineral assessed at different skeletal sites. *J Bone Miner Res* 1993;8:1227-1233.

39. Wasnich RD. *Fracture prediction with bone mass measurements*. In: Genant HK, ed. Osteoporosis update. Berkeley, CA: University Press, 1987: 95-101.

40. Wasnich RD, Ross PD, Davis JW, et al. A comparison of single and multi-site BMC measurements for assessment of spine fracture probability. *J Nucl Med* 1989;30:1166-1171.

41. Ross PD, Genant HK, Davis JW, et al. Predicting vertebral fracture incidence from prevalent fractures and bone density among non-black, osteoporotic women. *Osteoporos Int* 1993;3:120-126.

42. WHO Study Group. *Assessment of fracture risk and its application to screening for postmenopausal osteoporosis*. Geneva: World Health Organization, 1994 WHO Technical Report Series; vol 843).

43. Kanis JA, Melton LJ, Christiansen C, et al. The diagnosis of osteoporosis. *J Bone Miner Res* 1994;9:1137-1141.

44. Bouxsein ML, Coan BS, Lee SC. Prediction of the strength of the elderly proximal femur by bone mineral density and quantitative ultrasound measurements of the heel and tibia. *Bone* 1999; 25:49-54.

45. Mughal M, Lorenc R. *Assessment of bone status in children using QUS techniques in*. In: Njeh C HD, Fuerst T, Gluer C, Genant H, ed. Quantitative ultrasound: Assessment of osteoporosis and bone status. London: Martin Dunitz Ltd, 1999, 214-233.

46. McCarthy RN, Jeffcott LB, McCartney RN. Ultrasound speed in equine cortical bone: Effects of orientation, density, porosity and temperature. *J Biomech* 1990;23:1139-1143.

47. Taaffe DR, Duret C, Cooper CS, et al. Comparison of calcaneal ultrasound and DXA in young women. *Med Sci Sports Exerc* 1999;31:1484-1489.

48. Thompson P, Taylor J, Oliver R, et al. Quantitative ultrasound (QUS) of the heel predicts wrist and osteoporosis related fracture in women aged 45-75 years. *J Clin Densitom* 1998;1:219-225.

49. Huang C, Ross PD, Yates AJ, et al. Prediction of fracture risk by radiographic absorptiometry and quantitative ultrasound: a prospective study. *Calcif Tissue Int* 1998;63:380-4.

50. Pluijm SM, Graafmans WC, Bouter LM, et al. Ultrasound measurements for the prediction of osteoporotic fractures in elderly people. *Osteoporos Int* 1999;9:550-6.

51. Stegman MR, Heaney RP, Recker RR. Comparison of the speed of sound ultrasound with single photon absorptiometry for determining fracture odds ratios. *J Bone Miner Res* 1995;10:346-352.

52. Heaney RP, Avioli LV, Chesnut CH, et al. Osteoporotic bone fragility: Detection by ultrasound transmission velocity. *JAMA* 1989;261:2986-2990.

53. Nicholson PH, Lowet G, Cheng XG, et al. Assessment of the strength of the proximal femur in vitro: relationship with ultrasonic measurements of the calcaneus. *Bone* 1997;20:219-24.

54. van den Bergh JP, van Lenthe GH, Hermus AR, et al. Speed of sound reflects Young's modulus as assessed by microstructural finite element analysis. *Bone* 2000;26:519-24.

55. Frost ML, Blake GM, Fogelman I. Contact quantitative ultrasound: an evaluation of precision, fracture discrimination, age-related bone loss and applicability of the WHO criteria. *Osteoporos Int* 1999;10:441-9.

56. Nairus J, Ahmadi S, Baker S, et al. Quantitative ultrasound: An indicator of osteoporosis in perimenopausal women. *J Clin Densitom* 2000; 3:141-7.

57. Lang P, Steiger P, Faulkner K, et al. Osteoporosis: Current techniques and recent developments in

quantitative bone densitometry. *Radiol Clinics North America* 1991;29:49-76.

58. Adami S, Gatti D, Braga V, et al. Site-specific effects of strength training on structure and geometry of ultradistal radius in postmenopausal women. *J Bone Miner Res* 1999;14:120-124.
59. Jarvinen TLN, Kannus P, Sievanen H, et al. Randomized controlled study of effects of sudden impact loading on rat femur. *J Bone Miner Res* 1998;1475-1482.
60. Jarvinen TLN, Kannus P, Sievanen H. Have the DXA-based exercise studies seriously underestimated the effects of mechanical loading on bone? (Letter). *J Bone Miner Res* 1999;14:1634-1635.
61. Lindsay R. Clinicial utility of biochemical markers. *Osteoporos Int* 1999;9:S29-32.
62. Garnero P. Biochemical markers of bone turnover: recent data and avenues for the future. *Rev Rhum Engl Ed* 1999;66:538-42.
63. Eyre DR. Bone biomarkers as tools in osteoporosis management. *Spine* 1997;22:17S-24S.
64. Delmas PD. Biochemical markers of bone turnover. In: Theoretical considerations and clinical use in osteoporosis. *Am J Med* 1993;95:11S-16S.
65. Ebeling PR, Peterson JM, Riggs BL. Utility of type I procollagen propeptide assays for assessing abnormalities in metabolic bone diseases. *J Bone Miner Res* 1992;7:1243-1250.
66. Hansen MA, Overgaard K, Riis BJ, et al. Role of peak bone mass and bone loss in postmenopausal osteoporosis: 12-year study. *Br Med J* 1991;303:961-964.
67. Keen R, Nguyen T, Sobnack R, et al. Can biochemical markers predict bone loss at the hip and spine: a 4-year prospective study of 141 early postmenopausal women. *Osteoporos Int* 1996;6:399-406.
68. Brahm H, Mallmin H, Michaelsson K, et al. Relationships between bone mass measurements and lifetime physical activity in a Swedish population. *Calcif Tissue Int* 1998;62:400-412.
69. Garnero P, Sornay-Rendu E, Claustrat B, et al. Biochemical markers of bone turnover, endogenous hormones and the risk of fractures in postmenopausal women: the OFELY study. *J Bone Miner Res* 2000;15:1526-36.
70. Akesson K, Ljunghall S, Jonsson B, et al. Assessment of biochemical markers of bone metabolism in relation to the occurrence of fracture: a retrospective and prospective population-based study of women. *J Bone Miner Res* 1995;10:1823-9.
71. Bauer DC, Sklarin PM, Stone KL, et al. Biochemical markers of bone turnover and prediction of hip bone loss in older women: the study of osteoporotic fractures. *J Bone Miner Res* 1999;14:1404-10.
72. Bennell KL, Malcolm SA, Brukner PD, et al. A 12-month prospective study of the relationship between stress fractures and bone turnover in athletes. *Calcif Tissue Int* 1998;63:80-85.
73. Zanker CL, Swaine IL. Relation between bone turnover, oestradiol, and energy balance in women distance runners. *Br J Sports Med* 1998;32:167-71.
74. Nishiyama S, Tomoeda S, Ohta T, et al. Differences in basal and postexercise osteocalcin levels in athletic and nonathletic humans. *Calcif Tissue Int* 1988;43:150-154.
75. Virtanen P, Viitasalo JT, Vuori J, et al. Effect of concentric exercise on serum muscle and collagen markers. *J App Physiol* 1993;75:1272-1277.
76. Kristoffersson A, Hultdin J, Holmlund I, et al. Effects of short-term maximal work on plasma calcium, parathyroid hormone, osteocalcin, and biochemical markers of collagen metabolism. *Int J Sports Med* 1995;16:145-149.
77. Brahm H, Piehl-Aulin K, Ljunghall S. Biochemical markers of bone bone metabolism during distance running in healthy, regularly exercising men and women. *Scand Journal Med Sci Sports* 1996;6:26-30.
78. Malm HT, Ronni-Sivula HM, Viinikka LU, et al. Marathon running accompanied by transient decreases in urinary calcium and serum osteocalcin levels. *Calcif Tissue Int* 1993;52:209-211.
79. Thorsen K, Kristofferson A, Lorentzon R. The effects of brisk walking on markers of bone and calcium metabolism in postmenopausal women. *Calcif Tissue Int* 1996;58:221-225.
80. Salvesen H, Piehl-Aulin K, Ljunghall S. Change in levels of the carboxyterminal cross-linked telopeptide of type I procollagen, the carboxyterminal cross-linked telopeptide of type I collagen and osteocalcin in response to exercise in well-trained men and women. *Scand Journal Med Sci Sports* 1994;4:186-190.
81. Welsh L, Rutherford OM, James I, et al. The acute effects of exercise on bone turnover. *Int J Sports Med* 1997;18:247-251.
82. Hatori M, Hasegawa A, Adachi H, et al. The effects of walking at the anaerobic threshold level on vertebral bone loss in postmenopausal women. *Calcif Tissue Int* 1993;52:411-414.
83. Pruitt LA, Jackson RD, Bartels RL, et al. Weight-training effects on bone mineral density in early postmenopausal women. *J Bone Miner Res* 1992;7:179-185.
84. Ryan AS, Treuth MS, Rubin MA, et al. Effects of strength training on bone mineral density: Hormonal and bone turnover relationships. *J Appl Physiol* 1994;77:1678-1684.
85. Nelson ME, Fiatarone MA, Morganti CM, et al. Effects of high-intensity strength training on multiple risk factors for osteoporotic fractures. A randomized control trial. *JAMA* 1994;272:1909-1914.

86. Menkes A, Mazel S, Redmond RA, et al. Strength training increases regional bone mineral density and bone remodeling in middle-aged and older men. *J App Physiol* 1993;74:2478-2484.
87. Dalsky GP, Stocke KS, Ehsani AA, et al. Weight-bearing exercise training and lumbar bone mineral content in postmenopausal women. *Ann Int Med* 1988;108:824-828.
88. Bell NH, Godsen RN, Henry DP, et al. The effects of muscle-building exercise on vitamin D and mineral metabolism. *J Bone Miner Res* 1988;3:369-73.
89. Fiore CE, Cottini E, Fargetta C, et al. The effects of muscle building exercise on vitamin D and mineral metabolism. *J Bone Miner Res* 1991;3:369-373.
90. Karlsson MK, Vergnaud P, Delmas PD, et al. Indicators of bone formation in weight lifters. *Calcif Tissue Int* 1995;56:177-180.
91. Karlsson MK, Johnell O, Obrant KJ. Is bone mineral density advantage maintained long-term in previous weight lifters. *Calcif Tissue Int* 1995;57:325-328.
92. Beck TJ, Ruff CB, Warden KE, et al. Predicting femoral neck strength from bone mineral data. A structural approach. *Investigative Radiol* 1990;25:6-18.
93. Sievänen H, Koskue V, Rauhio A, et al. Peripheral quantitative computed tomography in human long bones: Evaluation of in vitro and in vivo precision. *J Bone Miner Res* 1998;13:871-882.

part II

Determinants of Bone Mineral Other Than Physical Activity

Before delving more deeply into the subject of physical activity and its influence on bone health, we will take the time to explore the many other determinants of bone health. Physical activity, after all, accounts for only 10% of bone mineral at the population level. The determinants that account for the other 90% are genetics, gender, age, soft tissue composition (lean mass, fat mass), lifestyle factors (smoking, alcohol intake), medications, hormones, and nutrition (table II.1). These factors may interact with one another, and their degree of influence varies at different stages of the life span and at different skeletal sites.

By studying the factors other than physical activity that influence bone health, you will have a better understanding of how bone behaves in general, as well as how it responds to physical activity. A knowledge of the concept of determinants, and the relative importance of each one, will help you develop realistic aspirations when prescribing exercise for bone health and when examining the literature in this field. If you are performing original bone health research, knowledge of the determinants of BMD will help you take account of (or control for) important potential confounds in your physical activity study.

The material in this part is deliberately aimed at the person with a physical activity bent. Physical activity influences some of the determinants discussed in this part (e.g., exercise can influence lean mass

TABLE II.1

Factors that may influence BMD	
Age	*Other environmental factors*
Genetics	Smoking (–)
Racial (risk lower in black people)	*Hormonal factors*
Familial history (–)	Delayed puberty (–)
Nutritional	Primary gonadal insufficiency (–)
Calcium (+/–)	Secondary gonadal insufficiency (–)
Vitamin D (+)	Use of oral contraceptives (+/–)
Malnutrition (–)	Multiparity (+/–)
Exercise	Lactation (+/–)
Daily physical activity (+)	Premenstrual tension (–)
Immobilization (–)	
Space flight (–)	

+ denotes a protective factor, – denotes a risk, and +/– denotes that the evidence for both exists.

or hormone levels), and so we will examine the influence it has on each.

Because BMD can be measured accurately and conveniently, it is used routinely as an outcome measure in bone research. Although BMD should not be considered a convenient synonym for bone strength (see chapter 3), it remains the best common denominator that measures a key property of bone. Therefore, this part tends to concentrate on BMD as the main outcome measure.

Bone mineral is age, sex, and race dependent. In chapter 5 we discuss the role of heredity, as this is thought to explain between 60% and 80% of the variation in bone mineral among individuals. We also outline the possible interaction between genetics and physical activity that may influence bone mineral.

Chapter 6 clarifies what appears at times to be a confusing area of research—soft tissue composition and bone mineral. Lean mass and fat mass, together with bone mineral, make up total body weight. It has long been known that body weight is positively associated with increased bone mineral. DXA scanning allowed researchers to more easily differentiate lean mass and fat mass while measuring bone mineral—the anthropometrist's dream come true! Clearly physical activity plays a role in influencing soft tissue composition. The chapter outlines the relative importance of lean mass and fat mass at various stages of the life span.

Chapter 7 provides essential background information about the endocrine system and bone. A proportion of endocrinologists specialize purely in bone, which illustrates the importance of the endocrine system in bone. We outline how estrogen influences bone in both women and men and discuss the influence the oral contraceptive pill, ovarian hormone therapy, and progesterone and growth hormone have on BMD. You will read about the deleterious effects of corticosteroids on bone, when this prescription medication is used to treat other significant conditions. If you are working with patients or research subjects other than the healthy young, you will likely come across the very common issues raised in this chapter in your day-to-day practice.

Chapter 8 reviews the role of nutrients, particularly calcium and vitamin D. It evaluates their role in BMD and summarizes the results of the small number of studies that have assessed the role of these nutrients in decreasing fracture risk. The chapter also examines whether physical activity and calcium have a synergistic effect on bone mineral.

The reader who takes the trouble to fully understand the role of the determinants of bone mineral will be much better placed to understand the exercise and physical activity data reported in part III of this book. Bone is a complex tissue and organ with many influences. A holistic perspective will lend credence to the work of exercise scientists studying the role of physical activity in bone health.

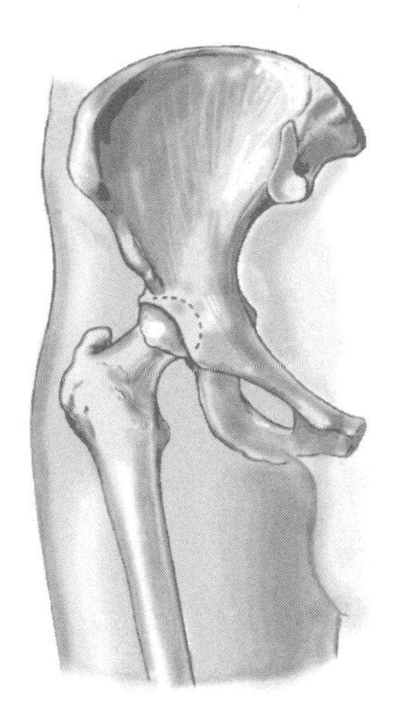

Chapter 5

Age, Sex, Genetics, and Race

When women who have osteoporosis (bone mineral density more than 2.5 standard deviation below the young normal mean, see chapter 15) are first diagnosed, their initial response may be, Can I build my bone strength back to normal with exercise or calcium or medication? Unfortunately, in virtually all cases, the answer to this question is no. The reason is that age and genetic factors are the greatest influence on bone mineral; other factors can alter it only to a relatively small degree. Nevertheless, strategies exist that can help *conserve* bone mineral and stay further loss.

In this chapter we will explore the normal pattern of bone mineral gain and loss in adults, as well as the role of genetics in determining bone mineral density. Having recognized the importance of these inborn determinants of bone mineral, the reader should not despair. The influence of all lifestyle factors, including physical activity, remains very important, as even a relatively small enhancement in bone mineral can substantially reduce the risk of osteoporotic fracture [1].

Age

Bone mineral is accrued at various rates during the childhood and adolescent years, remains largely constant during the young adult years, and then diminishes with aging. Figure 5.1 illustrates the trend for bone mineral loss with aging. In this chapter we focus on the adult data, as bone gain in childhood and adolescence is detailed in chapter 10.

The actual rates of loss vary among reports and depending on the sites measured. Ensrud and colleagues [2] examined 5689 community-dwelling white women who were 65 years or older at baseline. The rate of bone loss at the proximal femur and its subregions was 2.5 mg/cm^2 (95% confidence interval [CI] 2.0-2.9) in women 67 to 69 years old and 10.4 mg/cm^2 for those over 85 years old (95% CI 8.4-12.4). The average rate of femoral neck bone loss was −1.0% (the range was from −10% to +13%) in an age-stratified sample of 304 women (aged 30 to 94 years) from Rochester, Minnesota, and did not vary significantly with age [3]. Summarized by

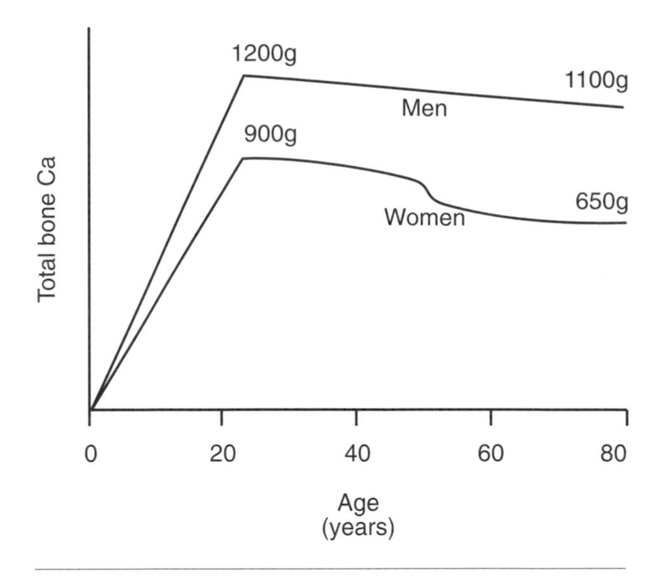

FIGURE 5.1 Schematic representation of bone mineral gain and loss across the life span.

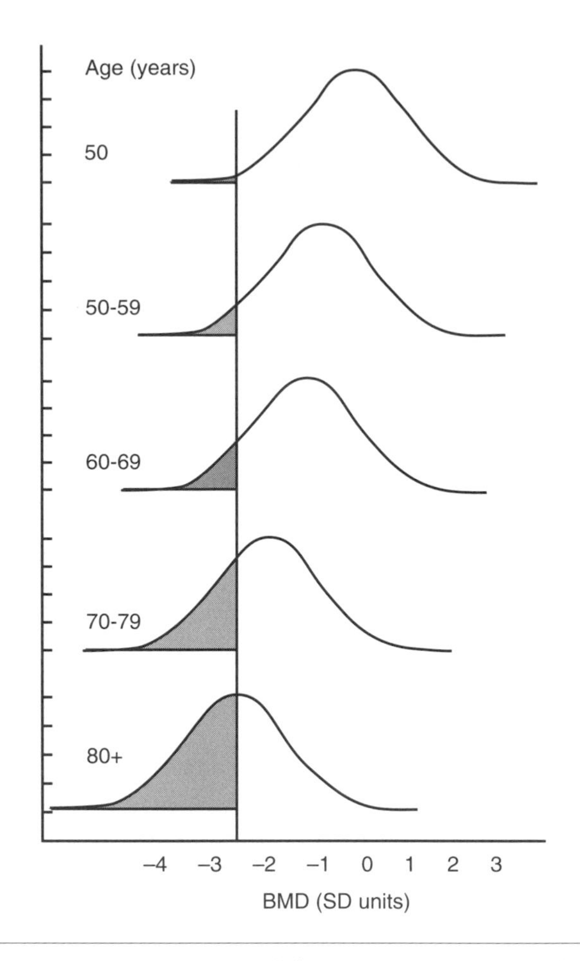

FIGURE 5.2 Bone mineral density (BMD) represented by standard deviations above or below the mean of each decade from 50 years. BMD remains normally distributed, but the mean decreases progressively with age. Note that the proportion of women with osteoporosis (see chapter 15) increases exponentially with age.
Modified from [90].

decade, menopausal loss seems to be about 10% per decade but levels off after age 75 to about 3% per decade. The rate of loss appears greater at the trabecular vertebral bone than at the more cortical femoral or radial sites. Figure 5.2 illustrates that the population mean of BMD diminishes, while individual variation remains wide.

In a large cross-sectional series of 965 Japanese men, bone loss per decade was 3.9% at the radius, 1.6% at the lumbar spine, and 3.3% at the hip [4]. Generally speaking, bone loss in men appears to be about 5% to 10% per decade (figure 5.3) [5-8]. At the proximal femur, cross-sectional studies reveal a negative relationship

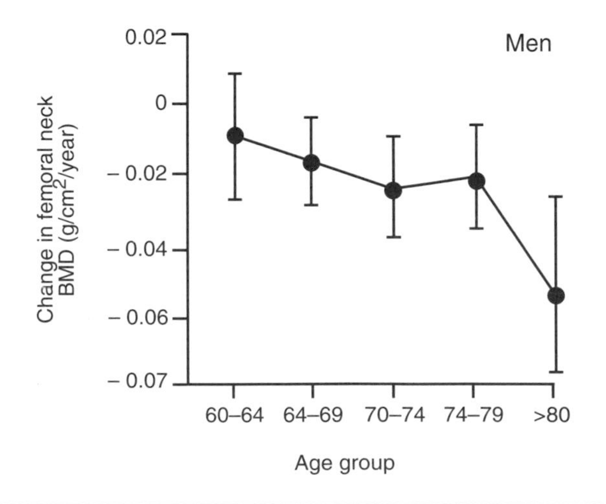

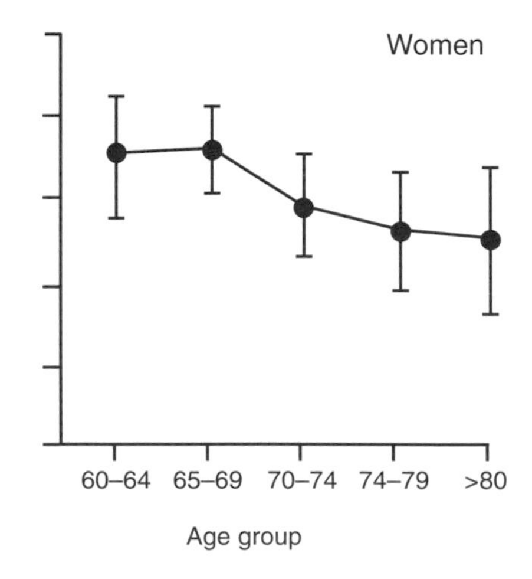

FIGURE 5.3 Longitudinal rates of bone loss in men (left panel) and women (right panel) at the femoral neck. Data points represent mean (with 95% confidence interval) for each age group.

between age and bone mineral, but the slope of this relationship is less than that in women [9].

Sex

The relative difference between men and women for peak hip and lumbar spine bone mineral density and rate of decline with aging (before menopause) is quite small. However, men are conferred other biomechanical advantages that account for their lower rate of fracture.

Early in life there is little discernible difference in bone size, mass, or structure between boys and girls. At puberty, linear growth and skeletal mass accumulation is greater for boys than for girls, resulting in larger, longer bones and higher bone mineral density [10], on average (see chapter 10). By early adulthood the bones of the appendicular skeleton are wider in men (20% in the second metacarpal) and have a thicker cortex [11]. Primarily because of this thicker cortex in men, radial and total body areal bone mineral density are also reported to be greater. It is important to note that there is no difference in the density of cortical bone between the sexes. The sex difference in cortical width, however, is apparent throughout the rest of life (figure 5.4).

Vertebral cross-sectional area is also 25% greater in men [12]. Men benefit from accentuated periosteal apposition with aging, resulting in increased bone width, thus garnering them a bone strength advantage. At adulthood total body bone mineral content is approximately 3100-3500 g for men as compared to 2300-2700 g for women [13, 103], most of which can be explained by the larger overall size of men. This size effect was highlighted by a QCT study that showed similar lumbar spine BMD between young men and women [14]. Areal and estimated volumetric BMD (by DXA) were also compared between men and women aged 58-79 years [15]. Although men had significantly greater areal BMD at the total body ($p < 0.001$), femoral neck ($p < 0.01$), and lumbar spine ($p < 0.05$), there was no gender difference in estimated bone volumes (femoral neck or lumbar spine).

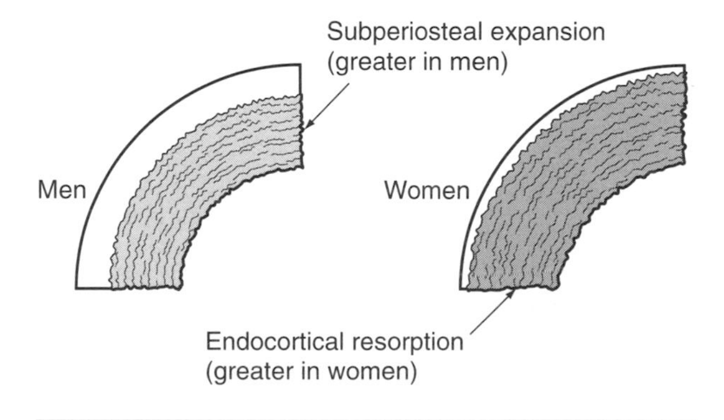

FIGURE 5.4 A sex difference that affects bone strength is the greater subperiosteal bone formation in men (strengthens bone) and greater endocortical resorption in women (weakens bone).

The way bone is lost also appears to differ between women and men as well as between the axial and appendicular skeleton. First of all, women lose more bone through life than men do due, for the most part, to the transient period of accelerated loss women experience at some sites at menopause [104]. A study from Finland of calcaneal bone loss in 161 men and 324 women (> 75 years at baseline) showed that the annual decrease in BMD was 2.5-2.7% greater in women than in men (0.8-1%) [105].

The cortex of women also becomes more porous as women lose more endosteal bone and gain relatively less periosteal bone than men do (figure 5.4) [106, 11]. Longitudinal studies have shown a more rapid decline in bone mass in men at both the appendicular and axial skeleton than was previously believed (see figure 5.3). However, men retain a skeletal advantage (and a lower rate of hip fracture incidence and vertebral fracture prevalence) [16] owed primarily to their higher rate of periosteal apposition and decreased rate of endosteal resorption in adult life [17].

At trabecular sites, the rates of bone loss are similar between sexes. The trabecular loss observed in women, however, is a result of both a decline in the number of trabecular elements and thinning of remaining trabeculae, while men experience generalized trabecular thinning [18]. Because of these sex differences in trabecular bone loss, men retain a greater load-bearing capacity as they age.

Genetics

The study of genetics has been turbulent in the past century. The abuse of genetic theories in the 1940s led to this field being almost taboo for several decades. However, the Human Genome Project and the search for the genes for osteoporosis and other diseases has ensured that genetic influences on biology is a household conversation topic. It is important to understand that genes do not act in isolation but interact with a large and uncertain number of other genes and environmental and physical factors, all of which influence their expression.

Bone Mineral Density and Genetics

How the genetic message governs bone density is not known. On theoretical grounds, both from heritability estimates and from assessments of candidate genes, BMD appears to be controlled by a panel of genes rather than one single locus (*L. Raisz, personal communication*) [19, 20]. This is termed polygenic inheritance. Candidate genes that may play a role include the vitamin D receptor gene (VDR) [21], the estrogen receptor (ER) gene [22, 23], the apolipoprotein E4 gene [24] (also associated with Alzheimer's disease), the interleukin genes, the parathyroid hormone receptor gene [25], and polymorphisms of the collagen gene (Collagen Ia1Sp1) [20, 26].

The VDR gene was claimed to be the "gene for osteoporosis" in 1994 when the journal *Nature* gave cover-story prominence to the observation that an allele defined by a BsmI polymorphism (a specific sequence of nucleic acids) in the gene encoding the 1,25-dihydroxyvitamin D receptor (VDR) was correlated with BMD of the spine and hip in adults [21]. This VDR genotype had earlier been shown to relate to osteocalcin levels [27]. Individuals homozygous for the presence of the BsmI site were designated as bb, and homozygotes without the particular site were labeled BB. The researchers found that the bb individuals had higher BMD and that this one genetic marker accounted for up to 75% of the variation in the heritable component of peak bone density. While a few other groups also reported finding significant associations [28, 29], the amount of heritability attributed to the VDR receptor was immediately questioned, even when using Morrison and colleagues' own published data (*John L Hopper, personal communication, 1994*) [21].

Since then, several groups have been unable to corroborate the finding that the VDR receptor polymorphism explained a large proportion of variation in BMD in adults [30-33]. Furthermore, Morrison and colleagues published a letter withdrawing part of their data that was published in *Nature* [107]. At present there is no final consensus on the relationship between VDR genotype and BMD, but it may explain at most several percents of population variation in BMD.

Krall and colleagues [31], however, despite finding no association between the BMD of a group of postmenopausal women and VDR polymorphism, found that women with the BB allele had most rapid bone loss at several sites. At the hip, this rate of loss was influenced by calcium intake [31]. Thus, it may be that VDR genotype plays some role in determining BMD, but it is clear that the initial estimates of the importance of this particular polymorphism were overly optimistic. Some important problems in a number of published studies include small sample sizes and different statistical approaches to the data.

Because of the likelihood that many genes contribute a small amount to the susceptibility to osteoporosis, researchers have begun to broaden their target to seek genetic markers for various bone-related functions likely to be contributing factors. For example, there may prove to be allelic variation in genetic control of the role of calcium (via the VDR genotypes), the role of estrogen (via the estrogen receptor genotypes), the role of bone matrix (via the type 1 collagen genotype), and the role of growth factors. Small effects in all of these genes could accumulate to a clinically relevant degree [20]. The occurence of a gene-gene interaction also is quite likely.

Genetic Influence on Peak Bone Mass

Researchers looking to examine the role of genetic factors in a human trait such as BMD generally measure the trait in related individuals and compare the correlation found with that in unrelated people. Such familial studies have shown that BMD is strongly influenced by parental bone mass [34-36]. Furthermore, the spinal and femoral BMDs of daughters of women with osteoporosis resemble those of their mothers [37, 38].

In women with a mean age of 31 years, inheritance accounted for between 41% and 61% of the variance of BMD at five skeletal sites [39]. These heritability estimates were made after adjusting BMD for age, height, weight, and significant lifestyle or environmental factors. Other workers have estimated the heritability to explain from 46 to 84% of BMD [40-44].

A more powerful strategy of family study is the twin study because identical (monozygotic) twins share all their genes and nonidentical (dizygotic) twins share, on average, only half their genes. Twin studies, both in adults [43, 45-48] and in children and young adults [19, 44, 49], have shown that pairs of monozygotic twins have comparative BMD values that differ less than the BMD values in dizygotic twins. These studies suggest a strong genetic contribution to BMD especially in the axial skeleton. These findings are, however, subject to certain assumptions inherent in the "classic twin model."

Although twin studies are powerful tools, they have some limitations. The results may slightly overestimate the influence of heritability because they may include some measure of what is referred to as "common environment" within twin pairs that cannot be adjusted [19, 39]. A common assumption is that both dizygotic pairs and monozygotic pairs share environmental effects equally. However, we know that this assumption is violated sometimes. It is not surprising, then, that estimates vary given (1) the inherent differences in populations studied, (2) the difficulty in adjusting precisely for environmental/lifestyle factors, and (3) the role of gene-environment interactions. Some investigators use sophisticated statistical (multivariate normal) modeling to estimate these additive genetic, common environmental, and unique environmental influences [43, 44].

Gene Expression and Interactions

Some studies attribute such large percentages of the variance in BMD to genetic or environmental factors that the sum of the determinants (across studies) is in excess of 100% [50, 51]. For example, heritability is believed to account for about 60% of the interindividual variation in BMD [39], but studies have also estimated physical activity [52] and dietary calcium [53] to each account for 40% of the variation. This apparent paradox could be explained by the presence of either one, or a combination, of three situations: (1) a gene-environment interaction, (2) an environment-environment interaction (e.g., physical activity and calcium), or (3) a common set of genes controlling both BMD and factors such as muscularity, which have been traditionally considered in the lifestyle category. Examples of each of these scenarios and evidence for them are discussed in the following sections.

Gene-Environment Interaction. A gene-environment interaction would be present if genetic factors regulated the response of bone to physical activity. Thus, some individuals would respond more and some less to identical doses of bone loading, because of genetic factors. Whether or not this occurs remains unknown, but if it does, it is not via common allelic variants of the vitamin D receptor (VDR) gene. In a prospective study that examined whether bone's response to physical activity differed among women with different alleles of the VDR gene, women with the BB gene allele (which may have a maximal adverse effect on BMD) had as good an osteogenic response to mechanical loading as subjects with other VDR genotypes [54]. Thus, irrespective of the VDR genotype, physical activity was beneficial for bones of premenopausal women [54].

Athletes may be genetically disposed to having high BMD, or they may respond more positively to exercise intervention than nonathletes. This might explain why cross-sectional BMD studies in athletes reveal substantial differences between subjects and controls, whereas intervention studies in general populations find a much smaller BMD difference between exercising subjects and controls (see part III).

Gene-environment interaction would be present if genetic factors determined an individual's choices of lifestyle (e.g., an active lifestyle, alcohol intake, smoking) or dietary habits (e.g., a high calcium diet) [55]. Gene-environment interaction would also arise if genes determined bone's responsiveness to these environmental factors. Twin studies, however, suggest that genetic factors play only a small role in physical fitness [56-61] (about 40% of variance in $\dot{V}O_2max$ per kilogram of body weight) and dietary intake [62].

Environment-Environment Interaction. Environment-environment interaction would be present if the level of dietary calcium regulated bone's response to physical activity. If mechanical loading were a totally independent predictor of bone mass, then a certain dose of mechanical loading should have the same influence on bone irrespective of dietary calcium intake. Whether this is the case remains unresolved [50] (see the discussion of calcium in chapter 8). Although the argument appears teleologically tempting, the only prospective study designed to answer it failed to show any additional benefit of calcium to physical activity intervention [63].

Another example of environment-environment interaction would be if individuals who undertook an active lifestyle (more accurately, a "loading" lifestyle) consequently ingested more dietary calcium. This has been proposed by several authors, but does not appear to be the case [64, 65].

Common Genes for Bone Mass. If physical-activity-related factors that are known to influence bone mass, such as lean mass and body size, had a genetic basis in common with BMD, then this would also permit the determinants of BMD to sum to more than 100% as described earlier. As mentioned, muscle mass and physical fitness have only a small genetic basis [57-61], but there is some evidence that there is a common genetic influence on muscle and bone [108].

It is feasible, therefore, that physical activity and calcium act in a permissive role (environment-environment) (see chapter 8). In addition, it is not inconceivable that the effects of both mechanical loading and dietary calcium are modulated by genetic factors.

Genetic Influence on Age-Related Loss

Studies that examined the genetic influence on rates of age-related or postmenopausal bone loss provide conflicting results. As a general statement, little evidence has shown that rates of bone loss differ markedly among races [66, 67]. For example, postmenopausal loss rates are similar between Japanese American and Caucasian women, although small differences have been reported. Rates of loss are slightly lower in African Americans as compared to Caucasian Americans across the life span [68], and differences also exist in rates of postmenopausal bone loss between women from Norway and Holland [69].

A study of female twins revealed a genetic contribution to change in BMD at the trabecular sites (spine and hip) but not the cortical femoral neck [47]. This could be explained either by cortical bone being less influenced by genetic factors, or by the slow rate of change of cortical bone requiring a longer period of observation to reveal correlations. Twin and family studies generally show that similarities in bone mass diminish with time, suggesting that genetics plays a greater role in the development of peak bone mass than it does in the rate of bone loss. However, even at older ages, heritability studies show that genetic factors still contribute substantially to the variation among individuals in BMC and BMD [43]. More research is required to identify the intricate role of genes in age-related bone loss.

Race and Ethnicity

The terms *race* and *ethnicity* are prevalent in the bone literature. *Race* refers to the belief that there are three main genetically characterized human groups: Caucasian (white), Negroid (black), and Mongoloid (Asian) [70]. Over time, however, a significant mixing of the races has occurred. *Ethnicity* refers to cultural, religious, dietary, geographic, and other differences among races [70]. Ethnicity can influence disease prevalence even within races. Note that bone mineral density (BMD) and fracture rates can vary within a race, according to ethnicity.

Race implies that a population has genetic similarities, and in many cases a common environment, which interact with the genetic template to establish the phenotypic expression of a trait or characteristic. Populations are characterized to some extent by their size and shape, which allows them to adapt to the geographical and climatic environment they live in. The rotund, short stature of the Inuit is well suited to conserve heat in northern climates, and the tallness of the Masai of Africa provides a greater surface area per unit of mass from which to dissipate heat. Although researchers have identified interracial differences in skeletal density, the independent contributions of genetic and environmental factors are unclear.

Many years ago Choi and Trotter [71] observed that black fetuses had longer and heavier bones even before birth than white fetuses. Trotter and colleagues [72] had earlier compared the weight per unit volume of the bones of dry, defatted black and white skeletons of different ages and found that the bones of black cadavers were denser than the bones of white cadavers, especially for females (13% greater). These differences were greater in non-weight-bearing bones where the genetic contribution to bone density would be less obscured by the effects of mechanical loading [72, 73]. Early studies using DPA reported higher BMD values in black girls than in white girls aged 7-12 years [74]. A more recent study of 423 Asian, black, and Hispanic subjects (9-25 years) replicated these findings and showed that black girls and boys had greater femoral neck BMD than their white counterparts in early puberty [75].

Studies have confirmed that compared with whites, black adults have greater BMD at both the proximal femur [68, 76, 77] and the lumbar spine [68, 77] and increased cortical thickness [78]. The lower incidence of especially proximal femur fracture in African Americans was reported over a quarter century ago [79], and more recent studies have confirmed these lower rates at both the lumbar spine and the proximal femur [68, 80, 81]. The age-adjusted hip fracture incidence is 50% lower in African American women than in Caucasian American women [80].

It is well accepted that bone loss is an inevitable consequence of aging in both men and women and in all races. Evidence suggests that there is either a similar rate of bone loss across black and white subjects [77] or a higher rate of loss for white early postmenopausal women as compared with blacks [68, 76, 82]. Findings varied depending on the menopausal status of the group and the site measured. Similar rates of premenopausal bone loss were documented in 122 Caucasian American women and 121 African American women followed prospectively for three to four years [82]. Bone loss at the forearm and the lumbar spine, however, did not differ by ethnic group five years after the menopause. The literature agrees that the pattern, if not the magnitude, of bone loss is similar between black and white postmenopausal women [68].

There is great variation in the occurrence of osteoporotic fracture within and among different racial and ethnic groups (table 5.1). For example, the difference in hip fracture rates among Caucasian ethnic groups approaches that between Caucasian and other racial groups [83].

Only some of the many factors that influence risk of osteoporosis have been well studied in a range of ethnic groups. The majority of studies have been performed with non-Hispanic Caucasian participants.

As this book is about physical activity and bone health, we review the role of race and ethnicity on both of those factors. Despite the very different cultural values and attitudes toward physical activity, its beneficial effect on bone appears to cross ethnic and racial boundaries. Studies from Sweden [84], Japan [85], and Hawaii [86] all reveal a similar association among bone mass, fracture risk, and physical activity habits. A cross-sectional study in children demonstrated a significant ethnic difference in baseline variables such as calcium intake and loaded physical activity and in BMD at the femoral neck in a group of 163 Asian and Caucasian prepubertal boys and girls [87]. The only other study that compared Asian and Caucasian children (n = 52) living in North America showed no difference in BMD for pre/early pubertal boys but a 20%, on average, greater femoral neck and whole body BMD for the Caucasian, as compared to the Asian, girls [88]. A longitudinal study that addressed the ethnic effect of a pediatric physical activity intervention on bone mineral accrual [89] found no difference between Asian and Caucasian children.

TABLE 5.1

Age-adjusted rates[a] of hip fracture per 100,000 population for females, males, and total (both age- and gender-adjusted), by ethnic group

Ethnic group	Site	Female	Male	Total	Female:Male
Blacks	USA, 1993 [91]	344	235	300	1.5
	USA, 1984 [92]	174	108	137	1.6
	Johannesburg, South Africa, 1968 [93]	26	20	23	1.3
Hispanics[b]	California, USA, 1988 [94]	219	97	165	2.3
Asians	California, USA, 1988 [94]	383	116	265	3.3
	Hong Kong, 1970 [95]	179	113	150	1.6
	Singapore, 1966 [96]	83	111	95	0.7
Caucasians[c]	USA, 1994 [97]	968	396	738	2.4
	Maryland, USA, 1991 [98]	950	358	712	2.7
	Alicante, Spain, 1987 [99]	90	57	75	1.6
	Finland, 1985 [100]	432	199	329	2.2
	Funen, Denmark, 1983 [110]	511	191	351	2.7
	Kuopio, Finland, 1973 [102]	280	107	204	2.6

[a]Rates were age- and gender-adjusted to the 1990 U.S. non-Hispanic Caucasian population, [b]Hispanic Caucasians, and [c]non-Hispanic Caucasians.

SUMMARY

- Age, sex, genetics, and race are major, nonmodifiable determinants of bone mineral. Bone mineral increases during the growing years, stays relatively constant until about age 20, and diminishes rapidly at about 2% per year in women around menopause. This rapid bone loss persists for 5 to 10 years, after which time it slows. Bone is lost at about 10% per decade until about age 75, at which time bone loss slows to about 3% per decade.
- Although men lose bone at a faster rate than was previously believed, and at the same approximate rate as women, they have greater bone strength as compared to women because of a thicker bone cortex. This results from a higher rate of periosteal apposition and decreased rate of endosteal resorption in adult life.
- The genetic influence on bone mineral may vary with age. There is a strong genetic influence on peak bone mass. Inheritance may account for up to 60% of the variance in BMD in women aged about 30 years. The contribution of genetic and environmental factors calculated in various studies can add to greater than 100% because (1) inherent differences exist in populations, (2) it is difficult to adjust precisely for environmental/lifestyle factors, and (3) a number of interactions may exist.
- Although data conflict, it appears that genetics plays less of a role in determining the rate of bone loss than it does in determining peak bone mass.
- BMD appears to be controlled by a panel of genes rather than one single locus [20]. Genes that may play a role include the vitamin D receptor gene (VDR), the estrogen receptor (ER) gene, the parathyroid hormone (PTH) receptor gene, the apolipoprotein E4 gene, and polymorphisms of the collagen gene (Collagen Ia1Sp1).
- Race is also an important determinant of BMD that needs to be accounted for when interpreting, and performing, physical activity and bone research. Blacks have greater BMD than whites, who in turn have greater BMD than Asians.

References

1. Cummings, S.R., D.M. Black, M.C. Nevitt et al. (1993). Bone density at various sites for prediction of hip fractures. *Lancet* 341: 72-75.
2. Ensrud, K.E., L. Palermo, D.M. Black et al. (1995). Hip and calcaneal bone loss increase with advancing age: longitudinal results from the study of osteoporotic fractures. *J Bone Miner Res* 10: 1778-1787.
3. Melton, L.J., 3rd, E.J. Atkinson, M.K. O'Connor et al. (2000). Determinants of bone loss from the femoral neck in women of different ages. *J Bone Miner Res* 15: 24-31.
4. Sone, T., M. Miyake, N. Takeda et al. (1996). Influence of exercise and degenerative vertebral changes on BMD: A cross-sectional study in Japanese men. *Gerontology* 42 Suppl 1: 57-66.
5. Davis, J.W., P.D. Ross, J.M. Vogel et al. (1991). Age-related changes in bone mass among Japanese-American men. *Bone Miner* 15: 227-236.
6. Orwoll, E.S., S.K. Oviatt, M.R. McClung et al. (1990). The rate of bone mineral loss in normal men and the effects of calcium and cholecalciferol supplementation. *Ann Int Med* 112: 29-34.
7. Slemenda, C.W., J.C. Christian, T. Reed et al. (1992). Long-term bone loss in men: Effects of genetic and environmental factors. *Ann Int Med* 117: 286-291.
8. Melton, L.J., S. Khosla, E.J. Atkinson, et al. (2000). Cross-sectional versus longitudinal evaluation of bone loss in men and women. *Osteoporos Int* 11:592-599.
9. Hannan, M.T., D.T. Felson, and J.J. Anderson. (1992). Bone mineral density in elderly men and women: Results from the Framingham osteoporosis study. *J Bone Miner Res* 7: 547-553.
10. Bailey, D.A., H.A. McKay, R.A. Mirwald et al. (1999). The University of Saskatchewan Bone Mineral Accrual Study: A six-year longitudinal study of the relationship of physical activity to bone mineral accrual in growing children. *J Bone Miner Res* 14: 1672-1679.
11. Garn, S.M. (1970). Bone loss as a general phenomenon in man. *Fed Proc* 26: 1729-1736.
12. Gilsanz, V., M.I. Boechat, R. Gilsanz et al. (1994). Gender differences in vertebral size in adults: Biomechanical implications. *Radiology* 190: 678-683.

13. Rico, H., M. Revilla, L.F. Villa et al. (1993). Age-related differences in total and regional bone mass: A cross-sectional study with DXA in 429 normal women. *Osteoporos Int* 3: 154-159.

14. Genant, H.K., C.Y. Wu, C. van Kuijk et al. (1993). Vertebral fracture assessment using a semiquantitative technique. *J Bone Miner Res* 8: 1137-1148.

15. Faulkner, R.A., R.G. McCulloch, S.L. Fyke et al. (1995). Comparison of areal and estimated volumetric bone mineral density values between older men and women. *Osteoporos Int* 5: 271-275.

16. Khosla, S., L.J. Melton, 3rd, and B.L. Riggs. (1999). Osteoporosis: Gender differences and similarities. *Lupus* 8: 393-396.

17. Garn, S.M. (1992). Role of set-point theory in regulation of body weight [letter; comment]. *Faseb J* 6: 794.

18. Mosekilde, L. (1990). Sex differences in age-related changes in vertebral body size, density and biomechanical competence in normal individuals. *Bone* 11: 67-73.

19. Slemenda, C.W., J.C. Christian, C.J. Williams et al. (1991). Genetic determinants of bone mass in adult women: A reevaluation of the twin model and the potential importance of gene interaction on heritability estimates. *J Bone Miner Res* 6: 561-567.

20. Sowers, M.F. (1998). Expanding the repertoire: The future of genetic studies (Editorial). *J Bone Miner Res* 113: 1657-1659.

21. Morrison, N.A., J.C. Qi, A. Tokita et al. (1994). Prediction of bone density from vitamin D receptor alleles. *Nature* 367: 284-287.

22. Mizunuma, H., T. Hosoi, H. Okano et al. (1997). Estrogen receptor gene polymorphism and bone mineral density at the lumbar spine of pre- and postmenopausal women. *Bone* 21: 379-383.

23. Kobayashi, S., S. Inoue, T. Hosoi et al. (1996). Association of bone mineral density with polymorphism of the estrogen receptor gene. *J Bone Miner Res* 11: 306-311.

24. Shiraki, M., Y. Shiraki, C. Aoki et al. (1997). Association of bone mineral density with apolipoprotein E phenotype. *J Bone Miner Res* 12: 1438-1445.

25. Duncan, E.L., M.A. Brown, J. Sinsheimer et al. (1999). Suggestive linkage of the parathyroid receptor type 1 to osteoporosis. *J Bone Miner Res* 14: 1993-1999.

26. Keen, R.W., K.L. Woodforde-Richens, S.F.A. Grant et al. (1997). Type 1 collagen gene polymorphism is associated with osteoporosis and fracture. *J Bone Miner Res* 12: S545.

27. Morrison, N., R. Yeoman, P.J. Kelly et al. (1992). Contribution of trans-acting factor alleles to normal physiological variability: Vitamin D receptor gene polymorphisms and circulating osteocalcin. *Proc Natl Acad Sci USA* 89: 6665-6669.

28. Yamagata, Z., T. Miyamura, S. Iljimara et al. (1994). Vitamin D receptor gene polymorphism and bone mineral density in healthy Japanese women. *Lancet* 344: 1027.

29. Spector, T.D., R.W. Keen, N.K. Arden et al. (1995). Influence of vitamin D receptor genotype on bone mineral density in postmenopausal women: A twin study in Britain. *Br Med J* 310: 1357-1360.

30. Gallagher, J.C., D. Goldgar, H. Kinyamu et al. (1994). Vitamin D receptor genotypes in type I osteoporosis. *J Bone Miner Res* 9 (Suppl 1): S90.

31. Krall, E.A., P. Parry, J.B. Lichter et al. (1995). Vitamin D receptor alleles and rates of bone loss: Influences of years since menopause and calcium intake. *J Bone Miner Res* 10: 978-984.

32. Looney, J., M. Fisher, H. Yoon et al. (1995). Lack of evidence for an increased prevalence of vitamin D receptor genotype in severe osteoporosis. *J Bone Miner Res* 9 (Suppl 1): S148.

33. Hustmeyer, F.G., M. Peacock, S. Hui et al. (1994). Bone mineral density in relation to polymorphism at the Vitamin D receptor gene locus. *J Clin Invest* 94: 2130-2134.

34. Tylavsky, F.A., A.D. Bortz, R.L. Hancock et al. (1989). Familial resemblance of radial bone mass between premenopausal mothers and their college-age daughters. *Calcif Tissue Int* 45: 265-272.

35. McKay, H.A., D.A. Bailey, A.A. Wilkinson et al. (1994). Familial comparison of bone mineral density at the proximal femur and lumbar spine. *Bone Miner* 24: 95-107.

36. Matkovic, V., D. Fontana, C. Tominac et al. (1990). Factors that influence peak bone mass formation: A study of calcium balance and the inheritance of bone mass in adolescent females. *Am J Clin Nutr* 52: 878-888.

37. Seeman, E., J.L. Hopper, L.A. Bach et al. (1989). Reduced bone mass in daughters of women with osteoporosis. *N Engl J Med* 320: 554-558.

38. Seeman, E., C. Tsalamandris, C. Formica et al. (1994). Reduced femoral neck bone density in the daughters of women with hip fractures: The role of low peak bone density in the pathogenesis of osteoporosis. *J Bone Miner Res* 9: 739-743.

39. Krall, E.A., and B. Dawson-Hughes. (1993). Heritable and life-style determinants of bone mineral density. *J Bone Miner Res* 8: 1-9.

40. Arden, N.K., J. Baker, C. Hogg et al. (1996). The heritability of bone mineral density, ultrasound of the calcaneus and hip axis length: A study of postmenopausal twins. *J Bone Miner Res* 11: 530-534.

41. Arden, N.K., and T.D. Spector. (1997). Genetic influences on muscle strength, lean body mass, and bone mineral density: A twin study. *J Bone Min Res* 12: 2076-2081.

42. Slemenda, C.W., C.H. Turner, M. Peacock et al. (1996). The genetics of proximal femur geometry,

distribution of bone mass and bone mineral density. *Osteoporos Int* 6: 178-182.

43. Flicker, L., J.L. Hopper, L. Rodgers et al. (1995). Bone density determinants in elderly women: A twin study. *J Bone Miner Res* 10: 1607-1613.
44. Hopper, J.L., R.M. Green, C.A. Nowson et al. (1998). Genetic, common environment, and individual specific components of variance for bone mineral density in 10- to 26-year-old females: A twin study. *Am J Epidemiol* 147: 17-29.
45. Smith, D.M., W.E. Nance, K.W. Kang et al. (1973). Genetic factors in determining bone mass. *J Clin Invest* 52: 2800-2808.
46. Dequeker, J., J. Nijs, A. Verstraeten et al. (1987). Genetic determination of bone mineral content at the spine and radius: A twin study. *Bone* 8: 207-209.
47. Kelly, P.J., J.L. Hopper, G.T. Macaskill et al. (1993). Genetic determination of changes in bone density with age: A twin study. *J Bone Miner Res* 8: 11-17.
48. Pocock, N.A., J.A. Eisman, J.L. Hopper et al. (1987). Genetic determinants of bone mass in adults. A twin study. *J Clin Invest* 80: 706-710.
49. Young, D., J.L. Hopper, C.A. Nowson et al. (1995). Determinants of bone mass in 10- to 26-year-old females: A twin study. *J Bone Miner Res* 10: 558-567.
50. Specker, B.L. (1996). Evidence for an interaction between calcium intake and physical activity on changes in bone mineral density. *J Bone Miner Res* 11: 1539-1544.
51. Kelly, P.J., J. Eisman, and P.N. Sambrook. (1990). Interaction of genetic and environmental influences on peak bone density. *Osteoporos Int* 1: 56-60.
52. Pocock, N.A., J.A. Eisman, M.G. Yeates et al. (1986). Physical fitness is a major determinant of femoral neck and lumbar spine bone mineral density. *J Clin Invest* 78: 618-621.
53. Kelly, P.J., N.A. Pocock, P.N. Sambrook et al. (1990). Dietary calcium, sex hormones and bone mineral density in man. *Br Med J* 1361-1364.
54. Jarvinen, T.L.N., T.A.H. Jarvinen, H. Sievanen et al. (1998). Vitamin D receptor alleles and bone's response to physical activity. *Calcif Tissue Int* 413-417.
55. Heaney, R.P. (1986). *Calcium, bone health and osteoporosis*. In W.A. Peck (Ed.), *Bone and mineral research* (vol. 4, pp. 255-301). Amsterdam: Elsevier.
56. Howard, H. (1976). Ultrastructure and biochemical function of skeletal muscle in twins. *Ann Hum Biol* 3: 455-462.
57. Komi, P.V., J.T. Viitasalo, M. Havu et al. (1976). Physiological and structural performance capacity: Effect of heredity. In P.V. Komi (Ed.), *Biomechanics V* (pp. 118-123). Baltimore: University Park Press.
58. Lortie, G., C. Bouchard, A. Le Blanc et al. (1982). Familial similarity in aerobic performance. *Hum Biol* 54: 801-812.
59. Lesage, R., J.A. Simoneau, J. Le Blanc et al. (1985). Familial resemblance in maximal heart rate, blood lactate and aerobic power. *Hum Hered* 35: 182-189.
60. Montoye, H.J., and R. Gayle. (1978). Familial relationships in maximal oxygen uptake. *Hum Biol* 50: 241-249.
61. Bouchard, C., R. Lesage, G. Lortie et al. (1986). Aerobic performance in brothers, dizygotic and monozygotic twins. *Med Sci Sports Exerc* 18: 639-646.
62. Corey, L.A., and W.E. Nance. (1980). A study of dietary intake in adult monozygotic twins. *Acta Genet Med Gamellol* 29: 263-271.
63. Friedlander, A.L., H.K. Genant, S. Sadowsky et al. (1995). A two-year program of aerobics and weight training enhances bone mineral density of young women. *J Bone Miner Res* 10: 574-585.
64. Orwoll, E.S., J. Ferar, S.K. Oviatt et al. (1989). The relationship of swimming exercise to bone mass in men and women. *Arch Intern Med* 149: 2197-2200.
65. Nelson, M.E., C.N. Meredith, B. Dawson-Hughes et al. (1988). Hormone and bone mineral status in endurance-trained and sedentary postmenopausal women. *J Clin Endocrinol Metab* 66: 927-933.
66. Garn, S.M. (1970). *The earlier gain and later loss of cortical bone*. Springfield, IL: Charles C Thomas.
67. Matkovic, V., K. Kostial, I. Simonovic et al. (1979). Bone status and fracture rates in two regions of Yugoslavia. *Am J Clin Nutr* 32: 540-549.
68. Aloia, J.F., A. Vaswani, J.K. Yeh et al. (1996). Risk for osteoporosis in black women. *Calcif Tissue Int* 59: 415-423.
69. Falch, J.A., L. Sandvik, and E.C. Van Beresteijn. (1992). Development and evaluation of an index to predict early postmenopausal bone loss. *Bone* 13: 337-341.
70. Villa, M.L. (1994). Cultural determinants of skeletal health: The need to consider both race and ethnicity in bone research [editorial]. *J Bone Miner Res* 9: 1329-1332.
71. Choi, S.C., and M. Trotter. (1970). A statistical study of the multivariate structure and race-sex differences of American white and Negro fetal skeletons. *Am J Phys Anthropol* 33: 307-312.
72. Trotter, M., G.E. Broman, and R.R. Peterson. (1960). Densities of bones of white and negro skeletons. *J Bone Joint Surg* 42A: 50-58.
73. Trotter, M., G.E. Broman, and R.R. Peterson. (1959). Density of cervical vertebrae and comparison with densities of other bones. *Am J Phys Anthropol* 17: 19-25.
74. Bell, N.H., J. Shary, J. Stevens et al. (1991). Demonstration that bone mass is greater in black than in white children. *J Bone Miner Res* 6: 719-723.

75. Wang, M.C. (1997). Bone mass and hip axis length in healthy Asian, black, Hispanic and white American youths. *J Bone Miner Res* 12: 1922-1935.

76. Daniels, E.D., J.M. Pettifor, C.M. Schnitzler et al. (1997). Differences in mineral homeostasis, volumetric bone mass and femoral neck axis length in black and white South African women. *Osteoporos Int* 7: 105-112.

77. Perry, H.M., 3rd, M. Horowitz, J.E. Morley et al. (1996). Aging and bone metabolism in African American and Caucasian women. *J Clin Endocrinol Metab* 81: 1108-1117.

78. Griffin, M.R., W.A. Ray, R.L. Fought et al. (1992). Black-white differences in fracture rates. *Am J Epidemiol* 136: 1378-1385.

79. Gyepes, M., H.Z. Mellins, and I. Katz. (1962). The lower incidence of fracture in the hip of the negro. *JAMA* 181: 1073-1074.

80. Bohannon, A.D. (1999). Osteoporosis and African American women. *J Womens Health Gend Based Med* 8: 609-615.

81. Heaney, R.P. (1995). Bone mass, the mechanostat, and ethnic differences (editorial). *J Clin Endocrinol Metab* 80: 2289-2290.

82. Luckey, M.M., S. Wallenstein, R. Lapinski et al. (1996). A prospective study of bone loss in African-American and white women—a clinical research center study. *J Clin Endocrinol Metab* 81: 2948-2956.

83. Maggi, S., J.L. Kelsey, J. Litvak et al. (1991). Incidence of hip fractures in the elderly: A cross-national analysis. *Osteoporos Int* 1: 232-241.

84. Jonsson, B., P. Gardsell, O. Johnell et al. (1993). Life-style and different fracture prevalence: A cross-sectional comparative population-based study. *Calcif Tissue Int* 52: 425-433.

85. Hirota, T., M. Nara, M. Ohguri et al. (1992). Effect of diet and lifestyle on bone mass in Asian young women. *Am J Clin Nutr* 55: 1168-1173.

86. Ross, P.D., H. Orimo, R.D. Wasnich et al. (1989). Methodological issues in comparing genetic and environmental influences on bone mass. *Bone Miner* 7: 67-77.

87. McKay, H.A., M. Petit, R. Schutz et al. (2000). Lifestyle determinants of bone mineral: A comparison between prepubertal Asian- and Caucasian-Canadian boys and girls. *Calcif Tissue Int* 66: 320-324.

88. Bhudhikanok, G.S., M.C. Wang, K. Eckert et al. (1996). Differences in bone mineral in young Asian and Caucasian Americans may reflect differences in bone size. *J Bone Miner Res* 11: 1545-1556.

89. McKay, H.A., M.A. Petit, R.W. Schutz et al. (2000). Augmented trochanteric bone mineral density after modified physical education classes: A randomized school-based exercise intervention in prepubertal and early-pubertal children. *J Pediatr* 136: 156-162.

90. Kanis, J.A. (1996). *Textbook of osteoporosis* Oxford: Blackwell Science.

91. Hinton, R.Y., and G.S. Smith. (1993). The association of age, race, and sex with the location of proximal femoral fractures in the elderly. *J Bone Joint Surg* 75A: 752-759.

92. Farmer, M.E., L.R. White, J.A. Brody et al. (1984). Race and sex differences in hip fracture incidence. *Am J Public Health* 74: 1374-1380.

93. Solomon, L. (1968). Osteoporosis and fracture of the femoral neck in the South African Bantu. *J Bone Joint Surg* 50B: 2-13.

94. Silverman, S.L., and R.E. Madison. (1988). Decreased incidence of hip fracture in Hispanics, Asians, and blacks: California Hospital Discharge Data. *Am J Public Health* 78: 1482-1483.

95. Chalmers, J., and K.C. Ho. (1970). Geographical variations in senile osteoporosis. The association with physical activity. *J Bone Joint Surg* 52B: 667-675.

96. Wong, P.C. (1966). Fracture epidemiology in a mixed southeastern Asian community (Singapore). *Clin Orthop* 45: 55-61.

97. Baron, J.A., J. Barrett, D. Malenka et al. (1994). Racial differences in fracture risk. *Epidemiology* 5: 42-47.

98. Fisher, E.S., J.A. Baron, D.J. Malenka et al. (1991). Hip fracture incidence and mortality in New England. *Epidemiology* 2: 116-122.

99. Lizaur-Utrilla, A., A. Puchades Orts, F. Sanchez del Campo et al. (1987). Epidemiology of trochanteric fractures of the femur in Alicante, Spain, 1974-1982. *Clin Orthop* 24-31.

100. Luthje, P. (1985). Incidence of hip fracture in Finland. A forecast for 1990. *Acta Orthop Scand* 56: 223-225.

101. Frandsen, P.A., and T. Kruse. (1983). Hip fractures in the county of Funen, Denmark. Implications of demographic aging and changes in incidence rates. *Acta Orthop Scand* 54: 681-686.

102. Alhava, E.M., and J. Puittinen. (1973). Fractures of the upper end of the femur as an index of senile osteoporosis in Finland. *Ann Clin Res* 5: 398-403.

103. Krall E, Dawson-Hughes B., K. Hirst, et al. (1997). Bone mineral density and biochemical markers of bone turnover in healthy elderly men and women. *J Gerontol A Biol Sci Med Sci* 52: M61-67.

104. Riggs, B.L. (2000). The mechanisms of estrogen regulation of bone resorption. *J Clin Invest* 106: 1203-1204.

105. Cheng, S., J.A. Toivanen, H. Suominen, et al. (1995). Estimation of structural and geometrical properties of cortical bone by computerized tomography in 78-year-old women. *J Bone Miner Res* 10: 139-148.

106. Brockstedt, H., M. Kassem, E.F. Eriksen, et al. (1993). Age- and sex related changes in iliac

cortical bone mass and remodeling. *Bone* 14: 681-691.

107. Morrison N. (1997). Letter. *Nature* 387:106.

108. Seeman E, J.L. Hooper, N.R. Young, et al. (1996). Do genetic factors explain associations between muscle strength, lean mass, and bone density? A twin study. *Am J Physiol* 207:E320-327.

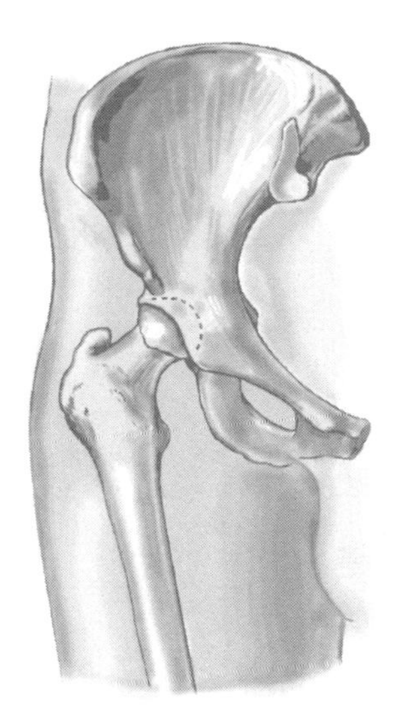

Chapter 6

Soft Tissue Determinants of Bone Mineral Density

The skeleton is a product of what we eat (or don't eat), what we do (or don't do), and who our parents and ancestors were. In this chapter we discuss how our skeleton is also influenced in part by soft tissue composition. We begin with a review of the evidence regarding total body mass and bone mineral, and then distinguish the influences of lean and fat mass as separate determinants of bone mineral.

Total body mass (weight) can be divided into a number of relevant compartments and tissue masses. This can be done anatomically (bone, muscle, fat, residual) or chemically (water, lipid, protein). The relative proportion of each of the tissue types that comprise total body weight is uniquely different among individuals, but surprisingly similar among athletes in the same sporting groups.

Consider the diversity of three athletic body types—the Sumo wrestler, the high jumper, and the gymnast. Intuitively we know that the Sumo wrestler will have a larger proportion of body fat than the gymnast, and that the gymnast will likely have a greater percentage of muscle mass than the high jumper. But we also know that there are remarkable similarities among all Sumo wrestlers and among all members of the women's Olympic gymnastics team. The sum of all body tissues, and the proportions of each, largely reflect genetics, body size, and past and current lifestyle.

Because DXA technology (see chapter 4) allows us to measure total body bone and soft tissue, it can compartmentalize bone mineral, fat, and (bone-mineral-free) lean tissue mass with reasonable accuracy. Note, however, that DXA soft tissue measurement assumes uniform hydration of each of the three compartments, yet lean mass in particular can vary in water content by up to 85% [1, 2]. This introduces the potential for error in DXA measurement of soft tissue.

Error is also introduced as DXA assumes that the soft tissue above and below the bone (where it cannot be measured) is equivalent to that on either side of the bone. If more fat is located above the spine than on either side of it, BMD will be underestimated. DXA also assumes that bone marrow is included within these soft tissue calculations, although this is not true in the strictest sense. A change in bone marrow fat of 50% will change BMD by 5-6% [3]. Aging causes a redistribution of body fat and an increase in yellow marrow. Taken together, these changes could result in erroneous estimations by DXA in older populations. It has been recommended that DXA assessments be limited to individuals whose abdominal thickness is between 15 and 30 cm, but these limitations are system specific [3, 4].

Despite these limitations, DXA has provided researchers with a better understanding of the role of soft tissue (fat and lean) in determining bone mass. Unfortunately, better understanding does not equal total clarity. The degree to which soft tissue mass influences bone remains an area of controversy since lean mass, fat mass, and total body weight are interrelated as well as being associated with lifestyle factors such as physical activity. Further, the specific and relative contribution of each tissue to bone varies across the life span and among diverse groups [5]. These issues are expanded in the following sections.

Total Body Mass and BMD

Studies in both children [6] and adults [7, 8] identified body weight as the primary predictor of bone mass at regional sites (hip, spine) and for the total body.

Nevertheless, the literature contains studies with contradictory findings. This may reflect variations in such methodological factors as the measures of body weight and fat mass used (e.g., BMI [body mass index], IDW [ideal body weight], mass), the age of the subjects, the study design (cross-sectional versus longitudinal), and the statistical approach (choice of control variables). Other factors such as the hormonal status of the subjects and the use of hormone replacement therapy may also have contributed to the contradictory findings.

Cross-sectional studies have demonstrated that body weight may be a more important predictor of bone mineral density (BMD) in older than in younger populations. Postmenopausal women who were at least 10% above their ideal body weight had significantly greater BMD at the spine, hip, and radius than did women of normal weight [9, 10]. A similar phenomenon was not found in premenopausal women, suggesting that the increased weight may be a larger factor in slowing postmenopausal bone loss than in maintaining bone mineral in the adult skeleton [9]. However, this hypothesis cannot be tested with a cross-sectional study design.

In large cohort studies, such as the Framingham study (1988-89), body weight explained up to 20% of the variance in BMD for both weight-bearing (lumbar spine, femoral neck) and non-weight-bearing sites (radius) [11]. The Rancho Bernado study found that body size explained a greater proportion of the variance of the weight-bearing hip and spine sites (17% and 12%, respectively) than of the non-weight-bearing radial sites (8% or less) [12].

For women in the four-year longitudinal Framingham Osteoporosis Study, lower baseline weight and weight loss were more strongly associated with BMD loss than factors such as age, smoking, caffeine, alcohol use, or (unfortunately) physical activity [13].

Weight loss generally appears to be associated with decreased BMD (figures 6.1 and 6.2) [14-

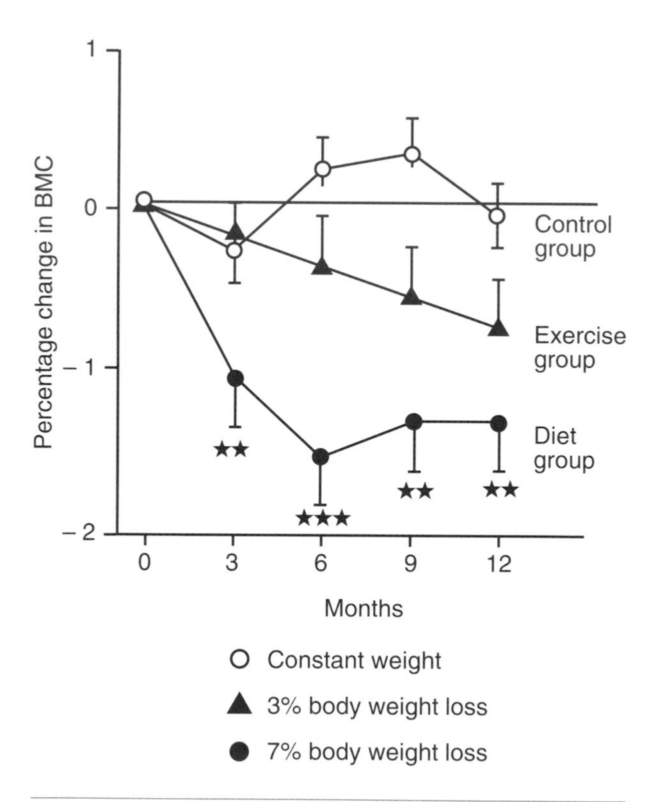

FIGURE 6.1 Bone mineral content (BMC) and weight loss over 12 months. BMC remained constant in control subjects whose weight remained constant. It tended to decrease in the exercising group, whose subjects lost on average 3% body weight. BMC decreased significantly in the subjects who dieted and lost on average 7% body weight.

Reprinted, by permission, from Pritchard JE, Nowson CA, Wark JD, 1996, "Bone loss accompanying diet-induced or exercise-induced weight loss: a randomised controlled study," *Int J Obesity* (20).

16], although inconsistencies do exist between studies [17]. Several mechanisms could explain this phenomenon. First, the fat tissue loss may decrease endogenous estrogen production. Also, weight loss may stimulate increased bone turnover and loss particularly at trabecular sites such as the spine and hip. Finally, changes in BMD may be artifactual as fat tissue itself may affect densitometric measurement of bone (see chapter 4) [18, 19]. It is particularly important for students and researchers in the field of physical activity and BMD to be aware of this, as weight loss that occurs during an exercise intervention can potentially confound results.

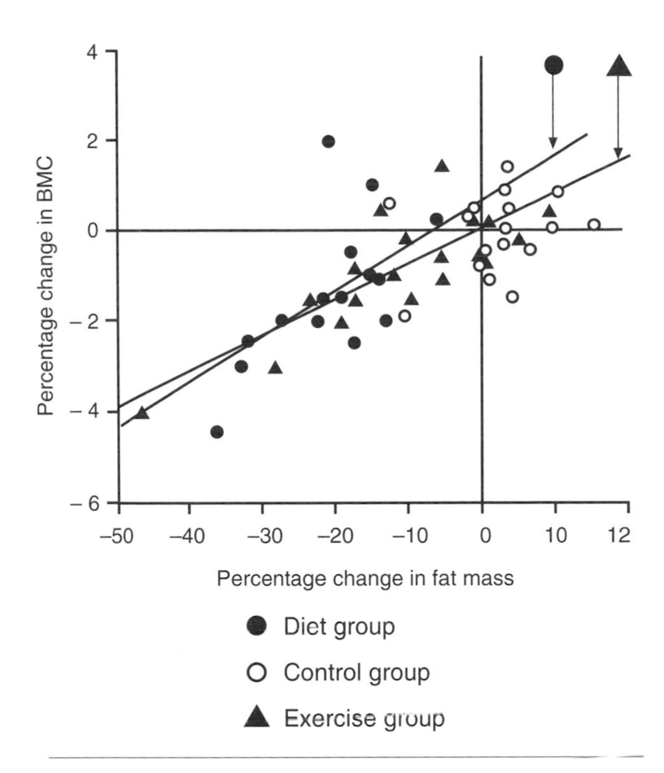

FIGURE 6.2 Change in BMC was significantly and positively correlated with change in fat mass in both the exercise group (triangles) and the diet group (filled circles) ($r = 0.8$, $p < 0.01$).

Reprinted, by permission, from Pritchard JE et al., 1996, "Bone loss accompanying diet-induced or exercise-induced weight loss: a randomised controlled study," *Int J Obesity* (20).

Lean Mass and BMD

With the advent of DXA technology, researchers were able to distinguish the specific tissues that comprise body weight. Of these, lean mass proved the most robust predictor of BMD in young [20-23], middle-aged [24, 25], and elderly [26-29] women. In a cross-sectional study of young women (aged 10-26), each kilogram of lean mass, with all other factors held constant, was associated with about 1% greater proximal femoral BMD [21].

There has been some consensus that lean mass may have different significance at different stages of life. Total lean mass may be the most powerful soft tissue determinant of lumbar and total body BMD in children (figures 6.3 and 6.4) [6, 21, 30, 31] and premenopausal women [32], whereas it may be less important during infancy [33] and after the menopause [24, 34, 35]. In elderly women (60-89 years) lean mass was independently associated with BMD at all sites when baseline data were analyzed cross-sectionally [26]. In longitudinal follow-up, however, change in lean mass did not predict change in BMD [36]. This may occur because fat mass accounts for a greater proportion of overall body mass in the very young and the elderly. Moreover, the association between lean mass and BMD may be established most strongly during growth and development, so that the associations seen cross-sectionally in later life may reflect early development. Further longitudinal studies on changes in BMD in the postmenopausal years are needed to determine whether changes in BMD are sensitive to changes in body composition and to delineate the mechanism of any such association [17].

Fat Mass and BMD

Fat mass is an independent predictor of BMD, particularly the weight-bearing axial and lower limb BMD, to a degree that is not accounted for merely by its mechanical load on the skeleton. During childhood fat mass probably has less influence on BMD than does lean mass [21]. It is an important determinant during the premenopausal years [38-40] and becomes increasingly significant after menopause [41]. Fat mass may predict change in BMD in postmenopausal women better than does lean mass [42, 43].

Potential mechanisms whereby fat could influence bone mass in addition to its loading effect [44] are shown in table 6.1.

Postmenopausal women convert 10-15% of circulating androstenedione (a hormone from the androgen family) to estrone in adipose tissue [45] and 25-30% in muscle [46]. Obese women appear to have greater supplies of androgens available for conversion [47]. Women of normal weight convert 1-2% of androstenedione to es-

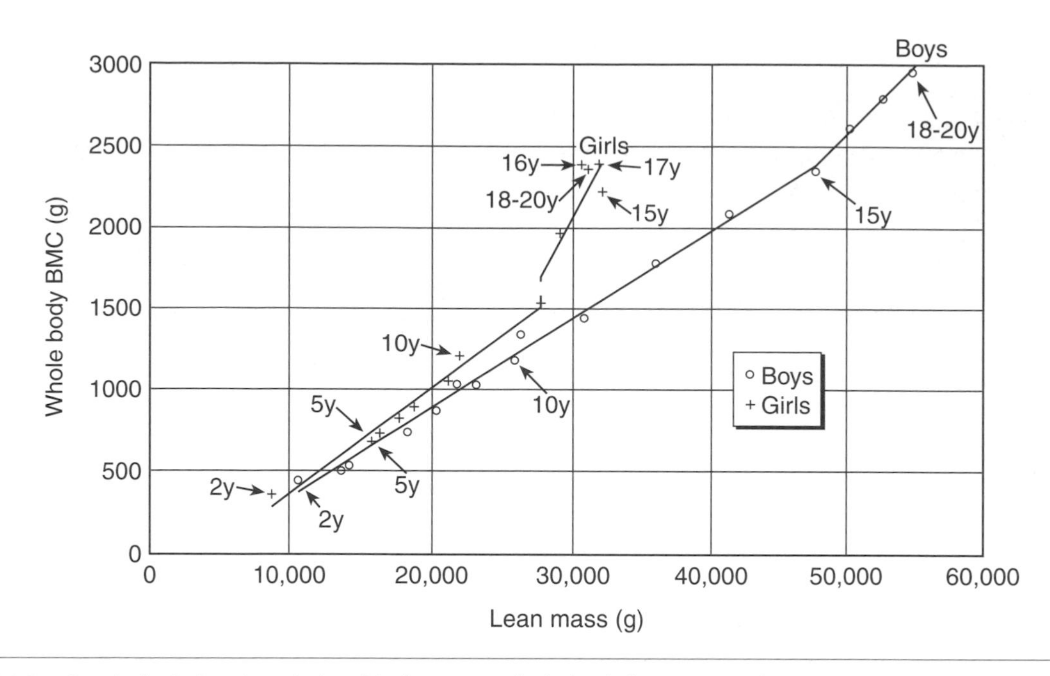

FIGURE 6.3 Graph depicting the relationship between whole body bone mineral content (BMC) and lean mass (crosses = girls; open circles = boys).

Reprinted from *Bone*, 22, Schiessl H, Frost HM, Jee WSS, "Estrogen and bone-muscle strength and mass relationship," 1-6, copyright 1998, with permission from Elsevier Science.

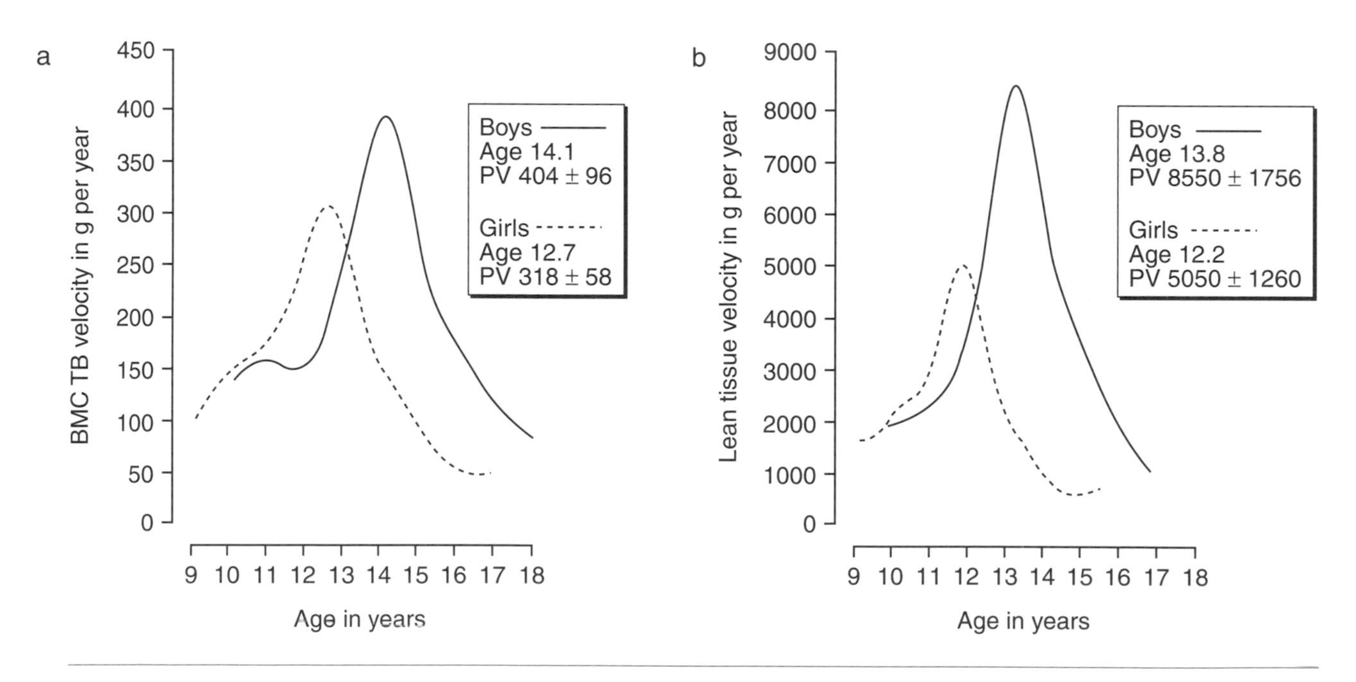

FIGURE 6.4 *(a)* The timing of peak gains in bone mineral is very closely associated with *(b)* timing of peak gains in lean mass [48].

trone, whereas women weighing between 300 and 400 lb (130 and 180 kg) convert 12-15% [45]. In addition, the conversion of androgens to estrogen increases with age [46].

Soft Tissue in Bone Research Studies

Absolute body weight, and the amount of fat or lean tissue mass that comprises this weight, has the potential to confound findings in cross-sectional and longitudinal studies. This is especially true for physical activity intervention studies in which a program of physical activity may change tissue composition. This is also true in studies of growing children in which size, weight, and tissue composition change as children mature. It is therefore crucial that these variables (absolute values and change in these values over the course of a study) be controlled for within study design (matching) or in the statistical analysis.

TABLE 6.1

Association between fat mass and bone mass
Potential mechanisms whereby fat mass may influence bone mass
• Aromatization of androgens
• Metabolism of estrogen from the more potent to the less potent catechol estrogens
• Alteration of the binding capacity of estrogen for sex hormone binding globulin
• Acting as a storage unit for steroid hormones
• Via development of obesity-related insulin resistance

SUMMARY

- Total body mass and its components, lean mass and soft tissue mass, are major determinants of bone mass.
- There is a positive correlation between lean mass and bone mass.
- Fat mass is also correlated with bone mass.
- The relative contributions of these soft tissues to bone mass varies throughout the life span and between the sexes. Because soft tissue mass can easily confound studies of bone and physical activity, it must be assessed and accounted for.

References

1. Heymsfield SB, Waki M. Body composition in humans: advances in the development of multi-compartment chemical models. *Nutr Rev* 1991; 49:97-108.
2. Roubenoff R, Kehayias JJ, Dawson-Hughes B, et al. Use of dual-energy x-ray absorptiometry in body-composition studies: not yet a "gold standard". *Am J Clin Nutr* 1993;58:589-91.
3. Sorenson J. Effects of nonmineral tissues on measurement of bone mineral content by dual-photon absorptiometry. *Med Phys* 1990;17:905-912.
4. Wahner HW. *Use of densitometry in management of osteoporosis*. In: Marcus R, Feldman D, Kelsey J, ed. Osteoporosis. San Diego, CA, USA: Academic Press, 1996: 1055-1074.
5. Frost HM. Why do marathon runners have less bone than weight-lifters? A vital-biomechanical view and explanation. *Bone* 1997; 20:183-189.
6. Faulkner RA, Bailey DA, Drinkwater DT, et al. Bone densitometry in Canadian children 8-17 years of age. *Calcif Tissue Int* 1996;59:344-351.
7. Flicker L, Hopper JL, Rodgers L, et al. Bone density determinants in elderly women: a twin study. *J Bone Miner Res* 1995;10:1607-13.

8. Salamone LM, Glynn N, Black D, et al. Body composition and bone mineral density in premenopausal and early perimenopausal women. *J Bone Miner Res* 1995;10:1762-8.
9. Ribot C, Tremollieres F, Pouilles JM, et al. Obesity and postmenopausal bone loss: the influence of obesity on vertebral density and bone turnover in postmenopausal women. *Bone* 1988;8:327-331.
10. Dawson-Hughes B, Shipp C, Sadowski L, et al. Bone density of the radius, spine and hip in relation to percent ideal body weight in postmenopausal women. *Calcif Tissue Int* 1987;40:310-314.
11. Felson DT, Zhange Y, Hannan MT, et al. Effects of weight and body mass index on bone mineral density in men and women: The Framingham study. *J Bone Miner Res* 1993;8:567-573.
12. Edelstein SL, Barrett-Connor E. Relation between body size and bone mineral density in elderly men and women. *Am J Epidemiol* 1993;138:160-169.
13. Hannan MT, Felson DT, Dawson-Hughes B, et al. Risk factors for longitudinal bone loss in elderly men and women: the Framingham Osteoporosis Study. *J Bone Miner Res* 2000;15:710-20.
14. Pritchard JE, Nowson CA, Wark JD. Bone loss accompanying diet-induced or exercise-induced weight loss: a randomised controlled study. *Int J Obesity* 1996;20:513-520.
15. Jensen LB, Quaade F, Sorensen OH. Bone loss accompanying voluntary weight loss in obese humans. *J Bone Miner Res* 1994;9:459-63.
16. Compston JE, Laskey MA, Croucher PI, et al. Effect of diet-induced weight loss on total body bone mass. *Clin Sci (Colch)* 1992;82:429-432.
17. Cauley J, Salamone LM, Lucas FL. *Postmenopausal endogenous and exogenous hormones, degree of obesity, thiazide diuretics and risk of osteoporosis.* In: Marcus R, Feldman D, Kelsey J, ed. Osteoporosis. San Diego, CA, USA: Academic Press, 1996: 551-576.
18. Parfitt AM, Miller MJ, Frame B, et al. Metabolic bone disease after intestinal bypass for treatment of obesity. *Ann Intern Med* 1978;89:193-9.
19. Sievanen H, Oja P, Vuori I. Scanner-induced variability and quality assurance in longitudinal dual energy x-ray-absorptiometry measurements. *Med Phys* 1994;21:1795-1805.
20. Henderson NK, Price RI, Cole JH, et al. Bone density in young women is associated with body weight and muscle strength but not dietary intakes. *J Bone Miner Res* 1995;10:384-393.
21. Young D, Hopper JL, Nowson CA, et al. Determinants of bone mass in 10- to 26-year-old females: A twin study. *J Bone Miner Res* 1995;10:558-567.
22. McKay HA, Petit M, Schutz R, et al. Lifestyle determinants of bone mineral: A comparison between prepubertal Asian- and Caucasian-Canadian boys and girls. *Calcif Tissue Int* 2000;66:320-324.
23. McKay HA, Petit MA, Schutz RW, et al. Augmented trochanteric bone mineral density after modified physical education classes: A randomized school-based exercise intervention in prepubertal and early-pubertal children. *J Pediatr* 2000;136:156-162.
24. Khosla S, Atkinson EJ, Riggs BL, et al. Relationship between body composition and bone mass in women. *J Bone Miner Res* 1996;11:857-863.
25. Ho SC, Wong E, Chan SG, et al. Determinants of peak bone mass in Chinese women aged 21-40 years. III. Physical activity and bone mineral density. *J Bone Miner Res* 1997;12:1262-1271.
26. Flicker L, Hopper JL, Rodgers L, et al. Bone density determinants in elderly women: a twin study. *J Bone Miner Res* 1995;10:1607-1613.
27. Bevier WC, Wiswall RA, Pyka G, et al. Relationship of body composition, muscle strength and aerobic capacity to bone mineral density in older men and women. *J Bone Miner Res* 1989;4:421-432.
28. Hoover PA, Webber CE, Beaumont LF, et al. Postmenopausal bone mineral density: relationship to calcium intake, calcium absorption, residual estrogen, body composition and physical activity. *Can J Physiol Pharmacol* 1996;74:911-917.
29. Taaffe DR, Pruitt L, Lewis B, et al. Dynamic muscle strength as a predictor of bone mineral density in elderly women. *J Sports Med Phys Fitness* 1995;35:136-142.
30. Morris FL, Naughton GA, Gibbs JL, et al. Prospective 10-month exercise intervention in premenarcheal girls: positive effects on bone and lean mass. *J Bone Miner Res* 1997;12:1453-1462.
31. Bonjour JP, Rizzoli R. *Bone acquisition in adolescence.* In: Marcus R, Feldman D, Kelsey J, ed. Osteoporosis. San Diego, CA, USA: Academic Press, 1996: 465-476.
32. Heinonen A, Kannus P, Sievänen H, et al. Randomised control trial of effect of high-impact exercise on selected risk factors of osteoporotic fractures. *Lancet* 1996;348:1343-1347.
33. Jones G, Couper D, Riley M, et al. Determinants of bone mass in prepubertal children: Antenatal, neonatal and current influences. *J Bone Miner Res* 1997;12 (Supplement 1):S145 (abstract #169).
34. Douchi T, Oki T, Nakamura S, et al. The effects of body composition on bone density in pre- and post-menopausal women. *Maturitas* 1997;27:55-60.
35. Reid IR, Legge M, Stapleton JP, et al. Regular exercise dissociates fat mass and bone density in premenopausal women. *J Clin Endocrinol Metab* 1995;80:1764-1768.
36. Paton LM, Flicker L, Hopper J, et al. Determinants of change in bone mass in elderly women - a longitudinal twin study [abs]. *Proceedings of the ANZBMS* 1996;6:36.
37. Schiessl H, Frost HM, Jee WSS. Estrogen and bone-muscle strength and mass relationship. *Bone* 1998;22:1-6.

38. Martini G, Valenti R, Giovani S, et al. Age-related changes in body composition of healthy and osteoporotic women. *Maturitas* 1997;27:25-33.
39. Reid IR, Plank LD, Evans MC. Fat mass is an important determinant of whole body bone density in premenopausal women but not in men. *J Clin Endocrinol Metab* 1992;75:779-782.
40. Sowers MF, Kshirsagar A, Crutchfield MM, et al. Joint influence of fat and lean body composition compartments on femoral bone mineral density in premenopausal women. *Am J Epidemiol* 1992;136: 257-265.
41. Reid IR, Ames R, Evans MC, et al. Determinants of total body and regional bone mineral density in normal postmenopausal women - a key role for fat mass. *J Clin Endocrinol Metab* 1992;75:45-51.
42. Chen Z, Lohman TG, Stini WA, et al. Fat or lean tissue mass: which one is the major determinant of bone mineral mass in healthy postmenopausal women? *J Bone Miner Res* 1997;12:144-151.
43. Wark JD, Cassar C, Hoang H, et al. Factors associated with bone mass in midlife twins. *Osteoporos Int* 1998;8:106.
44. Frost HM. Obesity, and bone strength and "Mass": A tutorial based on insights from a new paradigm. *Bone* 1997;21:211-214.
45. Siiteri P. Adipose tissue as a source of hormones. *Am J Clini Nutr* 1987;45:277-282.
46. Longcope C, Baker S. Androgen and estrogen dynamics: relationships with age, weight and menopausal status. *J Clin Endocrinol Metab* 1993;76:601-604.
47. MacDonald PC, Edman CD, Hemsell DL, et al. Effect of obesity on conversion of plasma androstenedione to estrone in postmenopausal women with endometrial cancer. *Am J Obstet Gynecol* 1978; 130:448-455.
48. McKay HA, Mirwald RL, Bailey DA. A 7-year longitudinal comparison of peak bone mineral, peak lean, and peak fat mass accrual velocities and peak height velocity [abstract]. Proceedings of the European Symposium on Calcified Tissues, Tampere, Finland. 2000;12.

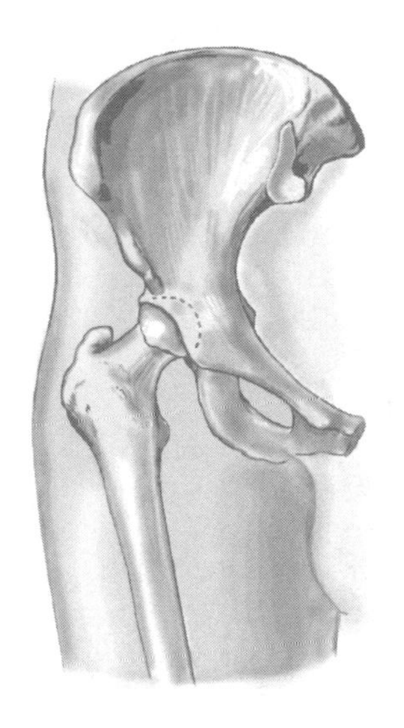

Chapter 7

Influence of Normal Endocrine Function on Bone Mineral

The endocrine system is intimately involved in the regulation of bone biology. In fact, the cumulative effect of the interaction between hormones and minerals results in many of the overall tissue features of the skeleton. The hormones that influence bone can be thought of as belonging to one of two classes, either "controlling" or "influencing" serum calcium levels and levels of other bone-related agents. The calciotropic, or "controlling," hormones—parathyroid hormone, vitamin D, and calcitonin—respond to changes in plasma-ionized calcium concentrations. How they regulate serum calcium was discussed in chapter 2. The "influencing hormones" also modify calcium metabolism, but their secretion is determined primarily by factors other than changes in plasma calcium.

In this chapter we summarize the effects of estrogen, progesterone, testosterone, growth hormone, insulin-like growth factor I (IGF-I), corticosteroids, and thyroid hormone on bone.

Estrogen

Estrogen affects bone both directly and indirectly. After discussing how it does this, we consider its relationship to bone in pregnancy, in women using the oral contraceptive pill, and in women on hormone replacement therapy. We then outline its pivotal role in the bone health of men.

Direct Actions

Estrogen's role in decreasing bone turnover is complex and involves interaction with multiple cell types and the regulation of multiple mediators. At least some actions are mediated via the estrogen receptor on the osteoblast [1]. When estrogen binds to the receptor, it increases production of type I collagen and transforming growth factor-beta (TGF-beta). Estrogen also acts by controlling production of locally produced factors: cytokines (e.g., IL-1, IL-6), growth factors, prostaglandins. Cytokines (IL-1,

IL-6), tumor necrosis factors, and TGF-beta are all potent stimulators of bone resorption and osteoclast formation that appear to be secreted by stromal cells and osteoclasts and whose levels are depressed by estrogen. In this way estrogen maintains bone mass by limiting bone resorption, and thus, acts to maintain bone mass [2].

Indirect Actions

Estrogen may also have an indirect effect on bone via the parathyroid gland, gut, or kidney [3]. For example, a change in the set-point at which parathyroid hormone (PTH) responds to serum calcium (see chapter 2) would promote bone mineralization by reducing bone turnover. Increased gastrointestinal absorption of calcium would have the same effect. Estrogen may increase renal calcium retention by stimulating an increase in renal synthesis of 1,25-dihydroxyvitamin D_3 [4]. Another indirect effect of estrogen on bone is via the stimulation of calcitonin release. Calcitonin, which limits bone turnover and thus bone loss, also affects renal calcium conservation.

Oral Contraceptive Pill and Bone

The effect of the oral contraceptive pill on BMD in normally menstruating women is unclear, as some studies report no effect, others a positive effect, and some even a negative effect on BMD [5, 60]. The oral contraceptive pill is often prescribed for treatment of menstrual disturbances in female athletes, but whether BMD improves in these situations remains unclear (see chapter 16). Furthermore, whether estrogen in the oral contraceptive pill helps prevent stress fractures in female athletes is also unclear (see chapter 17) [5]. The dearth of good quality data about important issues regarding the oral contraceptive pill highlights this as an excellent area for future research studies.

Estrogen Replacement Therapy and Bone

Hormone replacement therapy (HRT), either with estrogen alone (unopposed) or in combinations with progestins, is a commonly used medication, accounting for over 32 million prescriptions per year in the United States alone. This therapy has a moderate and consistent effect on bone mineral [6]. However, the severity of side effects (breast tenderness, spotting, pelvic discomfort, and mood changes) and problems associated with its long-term use make it an unpleasant choice for a large number of women [7]. In a randomized, double-blind, placebo-controlled trial of 128 healthy white women given relatively low doses of conjugated estrogen and medroxyproges-terone for 3.5 years [7], spinal BMD improved 3.5% ($p < 0.001$), as did BMD at the forearm and total body. The women were also supplemented with calcium to 1000 mg/d and vitamin D to maintain serum levels of 25-hydroxyvitamin D at 75 nmol/L.

Skeletal benefits of estrogens continue to accrue with duration of use for up to five years [8], and at least seven years of estrogen therapy is necessary for a persistent long-term effect on BMD [9]. Since few women take estrogen for such a long time, treatment may generally have been of insufficient duration to affect BMD in the long term (figure 7.1). Furthermore, the effect of estrogen is no longer evident many years after the discontinuation of therapy [9, 10]. Even 10 or more years of past estrogen therapy among women 75 years or older did not have a significant residual effect on bone density [9]. Nevertheless, hip fracture rates were lower in women taking hormone replacement than in those who were not (figure 7.2) [11]. This implies that current estrogen use may protect against fractures, but past use may not be effective.

The mechanism explaining estrogen's role in reducing fracture risk independently of BMD is unclear. Although some studies have suggested that women taking estrogen have improved balance [12], contradictory evidence exists [13].

Interaction Between Estrogen and Physical Activity in Women

Kohrt argues that weight-bearing exercise and HRT have independent and additive effects on BMD at the lumbar spine and hip sites and a synergistic effect on total body BMD [16, 17]. These results are consistent with experiments in ovariectomized rats where exercise and estrogen replacement had independent and additive effects on vertebral and femoral bone mass [18, 19].

It is proposed that the independent effects of exercise and HRT could be mediated in two ways. Estrogen replacement suppresses bone turnover, and exercise may act to increase bone formation. However, bone turnover markers in exercise studies have not shown this pattern consistently [16, 17] and further studies are required.

On the other hand, some have found that exercise plus HRT but not exercise alone increased BMD. Exercise did not add to the effect of HRT

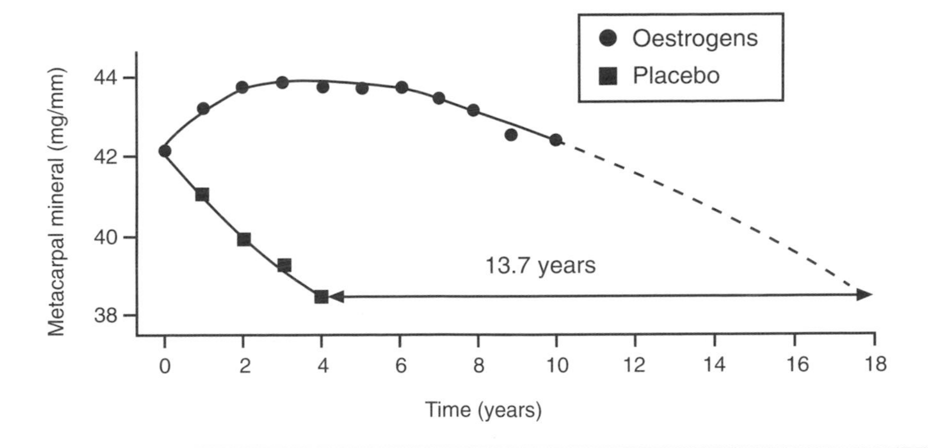

FIGURE 7.1 In untreated patients after oophorectomy, bone loss in 3-4 years is similar to that lost in almost 20 years by patients treated with hormone replacement therapy [14, 15].

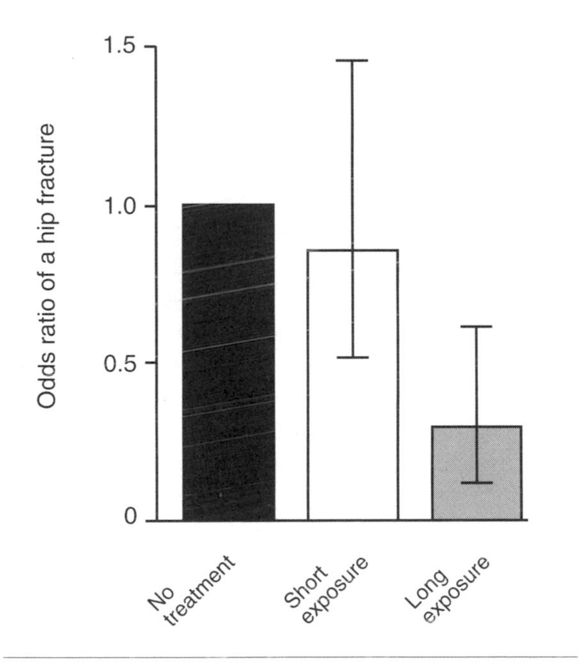

FIGURE 7.2 Hormone replacement therapy decreases the risk of hip fracture, but the benefit depends on the number of years that the patient has been taking the medication [11].

alone [20, 21]. However, the exercise level in these studies was poorly described and there were no data to support the effectiveness of exercise interventions (for example, no measures of increased strength) [22]. A more recent study that overcame these limitations also found no synergistic effect of exercise and HRT [23].

Estrogen in Men

Estrogen is important for bone maturation and mineralization in men as well as in women [24-26]. In 20- to 90-year-olds it is the sex steroid most strongly associated with BMD [13, 27]. This has been illustrated in the case of a 28-year-old man in whom an abnormal estrogen-receptor gene effectively deprived the skeleton of any estrogen. As a result, the patient was tall (80 in. [204 cm]) and had incomplete epiphyseal closure, with a history of continued linear growth into adulthood despite otherwise normal pubertal development. In addition, he had low BMD (t-score was –3.1). Thus, estrogen is important for bone maturation and mineralization in men as well as in women [24].

Progesterone

Progesterone has not been studied as thoroughly as estrogen, and whether or not it increases bone mineral remains a controversy. On the one hand, animal studies show that treatment with progestins (synthetic progesterone) decreases bone resorption and increases formation, whereas treatment with estrogen produced only decreased resorption [28].

Clinical studies indicate that progesterone treatment in postmenopausal women decreases the markers of bone resorption [29, 30] but does not appear to elevate the markers of bone formation. While existing data are inconclusive, the balance of evidence suggests that progesterone has relatively modest effects on bone in women who have normal menstrual cycles. In studies of combined estrogen and progestin treatment, however, there was both decreased resorption and increased formation [30, 31]. The role of progesterone treatment of the menstrual disturbance known as "short luteal phase" is discussed in chapter 16 [32, 33].

Effects of Pregnancy and Lactation

During the third trimester of pregnancy maternal estrogen levels are high, but during lactation, the mother becomes hypoestrogenic as prolactin dominates [34]. If maternal bone mineral were the sole source of calcium for the fetus, a mother's skeleton would lose about 3% of her BMD per pregnancy. However, bone mass is preserved, presumably in part because of higher circulating levels of 1,25-dihydroxyvitamin D and increased intestinal absorption efficiency [35]. Extensive lactation causes an early, perhaps physiologic, bone loss that recovers with weaning. Breast-feeding duration of less than four to six months does not generate enough calcium demand to create significant maternal bone loss [35]. A recent commentary in *The Lancet* concluded, "women can be assured that even with moderate calcium intakes, there is no evidence for any long-term harm to their skeleton or long-term osteoporosis risk from pregnancy and lactation" [36].

However, in women who might already have low BMD and other risk factors for osteoporosis before pregnancy, it would appear prudent to ensure that dietary calcium intake is adequate during pregnancy [37-40]. This will reduce the risk of pregnancy-associated osteoporosis, an uncommon, and often serious condition, which is usually reversible at least in part [39].

Testosterone

The effects of testosterone on the body are multiple and widespread. Testosterone may contribute to the larger peak bone size in men than in women, as well as contributing to periosteal bone apposition (cortical bone accrual of bone mineral) [41-46]. In the largest study reported to date (534 men aged 69± 8 years) using recently available technology, bioavailable (i.e., measuring free, biologically available testosterone) testosterone was positively associated with BMD at the ultradistal radius, lumbar spine, and hip [27]. This is consistent with findings in Australian [45] and British [46] studies.

The action mechanism of sex steroids is not well understood. Androgen receptors have been reported in the nucleus of the osteoblast [47], and they may exist in or around the osteoclast nucleus [48].

It is not known at what level serum testosterone becomes inadequate to contribute to skeletal health. A few studies have shown an association between bone mass and testosterone, but these studies are difficult to interpret since a single measurement of serum testosterone has a precision error of about 10-15% [27, 45, 46, 49, 50].

Although the predominant circulating sex steroid in men is testosterone and the predominant circulating sex steroids in women are estradiol and estrone, androgens and estrogens circulate in both sexes and contribute to bone health in both men and women [51].

Growth Hormone and Insulin-Like Growth Factor I

Growth hormone (GH) is known for its effect at the epiphyseal growth plate, an effect mediated via insulin-like growth factor I (IGF- I). An excess or deficiency of GH is associated with obvious abnormalities of skeletal growth, such as acromegaly with GH excess. In this condition of excess GH, exuberant periosteal apposition of bone produces the characteristic protruding jaw and prominent brow. Administration of GH to healthy humans and pituitary dwarfs increases net intestinal absorption of calcium.

GH and IGF-I decrease with advancing age, and these decrements may contribute to the reduced bone formation found during aging. GH may play a role in the accretion and/or maintenance of bone mass in either puberty [52] or young adulthood [53]. The release of GH is pulsatile and circadian, with the highest pulse amplitudes observed at night with the onset of sleep. The pattern of GH release is also a function of age and sex, with the frequency and magnitude of the pulses increasing as children pass through puberty.

It is important to remember that exercise, stress, and sleep, as well as levels of body fuels and hormones, increase the production of GH. For example, the low circulating levels are punctuated by four to eight bursts, which occur after meals or exercise, during slow-wave sleep, or without an obvious cause [54]. Thus, exercise may modulate levels of GH and influence bone mineral through that mechanism.

After about the age of 35, there is a decline in the secretion of GH and IGF-I, and this has been called the "somatopause" [41]. In a group of men with widely ranging ages, insulin-like growth factor binding protein (IGFBP-3) was strongly and significantly correlated with BMD at the total body, lumbar spine, and hip [55].

Corticosteroid Hormones

Corticosteroids rate a mention, not because of their physiological role in bone metabolism, but because excess glucocorticoid can have major adverse effects on bone (figure 7.3). While the textbook cause of glucocorticoid excess is a certain type of pituitary or adrenal tumors, the vast majority of glucocorticoid bone disease today is due to the hormone being used in its capacity as a powerful anti-inflammatory in immunosuppressive medication, used for conditions such as asthma, rheumatoid disease, and in transplant patients.

As osteoporosis associated with therapeutic glucocorticoid use is highly predictable, opportunities may arise for the exercise specialists or the physicians to prescribe appropriate exercise immediately after a patient starts on the medication. This adjunct therapy aims to improve balance in order to reduce fall and fracture risk as well as reduce the rate of bone loss that inevitably occurs with the use of corticosteroids. We emphasize that patients on corticosteroids are often not in a position to perform vigorous exercise (e.g., patients with severe asthma or inflammatory bowel disease) and that exercise prescription must be medically supervised [62]. In these cases, the use of HRT or other pharmacotherapy (e.g., bisphosphonates) should be prescribed to minimize bone loss.

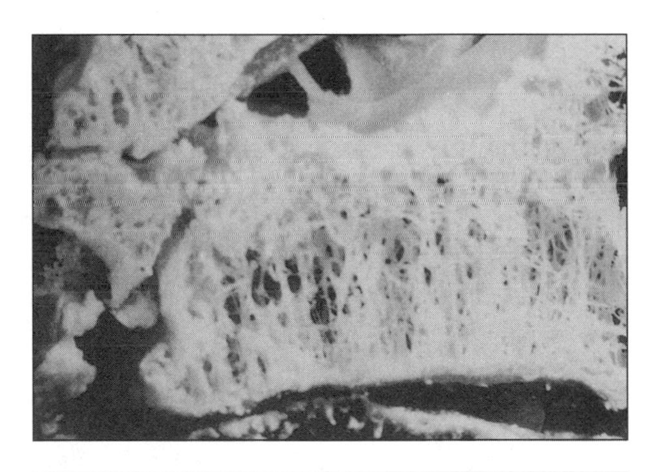

FIGURE 7.3 Example of a vertebral compression fracture such as can result from corticosteroid use.

Figure from "The Bone Organ System: Form and Function" by Thomas A. Einhorn, in OSTEOPOROSIS, edited by Robert Marcus, et al., copyright © 1996 by Academic Press, reproduced by permission of the publisher.

A substantial body of literature confirms the osteopenic effects of glucocorticoid medications [56, 57]. Glucocorticoids promote trabecular and cortical bone loss due to (1) a direct effect on osteoblasts and osteoclasts to decrease bone formation and possibly to increase bone resorption; (2) the inhibition of gastrointestinal absorption of calcium; (3) a reduction of tubular reabsorption of calcium with hypercalcuria; (4) secondary hyperparathyroidism, although that remains controversial; (5) myopathy and reduced physical activity; and (6) effects on pituitary and gonadal function in men and women.

Thyroid Hormone

Deficiencies of thyroid hormones early in life produce the well-known skeletal deformities of cretinism, but this is largely prevented in countries that perform neonatal screening. Before skeletal maturity, excess thyroid hormone may increase longitudinal skeletal growth. In adults, thyrotoxicosis increases bone turnover. Serum calcium tends to rise, reducing PTH and 1,25-dihydroxyvitamin D_3. Both hyperthyroidism and hypothyroidism are associated with osteoporosis [58]. Over-exuberant thyroid replacement treatment also contributes to bone loss due to increased bone turnover [59]. This situation can be avoided by monitoring serum TSH levels and adjusting the dosage of thyroxine to maintain the normal levels of TSH.

SUMMARY

- Bone is under complex endocrine control, and therefore many different endocrine abnormalities have a deleterious effect on bone.
- The calciotropic hormones, vitamin D, parathyroid hormones, and calcitonin regulate calcium metabolism (see chapter 2).
- Estrogen decreases bone turnover by a direct effect on the osteoblast to produce factors that inhibit bone resorption. The hormone also acts indirectly via the gut, kidney, and parathyroid gland to help the body to retain calcium and to change the set-point at which parathyroid hormone (PTH) responds to calcium.
- The effect of oral contraceptive use on bone mass is unclear, but it appears not to have major skeletal effects in women with pre-existing normal ovarian function. There are clear skeletal benefits from hormone replacement therapy with long-term use, but these benefits are lost when therapy supplementation ceases.
- There is no evidence for any long-term harm to the skeleton or long-term osteoporosis risk from pregnancy and lactation.
- Progesterone decreases bone resorption and increases formation in animal studies, but whether or not it does this in humans to a clinically-relevant extent remains controversial. Studies of combined estrogen and progestin treatment showed biochemical evidence of both decreased bone resorption and increased bone formation.
- Testosterone contributes to increased peak bone mass in men. Recently, it has been appreciated that estrogen is a key regulator of skeletal development and of bone mass in males as well as females. It appears that estrogen is a more important determinant of bone mineral than is testosterone in older men.
- Growth hormone (and IGF-I) is important throughout life but has been identified as a major determinant of bone mineral in men in their 20s. The decrease of this hormone and IGF-I with aging may contribute to age-associated bone loss.
- Glucocorticoids are a group of hormones important for mineral metabolism as they have a strong deleterious effect on bone, yet they are administered widely to patients with conditions for which little else can provide a benefit.
- Thyroid hormone imbalance also causes osteoporosis.

References

1. Eriksen EF, Colvard DS, Berg NJ, et al. Evidence of estrogen receptors in normal human osteoblast like cells. *Science* 1988;241:84-86.
2. Pacifici R. Estrogen, cytokines, and pathogenesis of postmenopausal osteoporosis. *J Bone Miner Res* 1996;11:1043-51.
3. Prince RL. Counterpoint: estrogen effects on calcitropic hormones and calcium homeostasis. *Endocr Rev* 1994;15:301-309.
4. Geusens P, ed. *Osteoporosis in clinical practice: a practical guide for diagnosis and treatment.* London: Springer-Verlag, 1998:188.
5. Bennell KL, White S, Crossley K. The oral contraceptive pill: a revolution for sportswomen? *Br J Sports Med* 1999;33:231-238.
6. Cauley J, Salamone LM, Lucas FL. *Postmenopausal endogenous and exogenous hormones, degree of obesity, thiazide diuretics and risk of osteoporosis.* In: Marcus R, Feldman D, Kelsey J, ed. Osteoporosis. San Diego, CA, USA: Academic Press, 1996: 551-576.
7. Recker RR, Davies KM, Dowd RM, et al. The effect of low-dose continuous estrogen and progesterone therapy with calcium and vitamin D on bone in elderly women. A randomized, controlled trial. *Ann Intern Med* 1999;130:897-904.
8. Marcus R, Greendale G, Blunt BA, et al. Correlates of bone mineral density in the postmenopausal estrogen/progestin interventions trial. *J Bone Miner Res* 1994;9:1467-1476.

9. Felson DT, Zhange Y, Hannan MT, et al. The effect of postmenopausal estrogen therapy on bone density in elderly women. *N Engl J Med* 1993;329:1141-1146.
10. Heaney RP. Estrogen-calcium interactions in the post-menopause: A quantitative description. *Bone Miner* 1990;11:67-84.
11. Kanis JA, Johnell O, Gullberg B, et al. Evidence for efficacy of drugs affecting bone metabolism in preventing hip fracture. *Br Med J* 1992;305:1124-1128.
12. Naessen T, Lindmark B, Larsen HC. Better postural balance in elderly women receiving estrogens. *Am J Obstet Gynecol* 1997;177:412-6.
13. Ekblad S, Lonnberg B, Berg G, et al. Estrogen effects on postural balance in postmenopausal women without vasomotor symptoms: a randomized masked trial. *Obstet Gynecol* 2000;95:278-83.
14. Stepan JJ, Pospichal J, Presl J, et al. Bone loss and biochemical indices of bone remodeling in surgically induced postmenopausal women. *Bone* 1987;8:279-284.
15. Lindsay R, Hart DM, Aitken JM, et al. Long-term prevention of postmenopausal osteoporosis by oestrogen. Evidence for an increased bone mass after delayed onset of oestrogen treatment. *Lancet* 1976;1:1038-41.
16. Kohrt WM, Snead DB, Slatopolsky E, et al. Additive effects of weight-bearing exercise and estrogen on bone mineral density in older women. *J Bone Miner Res* 1995;10:1303-11.
17. Kohrt WM, Ehsani AA, Birge SJ, Jr. HRT preserves increases in bone mineral density and reductions in body fat after a supervised exercise program. *J Appl Physiol* 1998;84:1506-12.
18. Yeh JK, Aloia JF, Barilla ML. Effects of 17 beta-estradiol replacement and treadmill exercise on vertebral and femoral bones of the ovariectomized rat. *Bone Miner* 1994;24:223-34.
19. Yeh JK, Liu CC, Aloia J. Additive effect of treadmill exercise and 17b-estradiol replacement on prevention of tibial bone loss in adult ovariectomized rat. *J Bone Miner Res* 1993;8:677-683.
20. Prince RL, Smith M, Dick IM, et al. Prevention of postmenopausal osteoporosis: a comparative study of exercise, calcium supplementation and hormone-replacement therapy. *N Engl J Med* 1991;325:1189-1195.
21. Heikkinen J, Kurttila-Matero E, Kyllonen E, et al. Moderate exercise does not enhance the positive effect of estrogen on bone mineral density in postmenopausal women. *Calcif Tissue Int* 1991;49:S83-4.
22. Kohrt WM, Snead DB, Slatopolsky E, et al. Additive effects of weight-bearing exercise and estrogen on bone mineral density in older women. *J Bone Miner Res* 1995;10:1303-1311.
23. Heikkinen J, Kyllonen E, Kurttila-Matero E, et al. HRT and exercise: effects on bone density, muscle strength and lipid metabolism. *Maturitas* 1997;26:139-149.
24. Smith EP, Boyd J, Frank GR, et al. Estrogen resistance caused by a mutation in the estrogen-receptor gene in man. *N Engl J Med* 1994;331:1056-1061.
25. Colvard DS, Eriksen EF, Keeting PE, et al. Identification of androgen receptors in normal human osteoblast-like cells. *Proc Natl Acad Sci USA* 1989;86:854-857.
26. Nawata H, Tanaka S, Tanaka S, et al. Aromatase in bone cell: Association with osteoporosis in postmenopausal women. *F Steroid Biochem Mol Biol* 1995;53:165-174.
27. Greendale GA, Edelstein S, Barrett-Connor E. Endogenous sex steroids and bone mineral density in older women and men: the Rancho Bernardo Study. *J Bone Miner Res* 1997;12:1833-1843.
28. Aitken JM, Armstrong E, Anderson JB. Osteoporosis after oophorectomy in the mature female rat and the effect of oestrogen and progestogen replacement therapy in its prevention. *J Endocrinol* 1972;55:79-87.
29. Abdalla HI, Hart DM, Lindsay R, et al. Prevention of bone mineral loss in postmenopausal women by norethisterone. *Obstet Gynecol* 1985;66:789-792.
30. Oursler MJ, Kassem M, Turner R, et al. *Regulation of bone cell function by gonadal steroids.* In: Marcus R, Feldman D, Kelsey J, ed. Osteoporosis. San Diego, CA, USA: Academic Press, 1996: 237-260.
31. Christiansen C, Nilas L, Riis BJ, et al. Uncoupling of bone formation and resorption by combined oestrogen and progestogen therapy in postmenopausal osteoporosis. *Lancet* 1985;2:800-801.
32. Prior JC, McKay DW, Vigna YM, et al. Medroxyprogesterone increases basal temperature: a placebo-controlled crossover trial in postmenopausal women. *Fertil Steril* 1995;63:1222-6.
33. Prior JC, Vigna YM, Barr SI, et al. Ovulatory premenopausal women lose cancellous spinal bone: a five year prospective study. *Bone* 1996;18:261-267.
34. Speroff L, Glass R, Kase N. *Clinical gynecology, endocrinology and infertility*. Baltimore: Williams & Wilkins, 1989.
35. Sowers M. *Premenopausal reproductive and hormonal characteristics and the risk for osteoporosis.* In: Marcus R, Feldman D, Kelsey J, ed. Osteoporosis. San Diego, CA, USA: Academic Press, 1996: 529-550.
36. Eisman J. Relevance of pregnancy and lactation to osteoporosis? *Lancet* 1998;352:504-5.
37. Funk JL, Shoback DM, Genant HK. Transient osteoporosis of the hip in pregnancy: natural history of changes in bone mineral density. *Clin Endocrinol (Oxf)* 1995;43:373-82.
38. Anai T, Tomiyasu T, Arima K, et al. Pregnancy-associated osteoporosis with elevated levels of circulating parathyroid hormone-related protein: a report of two cases. *J Obstet Gynaecol Res* 1999; 25:63-7.
39. Yamamoto N, Takahashi HE, Tanizawa T, et al. Bone mineral density and bone histomorphometric as-

sessments of postpregnancy osteoporosis: a report of five patients. *Calcif Tissue Int* 1994;54:20-5.

40. Ghannam NN, Hammami MM, Bakheet SM, et al. Bone mineral density of the spine and femur in healthy Saudi females: relation to vitamin D status, pregnancy, and lactation. *Calcif Tissue Int* 1999;65:23-8.
41. Cooper CS, Taaffe DR, Guido D, et al. Relationship of chronic endurance exercise to the somatotropic and sex hormone status of older men. *Eur J Endocrinol* 1998;138:517-523.
42. Drinka PJ, Olson J, Bauwens S, et al. Lack of association between free testosterone and bone density separate from age in elderly males. *Calcif Tissue Int* 1993;52:67-69.
43. Finkelstein BL, Klibanski A, Neer RM, et al. Osteoporosis in men with idiopathic hypogonadotropic hypogonadism. *Ann Int Med* 1987;106:354-361.
44. Greenspan SL, Oppenheim DS, Klibanski A. Importance of gonadal steroids to bone mass in men with hypoprolactinemic hypogonadism. *Ann Int Med* 1989;110:526-531.
45. Kelly PJ, Pocock NA, Sambrook PN, et al. Dietary calcium, sex hormones and bone mineral density in man. *Br Med J* 1990;300:1361-1364.
46. Murphy S, Khaw K, Cassidy A, et al. Sex hormones and bone mineral density in elderly men. *Bone Miner* 1993;20:133-140.
47. Nakano Y, Morimoto I, Ishada O, et al. The receptor, metabolism and effects of androgen in osteoblastic MC3T3-E1 cells. *Bone Miner* 1994;26:245-59.
48. Mizunuma H, Hosoi T, Okano H, et al. Estrogen receptor gene polymorphism and bone mineral density at the lumbar spine of pre- and postmenopausal women. *Bone* 1997;21:379-383.
49. Rudman D, Drinka PJ, Wilson CR, et al. Relations of endogenous anabolic hormones and physical activity to bone mineral density and lean body mass in adult men. *Clin Endocrinol* 1994;40:653-661.
50. Stanley HL, Schmitt BP, Poses RM, et al. Does hypogonadism contribute to the occurrence of a minimal trauma hip fracture in elderly men? *J Am Geriatr Soc* 1991;39:766-771.
51. Dequeker J, Geusens P. Anabolic steroids and osteoporosis. *Endocrinologica* 1985;271:45-52.
52. Martha PM, Rogol AD, Veldhuis JD, et al. A longitudinal assessment of hormonal and physical alterations during normal puberty in boys. III. The neuroendocrine growth hormone axis during later prepuberty. *J Clin Endocrinol Metab* 1996;81:4068-4074.
53. Russell-Aulet M, Shapiro B, Jaffe C, et al. Peak bone mass in young healthy men is correlated with the magnitude of endogenous growth hormone secretion. *J Clin Endocrinol Metab* 1998;83: 3463-3468.
54. Yarasheski KE, Campbell JA, Kohrt WM. Effect of resistance exercise and growth hormone on bone density in older men. *Clin Endocrinol* 1997;47:223-239.
55. Johansson AG, Forslund A, Hambraeus L, et al. Growth hormone-dependent insulin-like growth factor binding protein is a major determinant of bone mineral density in healthy men. *J Bone Miner Res* 1994;9:915-921.
56. Reid IR. Editorial: Glucocorticoid effects on bone. *J Clin Endocrinol Metab* 1998;83:1860-1862.
57. Reid IR. Steroid-induced osteoporosis. *Osteoporos Int* 1997;7:S213-6.
58. Suwanwalaikorn S, Baran D. *Thyroid hormone and the skeleton*. In: Marcus R, Feldman D, Kelsey J, ed. Osteoporosis. San Diego, CA, USA: Academic Press, 1996: 855-861.
59. Krakauer JC, Kleerekoper M. Borderline-low serum thyrotropin level is correlated with increased fasting urinary hydroxyproline excretion. *Arch Intern Med* 1992;152:360-364.
60. Burr DB, Yoshikawa T, Teegarden D, et al. Exercise and oral contraceptive use suppress the normal age-related increase in bone mass and strength of the femoral neck in women 18-31 years of age. *Bone* 2000; 27:855-863.
61. Einhorn TA. *The bone organ system: Form and function*. In Marcus R, Feldman D, Kelsey J, ed. Osteoporosis. San Diego, CA, USA: Academic Press, 1996: 3-22.
62. Robinson RJ, Krzywick T, Almond L, et al. Effect of a low-impact exercise program on bone mineral density in Chron's disease: A randomized controlled trial. *Gastroenterology* 1998;115:36-41.

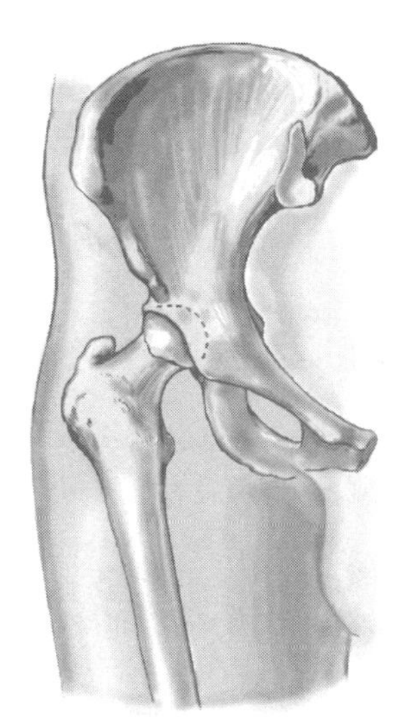

Chapter 8

Dietary Intake and Bone Mineral

Although the association between dietary calcium and bone health is widely accepted, the true relationship between calcium and both BMD and fracture is not as straightforward as may generally be presumed. Furthermore, a great deal of what is often called "disinformation" (misleading information) surrounds nutrition and bone.

Disinformation on the World Wide Web

On their Internet Web site a company that sells calcium tablets states that their product is formulated to "increase uptake of calcium and magnesium," and to "keep good bone density." They also claim that their "suppliment [*sic*] is the choice for anyone who is concerned about getting enough calcium." They then advocate taking 1400 mg/day of elemental calcium supplementation, which is 140% of the RDI. There is no mention of dietary calcium! The company running this site ignores the potential role of dietary calcium and makes claims about calcium and bone density that are not evidence based. Furthermore, they advocate excessive calcium intake in the name of "good bone density." As this chapter points out, the skeleton does not retain ever-increasing amounts of calcium merely because more calcium is ingested.

A Medline computer search performed on August 25, 2000, found 1542 papers with both *calcium* and *bone mineral density* as keywords MeSH (Medline Sub Heading). Some pivotal papers on this subject were published in the late 1980s and early 1990s [1-8], and since then, a number of randomized, controlled studies have provided a great deal of useful data [9-11].

In this chapter we summarize studies of the association between calcium intake and bone mineral density according to life stage, distinguishing between cross-sectional descriptive studies of intake and the prospective studies of calcium supplementation. We then discuss the association between vitamin D intake and bone mineral density before examining the role of calcium and vitamin D intake in reducing fracture risk. Finally, we review the associations of alcohol and smoking, respectively, with bone. Alcohol and smoking are defined as drugs, not nutrition, but the former is generally measured via food questionnaires. The latter, although hopefully soon to disappear from the planet, remains a major insult to bone health.

Calcium Intake and Bone Mineral Density at Various Life Stages

The human skeleton serves as a storage site for excess calcium and also as a supply of calcium during times of deprivation. This portable supply of calcium, however, is a double-edged sword. When the calcium reserve, our skeleton, is called on to meet dietary insufficiencies for long periods, bone strength becomes compromised.

Childhood and Adolescence

To help make sense of the various studies that we are about to summarize, it is useful to review one aspect of gastrointestinal physiology. A complex homeostatic control regulates the amount of calcium ingested, the amount retained after obligatory losses from the various sites (digestive tract, skin, nails, hair, sweat, and urine), and the amount that is finally incorporated into the skeleton (figure 8.1).

The amount of calcium that is retained in bone from dietary intake is called calcium retention. When this is expressed as a percentage of calcium intake, it is called calcium retention efficiency. The inverse relationship between calcium intake and retention efficiency is one mechanism that has evolved to compensate for low dietary intakes of calcium. Thus, even if dietary intake is relatively poor, it may provide sufficient calcium for bone health if retention efficiency can be increased.

In children, apparent calcium retention efficiency at the age of peak growth appears to be about 30% for girls and 36% for boys, although individuals can vary widely [12]. The hyperbolic inverse relationship between calcium intake and retention efficiency as shown in figure 8.2 reflects a compensatory mechanism that increases calcium retention when dietary intake is low. When calcium intakes dropped to 400 mg/day

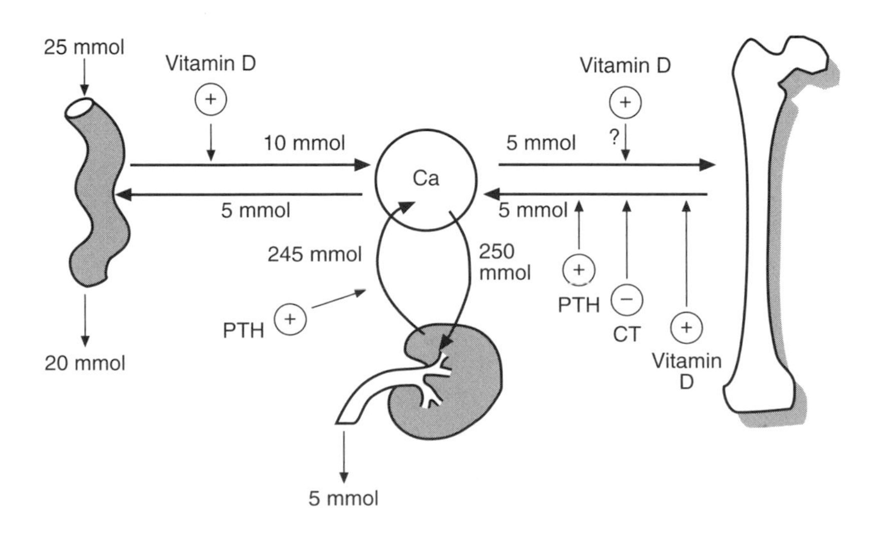

FIGURE 8.1 Schematic representation of the pathways linking the intestines, kidney, and bone. Chapter 2 contains a more detailed discussion of calcium physiology.

(just over a quarter of the recommended daily intake), absorption efficiency reached almost twice the usual mean [13]. Calcium absorption efficiency as high as 80% has been reported [1, 14], suggesting that some children have almost three times the average efficiency. It has been postulated that low calcium intake may also be compensated for temporarily during growth by a "borrowing" of calcium from the cortical shell [15, 16] to maintain serum calcium in the physiological level (see chapter 2). A precautionary note is needed to acknowledge that the long-term repercussions of low calcium intakes on bone geometry are not known.

Against this background of physiology emphasizing that calcium does not pass directly from the mouth into the bone reservoir, we report that randomized, double-blind, placebo-controlled studies in children found that calcium supplementation during the growing years increased bone mineral by about 1-3% independently of energy intake or other nutritional factors [18-24]. A randomized, controlled intervention with 82 white girls (mean age 12.2 ± 0.3) who were supplemented with a pint of milk for 18 months demonstrated a significant ($p < 0.02$) difference in total body bone mineral accrual for the treatment group (9.6%) as compared with controls (8.5%) [25].

In most follow-up studies, however, after withdrawal of supplementation, the benefit achieved by the formerly supplemented group decreased or disappeared [24, 26-28]. In one of these calcium supplementation studies, serum osteocalcin was 15% lower in the supplemented than in the placebo group, reflecting lower bone turnover. After the intervention ceased, however, this difference disappeared [18].

In the one exception to the studies that demonstrated only temporary benefits, Swiss researchers supplemented prepubertal girls with calcium as milk extract added to food products [22]. One year after supplementation was withdrawn, girls with habitually low dietary calcium intakes who had previously been supplemented still had greater increments in BMC and BMD in the femoral shaft than the previously unsupplemented children [22].

In adolescent girls, calcium supplementation also caused a moderate increase in BMD. Girls with a mean age of 14 years (range 10-17) who were supplemented with 1000 mg of calcium for 18 months had a 1.5% increase in BMD at both the spine and total hip. The greatest effect occurred during the first six months of supplementation [21].

You may be surprised to learn that high calcium intake, even for three years prior to puberty, has no long-term effect on bone mineral. This finding, however, is consistent with the hypothesis that bone changes following supplementation with dietary calcium are due to reversible changes in bone remodeling—the bone-remodelling transient (see chapter 2) [7]

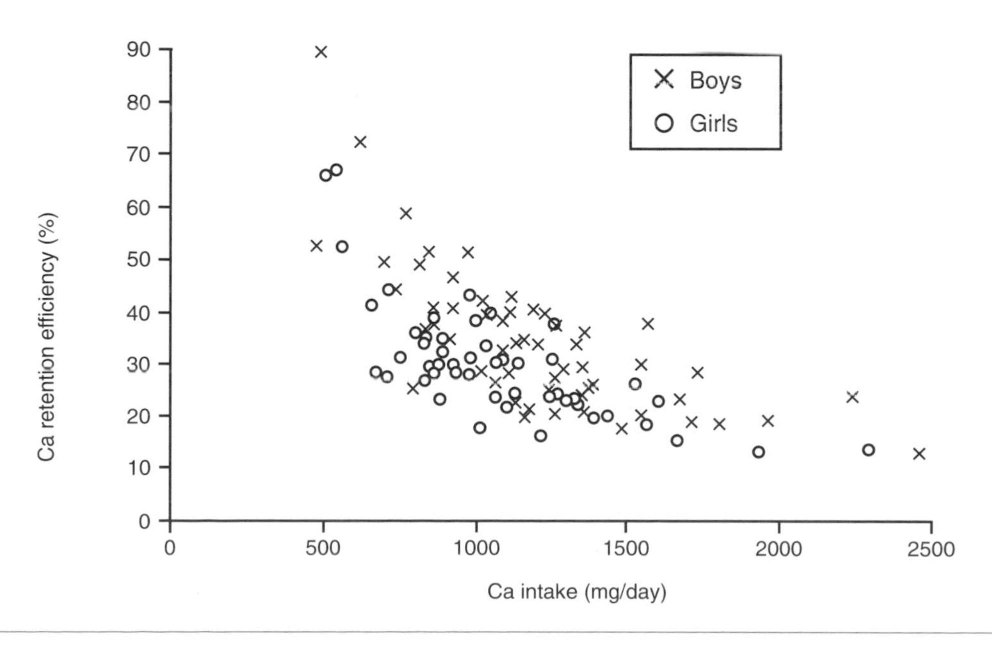

FIGURE 8.2 Graph of calcium retention calculated from food diaries and BMD by DXA.
Data from [17].

As an extension of this argument, Kanis contends that the RDA (recommended daily allowance) of calcium for young healthy individuals is unknown and that "there is no good evidence to suggest that variations in dietary intake of calcium are prejudicial for skeletal health in young individuals who take a reasonably balanced diet" [7]. This position is substantiated by the work we previously discussed by Bailey and colleagues that showed no relationship between dietary calcium intake and bone mineral accretion [12]. It appears that once peak bone mass is achieved, the mechanical forces acting on the skeleton maintain bone mass. As the adage suggests, we are what we eat, but children must be reminded that what we do is also important!

Premenopausal Years

Healthy premenopausal women are subjected to a barrage of advertising that promotes calcium supplementation to "prevent osteoporosis." Yet at this time of life they are estrogen replete with peak adult calcium absorption and retention. Is there any evidence that large calcium intakes are warranted in women of this age?

Some epidemiological studies have demonstrated that lifetime calcium intake is associated with increased BMD in women [29-38]. On the other hand, many similarly designed studies failed to show an association over a range of dietary intakes [39-48].

An oft-quoted study of calcium intake in two regions of Yugoslavia [30] reported an association between calcium intake and both increased bone mineral and decreased fracture rate. The study has been criticized on both statistical and biological grounds. Cumming [6] excluded this paper from his meta-analysis because dietary calcium was only measured in a subsample of the subjects whose BMD was measured. Thus, the comparison in this paper was between bone mass in a region with high average calcium intake and in a region with low average calcium intake, which leaves much room for other regional differences to confound the results.

The study conclusions may be invalid even if the calcium intake in the subsamples measured truly represented the entire populations. The high-calcium group had a 20% greater nutrient intake than the low-calcium groups despite their having similar body mass. Lower nutrient intake in populations with similar weight indicates less physical activity, so the difference attributed to calcium intake may be explained, at least partially, by differences in physical activity [1].

These cross-sectional data suggesting that dietary calcium intake does not substantially alter BMD in adult women are supported by large, longitudinal studies that also found no effect of high or low levels of calcium intake on BMD or rates of bone loss [48, 49].

A few studies have addressed the question of how calcium supplementation, as distinct from normal dietary intake, affects BMD in premenopausal women. The first such study found no effect after four years of calcium supplementation on wrist BMC [50]. Another study, which found that increasing calcium intake from 960 to 1600 mg prevented bone loss over a three-year period [51], has been criticized for its small numbers [49].

Taken together, these cross-sectional, longitudinal, and intervention studies suggest that dietary calcium in adulthood may have a smaller influence on BMD than genetic and other environmental factors [7]. Even the strongest proponents of calcium for bone health concede that "the evidence for a relation between bone density, bone loss, and estimated calcium intake in individuals is somewhat inconclusive" [5]. They attribute this to the imprecision of the diet history in measuring current and lifetime calcium intake, the confounding effect of other nutrients (phosphorus, protein, sodium), and differences among individuals in obligatory calcium losses [52].

Early Postmenopausal Years

As the biological response to calcium differs among women of early and late postmenopausal years, we summarize the results of intervention studies in women in each of these life stages separately.

During the early postmenopausal years, there may be a transient, rapid downward adjustment of bone mass of as much as 10-15% of premenopausal levels [53]. Whether additional calcium is beneficial during the perimenopausal and early postmenopausal years remains unresolved. At this stage of life, decreases in BMD are not related to nutrient deficiency but mainly to ovarian hormone withdrawal and thus are not substantially influenced by even high doses of calcium.

Researchers have conducted several key studies of dietary calcium supplementation and BMD in this age group. A daily supplement of 500 mg of calcium did not retard early postmenopausal lumbar spine bone loss in either Danish women with mean dietary intakes of 1000 mg/day [44] or American women with mean dietary intakes of around 700 mg/day [54, 55]. Although calcium supplementation to 2000 mg/day slowed forearm and total body cortical bone loss compared with placebo, it did not slow cancellous bone loss at the spine or ultradistal radius [44].

Cumming [6] reviewed six intervention studies [50, 54, 56-59] conducted in "healthy" women with a mean age of around 50 years and found a positive effect of calcium that ranged from 0% to 1.7% with a mean of 0.8% per year. These and other studies [55, 60] suggest that "a calcium supplement of around 1000 mg per day in early postmenopausal women can prevent loss of just under 1% of bone mass per year at all bone sites except at the vertebrae"[6]. This means that calcium has more of an effect on BMD than no treatment, but less of an effect than ovarian hormone therapy (also known as hormone replacement thrapy [HRT] or estrogen replacement therapy [ERT]).

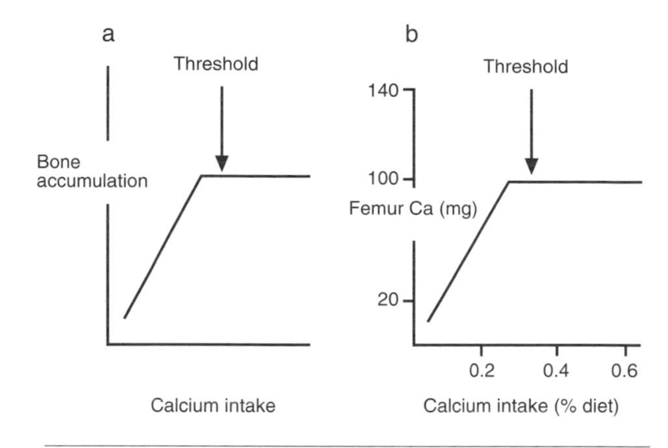

FIGURE 8.3 Threshold behavior of calcium intake. *(a)* Theoretical relationship of bone mineral accrual to calcium intake. Below the threshold value, bone mineral accrual and calcium intake form a linear relationship. Above the threshold, bone mineral is controlled by other factors and is no longer related to dietary calcium intake. *(b)* Data from growing rats showing bone mineral accrual to a threshold based on dietary intakes of calcium.

Figure adapted from "Nutrition and Risk for Osteoporosis" by RJ Heaney in OSTEOPOROSIS, edited by Robert Marcus et al. Copyright 1996 by Academic Press, reproduced by permission of the publisher.

Later Postmenopausal Years

In the later postmenopausal years, accelerated remodeling and accelerated bone loss may cause postmenopausal osteoporosis [61, 62]. Calcium supplementation in women of this age is associated with the maintenance, but not gain, of skeletal mass [3, 56, 58, 63-68]. For example, in healthy women who were more than five years postmenopausal and had dietary calcium intake of less than 400 mg/day, supplementation to 800 mg/day significantly reduced bone loss [55]. However, in women who had more substantial, but still very moderate, calcium intake (e.g., > 400 mg/day), calcium supplementation provided no advantage over placebo. This supports the argument that calcium is a "threshold nutrient" (figure 8.3), with 400 mg representing the threshold in this particular population. Because levels are likely to differ among individuals and populations, it is not advisable to extend this threshold indiscriminately to other populations.

While calcium supplementation slowed both appendicular and axial bone loss in women who were, on average, 10 years postmenopausal [50, 56, 58, 70-72], identical supplementation was ineffective in women who had just reached menopause [59, 73]. Studies that have found calcium supplementation to be effective [51, 55, 72, 74] used soluble calcium salts or dairy products as the supplement; calcium carbonate has not proven to be an effective supplement [55, 59, 73]. Differences in the short-term bioavailability of different calcium preparations [75] may influence their effect on bone loss. Long-term (four-year) studies in women who were 10 to 15 years postmenopausal showed a protective effect of calcium supplementation on BMD [76, 77].

In postmenopausal women, calcium is likely to attenuate bone loss by decreasing activation of new remodeling sites and promoting bone formation at the previously existing bone remodeling sites [7]. This phenomenon is known as the bone remodeling transient [78] (see chapter 2, page 18). Because there is still a deficit between the calcium lost in resorption and the calcium replaced in formation, bone continues to be lost, albeit at a slower rate, despite calcium supplementation.

Calcium Nutrition and Bone Mass in Men

Two papers from one research center reported a significant, and substantial, association between dietary calcium intake and BMD at both the lumbar spine and proximal femur in men ranging in age from 20 to 80 years [79, 80]. Kanis [1] argues that this association may merely reflect a true association between physical activity (and thus greater dietary intake and greater dietary calcium) and bone mineral density [1]. Both of the previously mentioned studies of calcium and BMD [79, 80] addressed this possibility and discounted it. The authors reported that calcium intake was independent of total energy intake [80] and that calcium intake was independent of weight or muscle strength and therefore unlikely to merely reflect better nutrition in the more active elderly [79]. It has been hypothesized that current calcium intake might be particularly important at trabecular sites, as they have higher bone turnover [81].

On the other hand, Holbrook and Barrett-Connor [82] could not find any relationship between dietary calcium intake (24-hour recall) and femoral BMD 18 years later. Given that the single questionnaire measuring recalled dietary intake over the previous 24 hours may not accurately represent calcium intake, this result is not surprising. It does, however, contrast with the finding in women that age-adjusted mean proximal femur BMD levels increased significantly with increasing tertile of calcium intake, associations that persisted after adjusting for potential confounders [82].

Vitamin D and Bone Mineralization

Vitamin D_3 (cholecalciferol) and its metabolites are critically important for the normal mineralization of bone [83] (see chapter 2). In this section, we summarize data on vitamin D and bone mineralization in adolescence, adulthood (skeletal maturity to menopause), and the postmenopausal years. We use the chemical name 1,25-dihydroxyvitamin D_3 unless referring to the pharmaceutical compound administered to subjects or animals, which is referred to as calcitriol.

Children and Young Adults

Winter sunlight exposure (and thus, vitamin D) was associated with BMD at the proximal femur in pubertal girls but not boys [84]. Vitamin D may facilitate calcium retention during the growth spurt [85]. In a cross-sectional (n = 178) and one-year longitudinal (n = 57) study, serum 1,25-dihydroxyvitamin D levels were greatest during the pubertal growth spurt (Tanner stage III-IV; age 11-13 years), which correlated with the time of peak bone mineral accrual. In the longitudinal study, annual changes in 1,25-dihydroxyvitamin D were associated with measures of BMD at both the total body and the proximal radial sites. Thus, 1,25-dihydroxyvitamin D correlated significantly with bone mineral acquisition during pubertal growth. This is presumably in response to the high requirement for calcium during that period [85].

On the other hand, in a population of slightly older Icelandic girls (16-20 years), almost one fifth of the subjects had serum levels of 25-hydroxyvitamin D below the lower level of normal in adults, but there was no significant association between serum vitamin D levels and BMD [86]. In a study of 109 boys and 119 girls (aged 9-18 years) whose heights were measured every six months over five years, summer growth velocities accounted for 67% of total yearly growth in boys and 60% in girls [87].

In 20- to 23-year-old women and men, a cross-sectional study found that dietary intake of vitamin D was positively associated with BMD among women [88]. The association may not be causal because body weight and sports activity during adolescence were stronger determinants of BMD than vitamin D intake, and these variables are both correlated with vitamin D intake [89].

Perimenopausal Women

Lukert and colleagues [90] reported a positive association between dietary intake of vitamin D and bone density in perimenopausal women. They found no association, however, between the circulating level of 1,25-hydroxyvitamin D_3 and bone density despite the association with dietary intake. As discussed earlier, dietary intake of vitamin D correlates closely with that of a number of other nutrients including calcium, indicating that vitamin D nutrition may not exclusively influence BMD in this age group [91].

Older Adults

Advancing age is associated with impaired levels of serum vitamin D, particularly in those not living independently [92]. The reasons for this are not clear but are thought to relate to sunlight intensity, sun exposure, and dietary practices [93]. Moreover, decreasing renal function reduces 1,25-dihydroxyvitamin D_3 production [91]. This relative deficiency of vitamin D and its active metabolites in some elderly populations explains the rationale for vitamin D supplementation in elderly women [91, 94]. However, when vitamin D (vitamin D_2/vitamin D_3) has been used as a pharmacological treatment of osteoporosis, the results have proven disappointing [71, 95, 96]. Therefore, at present, physiological doses of vitamin D are used to correct subclinical deficiency with the aim of maintaining BMD [94, 97].

The most prominent "vitamin D studies" evaluated the effect of vitamin D_3 on BMD in 27 elderly women and 29 controls (mean age 84 years) who were part of a much larger study of hip fractures among women living in nursing homes in France [98]. After 18 months of treatment using 800 IU (20 micrograms) of vitamin D_3 and also 1.2 g of calcium daily, BMD at the proximal femur increased 2.7% in the intervention group and decreased 4.6% in the placebo group ($p < 0.001$) [98]. Similar, but smaller, BMD changes favorable to the intervention group were found at the femoral neck and trochanteric sites. Bone loss was likely due to low calcium intake and decreased calcium absorption as well as vitamin D deficiency due to insufficient sunlight. (The fracture data from that study are discussed in the section, "Dietary Supplementation and Fracture Risk").

Mechanism of Action of Vitamin D on Bone Mineral

The risk of vitamin D deficiency is greatest in inhabitants of countries with little sunlight (e.g., Scandinavian countries) and whose dairy products are not fortified with vitamin D (e.g., France) [98]. Institutionalized individuals have poor sunlight exposure no matter where they live [92]. A deficit in calcium and vitamin D intake results in a negative calcium balance that stimulates secretion of PTH. This secondary hyperparathyroidism increases bone turnover, leading to diminished BMD compared with subjects who have adequate vitamin D and calcium intake [91,98]. Since Chapuy and colleagues supplemented participants with both vitamin D_3 and calcium, it is impossible to be certain which supplement contributed to the therapeutic effect on BMD or whether it was due to the combination of agents [91].

Theoretically, several mechanisms could explain vitamin D's role as a determinant of bone mass. These include (1) an age-related decline in vitamin D production in the skin, (2) an age-related decline in renal 1,25-dihydroxyvitamin D production contributing to senile osteoporosis [91, 100], or (3) an age related decline in the effect of vitamin D on intestinal absorption of calcium [101]. All of these factors can contribute to age-related bone loss that, according to some studies, can be countered by various forms of vitamin D and calcium supplements [98, 102, 103]. However, not all studies agree that such treatment is beneficial [91].

Dietary Supplementation and Fracture Risk

Far fewer studies have evaluated the effect of diet on fracture than on BMD. Fracture is a relatively uncommon outcome and therefore more difficult to study. A systematic review found that increased calcium intake among postmenopausal women was associated with a small decrease in fracture risk [104]. This conclusion was based on the results of randomized trials of calcium supplements that measured fracture outcomes and a meta-analysis of 16 observational epidemiological studies of dietary calcium and hip fractures. In biological terms, 1000 mg/day of dietary calcium was associated with a 24% reduction in risk of hip fracture [104].

A limitation of the available data is that only three randomized studies of calcium intervention have used fracture as the outcome measure (table 8.1) [77, 105, 106]. The randomized intervention studies produced greater decreases in fracture risk than could be explained purely by the differences in BMD between groups. As calcium supplementation reduces bone turnover, it may reduce cortical porosity and trabecular perforation and, in turn, reduce fractures. This would be consistent with findings of decreased fracture rates even in the first year of the studies, before differences in BMD (as measured by DXA) developed [77].

Three other studies examined fracture as an outcome after an intervention with calcium and vitamin D (see table 8.2) [98, 107, 108, 109]. In women who completed 18 months of calcium and vitamin D supplementation, in the French nursing home study [98], the number of hip fractures was 43% lower ($p = 0.04$) and the total number of nonvertebral fractures was 32% lower ($p = 0.02$) than among those who received placebo [98]. A letter to the editor after 18 additional months of intervention (36 months total) stated that the probability of hip fractures (-29%; $p < 0.01$) was reduced in the treatment group [109]. Fewer treatment subjects (17%) had one or more nonvertebral fractures (255 treated

TABLE 8.1

Randomized trials of the effect of calcium supplements on risk of fractures in postmenopausal women

Author	No. of subjects	Setting	Mean age (yr)	Mean dietary calcium (mg)	Calcium dose (mg)	Follow-up (yr)	Fracture outcome (RR and 95% CI)
Chevalley 1994 [105]	93	Healthy volunteers	72	619	800	1.5	Vertebral fractures, RR = 0.7 (p > 0.05)
Reid 1995 [77]	78	Healthy volunteers	58	734	1000	4.0	Symptomatic fractures, RR = 0.3 (0.1-1.4)
Recker 1996 [106]	197	Healthy volunteers	73	433	1200	4.3	Vertebral fractures, RR = 0.7 (0.5-1.1)

RR = relative risk; CI = confidence interval.

TABLE 8.2

Randomized trials of the effect of calcium and vitamin D supplementation, or vitamin D supplementation alone, on risk of fractures in postmenopausal women

Author	No. of subjects	Setting	Mean age (yrs)	Mean dietary calcium (mg)	Calcium (mg) and vitamin D (IU) dose (mg)	Follow-up (yrs)	Fracture outcome (RR and 95% CI)
Chapuy 1992 [98]	3270	Nursing homes	84	513	1200 mg; 800 IU	1.5	Nonspine fractures, RR = 0.7 (p < 0.05)
Chapuy 1994 [109]	3270	Nursing homes	84	513	1200 mg; 800 IU	3.0	Nonspine fractures, RR = 0.76 (p < 0.01); Hip fractures RR = 0.71 (p < 0.01)
Lips 1996 [108]	1916[1]	Independently living	80	868	0 calcium, 400 IU Vitamin D	3.5	No difference in hip fractures or other peripheral fractures
Dawson-Hughes 1997 [107]	213[2]	Healthy volunteers	72	746; 700 IU	500; 700 IU	3.0	Nonvertebral fractures, (men and women combined) RR = 0.4 (0.2-1.1)

[1]This study included a further 662 men, whose results are not tabled here. [2]This study included a further 176 men, whose results are not tabled here. RR = relative risk; CI = confidence level.

compared with 308 untreated, p < 0.02), and 23% fewer subjects had one hip fracture (137 compared with 178, p < 0.02). It is estimated that up to 60% of patients with hip fractures may have vitamin D deficiency [110]. Table 8.2 provides a summary of important prospective trials performed in postmenopausal groups.

Vitamin D therapy for osteoporosis remains an area of substantial controversy [91]. There is little convincing evidence for its use in younger individuals. However, in 70- to 80-year-old individuals, vitamin D deficiency results in secondary elevation of parathyroid hormone (PTH) with increased bone turnover and thus bone loss. Vitamin D and calcium reverse and prevent the effects of hyperparathyroidism on bone and should be increased in older adults with low femoral bone mass and high serum PTH concentrations or low serum 25 hydroxycholecalciferol concentrations [109]. Treatment of vitamin D deficiency would be predicted to benefit bone mass and reduce fracture incidence [91, 98, 109]. However, vitamin D_3 supplementation (400 IU) in 80-year-old men and women did not alter fracture rates in the treatment group as compared with placebo [130].

Interaction of Calcium and Physical Activity

Designing a study that adequately measures the interaction between calcium and physical activity is extremely difficult and, in our minds, has not yet been accomplished in young or older individuals. In premenopausal women, however, Friedlander and colleagues in San Francisco [111] used the classical 2 (calcium/no calcium) × 2 (exercise/no exercise) factorial design to test for interaction between physical activity and calcium intake. They found no effect of high or low calcium intake on the BMD changes achieved with an exercise intervention program. However, the calcium intake in the unsupplemented group was already relatively high and perhaps above the threshold required for calcium to have an effect. Whether calcium intake and physical activity interact is one of the most important questions unanswered in the arena of lifestyle-related bone health research. We suspect it will not be long before substantial studies provide very valuable and practical data.

Other Lifestyle Factors and Bone Mineral Density

A number of other nutritional and nonnutritional lifestyle factors have been linked to bone health, including caffeine, sodium, protein, various vitamins, medications, and alcohol. In addition, smoking has a significant cumulative negative effect on bone. A thorough discussion of all of these factors is beyond the scope of this book, but we provide a brief discussion of the links between alcohol and smoking and bone because they can be important determinants.

Alcohol abuse is associated with numerous factors that contribute to low bone mineral, including poor nutrition, leanness, liver disease, malabsorption, vitamin D deficiency, hypogonadism, parathyroid dysfunction, and tobacco use [112]. At the cellular level, alcohol is known to depress osteoblast function [99,113].

On the other hand, moderate alcohol consumption is unlikely to be associated with lower bone density [114]. In a prospective study of 220 U.S. women, alcohol intake was associated with higher BMD at the spine and radius [115]. Australian and Scandinavian researchers have also shown a beneficial effect of moderate alcohol intake in long-term observational studies [116-118]. However, it must be remembered that a potential confounder in this association is that higher socioeconomic status (and thus perhaps better nutrition) may be associated with moderate alcohol intake.

There is no doubt, however, that smoking has a significant cumulative negative effect on bone mineral [119, 120]. In a prospective study, bone loss from the radius, calcaneum, spine, and femoral neck was greater in 34 smoking than in 278 nonsmoking postmenopausal women [121]. Cross-sectional studies have shown no negative association of smoking in pre- and perimenopausal women [122] but reported deficits of 0.5 to 1.0 standard deviations in the postmenopausal years. The statistically powerful co-twin model has shown that for each 10 years of smoking one pack per day (10 pack years) BMD is decreased at least 2% at the lumbar spine and about 1% at the femoral neck [123] and possibly as much as 5% [131].

Smoking increases bone resorption and decreases formation. This may be due to the direct effects of nicotine on osteoblasts [124] as well as to decreased production and increased degradation of estrogen [125]. A meta-analysis of smoking, BMD, and risk of hip fracture found that smoking was a significant risk factor for decreased bone density and hip fracture [120].

When lifestyle influences on bone health are viewed across the life span, their positive effects are clear. Because lifestyle influences are easily modifiable, unlike genetic factors, we believe that public health promotion initiatives targeting lifestyle factors are vital to bone health at all ages.

SUMMARY

- Humans appear to be capable of adapting to large changes in the dietary intake of calcium. Children and adolescents have tremendous ability to adapt to low levels of dietary calcium [12]. However, the long-term repercussions of low calcium intakes on bone geometry are not known.
- Women who are at least five years postmenopausal may be able to attenuate, but not reverse, bone loss with dietary calcium intakes of at least 1200 mg/day [10, 126]. Healthy postmenopausal women who are most likely to benefit from calcium supplements are those for whom usual dietary intakes of calcium are low [127, 128].
- Physiological levels of vitamin D are necessary for normal bone development in childhood and bone maintenance in adulthood. In older subjects, particularly the frail elderly in institutions and those living in northern climates, vitamin D deficiency may result in secondary hyperparathyroidism and thus accelerate bone loss. For women in such situations, treatment with physiological doses of vitamin D may prevent bone loss.
- Increased calcium intake among postmenopausal women is probably associated with a small reduction in fracture risk [104].
- Interventions combining calcium and vitamin D reduced fracture rates in older populations [98, 107, 108, 109].
- Whether exercise and calcium intake have a synergistic role remains unknown. The only study that investigated the calcium and physical activity interaction with a 2 × 2 randomized, prospective factorial design [111] found no difference between the calcium supplemented and the nonsupplemented groups.
- Smoking has an undoubted substantial deleterious effect on bone health.

References

1. Kanis JA, Passmore R. Calcium supplementation of the diet. *Br Med J* 1989;298:137-140, 205-208, 673-674.
2. Kanis JA. Requirements of calcium for optimal skeletal health. *Calcif Tissue Int* 1991;49:Suppl, S33-S41.
3. Heaney RP. Nutritional factors in osteoporosis. *Annu Rev Nutr* 1993;13:287-316.
4. Nordin BEC. The calcium debate. *Med J Aust* 1988;148:608.
5. Nordin BEC, Heaney RP. Calcium supplementation of the diet: justified by the present evidence. *Br Med J* 1990;300:1056-1060.
6. Cumming RG. Calcium intake and bone mass: A quantitative review of the evidence. *Calcif Tissue Int* 1990;47:194-201.
7. Kanis JA. Calcium nutrition and its implications for osteoporosis. *Eur J Clin Nutr* 1994;48:757-767, 833-841.
8. Heaney RP. The calcium controversy: a middle ground between the extremes. *Public Health Rep* 1989;S104:36-46.
9. Heaney RP, Abrams S, Dawson-Hughes B, et al. Peak bone mass. *Osteoporosis Int* 2000;11:985-1009.
10. Prince RL. Diet and the prevention of osteoporotic fracture. *N Engl J Med* 1997;337:701-702.
11. Prince RL, Devine A, Dick I, et al. The effects of calcium supplementation (milk powder or tablets) and exercise on bone density in postmenopausal women. *J Bone Miner Res* 1995;10:1068-1075.
12. Bailey DA, Martin AD, McKay HA, et al. Calcium accretion in girls and boys during puberty: a longitudinal analysis. *J Bone Miner Res* 2000;15:2245-2250.
13. Abrams SA, Grusak MA, Stuff J, et al. Calcium and magnesium balance in 9-14-y-old children. *Am J Clin Nutr* 1997;66:1172-7.
14. Martin AD, Bailey DA, McKay HA, et al. Bone mineral accretion during puberty. *Am J Clin Nutr* 1997;66:611-615.
15. Bailey DA, Wedge JH, McCulloch RG, et al. Epidemiology of fractures at the distal end of the radius in children associated with growth. *J Bone Joint Surg* 1989;71-A:1225-1231.
16. Parfitt AM. The two faces of growth: benefits and risks to bone integrity. *Osteoporos Int* 1994;4:382-398.
17. Bailey DA. The Saskatchewan bone mineral accrual study: Bone mineral acquisition during the growing years. *Int J Sports Med* 1997;18 (Suppl3):S191-194.

18. Johnston CC, Miller JZ, Slemenda CW, et al. Calcium supplementation and increases in bone mineral density in children. *N Engl J Med* 1992; 327:82-87.

19. Lloyd T, Andon MB, Rollings N, et al. Calcium supplementation and bone mineral density in adolescent girls. *JAMA* 1993;270:841-844.

20. Lee WTK, Leung SSF, Wang S-H, et al. Double-blind controlled calcium supplementation and bone mineral accretion in children accustomed to a low-calcium diet. *Am J Clin Nutr* 1994;60:744-750.

21. Nowson CA, Green RM, Hopper JL, et al. A co-twin study of the effect of calcium supplementation on bone density during adolescence. *Osteoporos Int* 1997;7:219-225.

22. Bonjour JP, Carrie AL, Ferrari S, et al. Calcium-enriched foods and bone mass growth in prepubertal girls: a randomized double-blind, placebo-controlled trial. *J Clin Invest* 1997;99:1287-1294.

23. Lee WTK, Leung SSF, Leung DMY, et al. A randomized double-blind controlled calcium supplementation trial, and bone and height acquisition in children. *Brit J Nutr* 1995;74:125-139.

24. Nowson CA, Green RM, Guest CS, et al. *The effect of calcium supplementation on bone mass in adolescent female twins*. In: Burkhardt P, Heaney RP, ed. Challenges of modern medicine. Ares-Serono Symposia Publications, 1995: 169-175. vol 7.

25. Cadogan J, Eastell R, Jones N, et al. Milk intake and bone mineral acquisition in adolescent girls: randomised, controlled intervention trial. *Br Med J* 1997;315:1255-60.

26. Lee WTK, Leung SSF, Leung DMY, et al. A follow-up study on the effects of calcium-supplement withdrawl and puberty on bone acquisition of children. *Am J Clin Nutr* 1996;64:71-77.

27. Slemenda CW, Peacock M, Hui SL, et al. Reduced rates of skeletal remodelling are associated with increased bone mineral density during the development of peak skeletal mass. *J Bone Miner Res* 1997;12:676-682.

28. Slemenda CW, Reister TK, Peacock M, et al. Bone growth in children following the cessation of calcium supplementation. *J Bone Miner Res* 1993;8:Suppl. 1. 149 (Abstract 151).

29. Nordin BEC. International patterns of osteoporosis. *Clin Orthop* 1966;45:17-30.

30. Matkovic V, Kostial K, Simonovic I, et al. Bone status and fracture rates in two regions of Yugoslavia. *Am J Clin Nutr* 1979;32:540-549.

31. Kamiyama S, Kobayashi S, Abe S, et al. Osteoporosis prevalence and nutritional intake among the people in farm, fishing and urban districts. *Tohoku J Exp Med* 1972;107:387-394.

32. Kanders B, Dempster DW, Lindsay R. Interaction of calcium nutrition and physical activity on bone mass in young women. *J Bone Miner Res* 1988;3:145-149.

33. Lau E, Donnan S, Barker DJP, et al. Physical activity and calcium intake in fracture of the proximal femur in Hong Kong. *Br Med J* 1988;297:1441-1443.

34. Sandler RB, Slemenda CW, LaPorte RE, et al. Postmenopausal bone density and milk consumption in childhood and adolescence. *Am J Clin Nutr* 1985;42:270-274.

35. Picard D, Ste-Marie LG, Coutu D, et al. Premenopausal bone mineral content relates to height, weight, and calcium intake during early adulthood. *Bone Miner* 1988;4:299-309.

36. Yano K, Heilbrun LK, Wasnich RD, et al. The relationship between diet and bone mineral content of multiple skeletal sites in elderly Japanese-American men and women living in Hawaii. *Am J Clin Nutr* 1985;42:877-888.

37. Holbrook TL, Barrett-Connor E, Wingard DL. Dietary calcium and risk of hip fracture: a 14-year prospective population study. *Lancet* 1988;ii:1046-1049.

38. Uusi-Rasi K, Sievanen H, Vuori I, et al. Associations of physical activity and calcium intake with bone mass and size in healthy women at different ages. *J Bone Miner Res* 1999;13:133-142.

39. Smith RW, Frame B. Concurrent axial and appendicular osteoporosis. Its relation to calcium consumption. *N Engl J Med* 1965;273:73-78.

40. Smith RW, Rizek J. Epidemiological studies of osteoporosis in women of Puerto Rico and Southwest Michigan with special reference to age, race, nationality and other associated findings. *Clin Orthop* 1966;45:31-48.

41. Stevenson JC, Lees B, Devonport M, et al. Determinants of bone density in normal women: Risk factors for future osteoporosis? *Br Med J* 1989;298:924-928.

42. van Beresteijn ECH, van t'Hof MA, de Waard H, et al. Relation of axial bone mass to habitual calcium intake and cortical bone loss in healthy early postmenopausal women. *Bone* 1990;11:7-13.

43. Wickham CAC, Walsh K, Cooper C, et al. Dietary calcium, physical activity, and risk of hip fracture: a prospective study. *Br Med J* 1989;299:889-992.

44. Nilas L, Christiansen C, Rodbro P. Calcium supplementation and postmenopausal bone loss. *Br Med J* 1984;289:1103-1106.

45. Hegsted DM. Mineral intake and bone loss. *Fed Proc* 1967;26:1747-1754.

46. Donath A, Indermuhle P, Baud R. *Influence of the national calcium and fluoride supply and of a calcium supplementation on bone mineral content of healthy population in Switzerland*. In: Proceedings International Conference on bone mineral measurement. Department of Health, Education and Welfare Publication, 1975.

47. Garn SM, Robinson CG, Wagner B, et al. Population similarities in the onset and rate of adult endosteal bone loss. *Clin Orthop* 1969;65:51-60.
48. Riggs BL, Wahner HW, Melton LJ, et al. Dietary calcium intake and rates of bone loss in women. *J Clin Invest* 1987;80:979-982.
49. Mazess RB, Barden HS. Bone density in premenopausal women: Effects of age, dietary intake, physical activity, smoking and birth-control pills. *Am J Clin Nutr* 1991;53:132-142.
50. Smith EL, Gilligan C, Smith PE, et al. Calcium supplementation and bone loss in middle-aged women. *Am J Clin Nutr* 1989;50:833-842.
51. Baran D, Sorensen A, Grimes J, et al. Dietary modification with dairy products for preventing vertebral bone loss in premenopausal women: a three-year prospective study. *J Clin Endocrinol Metab* 1989;70:264-270.
52. Avioli L, Heaney RP. Calcium intake and bone health. *Calcif Tissue Int* 1991;48:221-223.
53. Heaney RP. Estrogen-calcium interactions in the post-menopause: A quantitative description. *Bone Miner* 1990;11:67-84.
54. Ettinger B, Genant HK, Cann CE. Postmenopausal bone loss is prevented by treatment with low-dosage estrogen with calcium. *Ann Int Med* 1987;106:40-45.
55. Dawson-Hughes B, Dallai GE, Krall EA, et al. A controlled trial of the effect of calcium supplementation on bone density in postmenopausal women. *N Engl J Med* 1990;323:878-883.
56. Horsman A, Gallagher JC, Simpson M, et al. Prospective trial of oestrogen and calcium in postmenopausal women. *Br Med J* 1977;ii:789-792.
57. Polley KJ, Nordin BEC, Baghurst PA, et al. Effect of calcium supplementation on forearm bone mineral content in postmenopausal women: a prospective, sequential controlled trial. *J Nutr* 1987;117:1929-1932.
58. Recker RR, Saville PD, Heaney RP. Effect of estrogens and calcium carbonate on bone loss in postmenopausal women. *Ann Int Med* 1977;87:649-655.
59. Riis B, Thomsen K, Christiansen C. Does calcium supplementation prevent postmenopausal bone loss? *N Engl J Med* 1987;316:173-177.
60. Elders PJM, Lips P, Netelenbos JC, et al. Long-term effect of calcium supplementation on bone loss in perimenopausal women. *J Bone Miner Res* 1994;9:963-970.
61. Heaney RP, Recker RR, Saville PD. Menopausal changes in bone remodelling. *J Lab Clin Med* 1978;92:964-970.
62. Stepan JJ, Pospichal J, Presl J, et al. Bone loss and biochemical indices of bone remodeling in surgically induced postmenopausal women. *Bone* 1987;8:279-284.
63. Albanese AA, Edelson AH, Lorenze EJ, et al. Problems of bone health in the elderly: ten year study. *NY State J Med* 1975;75:326-336.
64. Dawson-Hughes B. Calcium supplementation and bone loss: A review of controlled clinical trials. *Am J Clin Nutr* 1991;54:274-280S.
65. Lamke B, Sjoberg HE, Sylven M. Bone mineral content in women with Colles' fracture: effect of calcium supplementation. *Acta Orthop Scand* 1978;49:143-149.
66. Smith EL, Reddan W, Smith PE. Physical activity and calcium modalities for bone mineral increase in aged women. *Med Sci Sports Exerc* 1981;13:60-64.
67. Recker RR, Heaney RP. The effect of milk supplements on calcium metabolism, bone metabolism and calcium balance. *Am J Clin Nutr* 1985;41:254-263.
68. Smith DA, Anderson JJB, Aitken JM, et al. *The effects of calcium supplements of the diet on bone mass measurements*. In: Kuhlencordt F, Kruse HP, ed. Calcium metabolism, bone and metabolic bone disease. Berlin: Springer, 1975: 278-282.
69. Heaney RJ. *Nutrition and risk for osteoporosis*. In: Marcus R et al. eds. Osteoporosis. 1st ed. San Diego: Academic Press, 1996:492.
70. Reid IR, Ames R, Evans MC, et al. Effect of calcium supplementation on bone loss in postmenopausal women. *N Engl J Med* 1993;328:460-464.
71. Nordin BEC, Horsman A, Crilly AG, et al. Treatment of spinal osteoporosis in postmenopausal women. *Br Med J* 1980;280:451-454.
72. Prince RL, Smith M, Dick IM, et al. Prevention of postmenopausal osteoporosis: a comparative study of exercise, calcium supplementation and hormone-replacement therapy. *N Engl J Med* 1991;325:1189-1195.
73. Ettinger B, Genant HK, Cann CE. Long-term estrogen replacement therapy prevents bone loss and fractures. *Ann Int Med* 1985;102:319-324.
74. Elders PJM, Netelenbos JC, Lips P, et al. Long-term effect of calcium supplementation on bone loss in perimenopausal women. *J Bone Miner Res* 1991;9:963-970.
75. Reid IR, Schooler BA, Hannon SF, et al. The acute biochemical effects of four proprietary calcium preparations. *Aust NZ J Med* 1986;16:193-197.
76. Devine A, Dick IM, Heal SJ, et al. A 4-year follow-up study of the effect of calcium supplementation on bone density in early postmenopausal women. *Osteoporos Int* 1997;7:23-28.
77. Reid IR, Ames R, Evans MC, et al. Long-term effect of calcium supplementation in women: a randomized controlled trial. *Am J Med* 1995;98:331-335.
78. Heaney RP. The bone-remodeling transient: Implications for the interpretation of clinical studies of

bone mass change. *J Bone Miner Res* 1994;9:1515-1523.

79. Nguyen TV, Kelly PJ, Sambrook PN, et al. Lifestyle factors and bone density in the elderly: Implications for osteoporosis prevention. *J Bone Miner Res* 1994;9:1339-1346.

80. Kelly PJ, Pocock NA, Sambrook PN, et al. Dietary calcium, sex hormones and bone mineral density in man. *Br Med J* 1990;300:1361-1364.

81. Bendavid EJ, Shan J, Barrett-Connor E. Factors associated with bone mineral density in middle-aged men. *J Bone Miner Res* 1996;1185-1190.

82. Holbrook TL, Barrett-Connor E. An 18-year prospective study of dietary calcium and bone mineral density in the hip. *Calcif Tissue Int* 1995;56:364-367.

83. Feldman D, Malloy PJ, Gross C. *Vitamin D: metabolism and action*. In: Marcus R, Feldman D, Kelsey J, ed. Osteoporosis. San Diego, CA, USA: Academic Press, 1996: 205-227.

84. Jones G, Dwyer T. Bone mass in prepubertal children: Gender differences and the role of physical activity and sunlight exposure. *J Clin Endocrinol Metab* 1998;83:4274-4279.

85. Ilich JZ, Badenhop NE, Jelic T, et al. Calcitriol and bone mass accumulation in females during puberty. *Calcif Tissue Int* 1997;61:104-9.

86. Kristinsson JO, Valdimarsson O, Sigurdsson G, et al. Serum 25-hydroxyvitamin D levels and bone mineral density in 16-20 years-old girls: lack of association. *J Intern Med* 1998;243:381-8.

87. Mirwald RL, Bailey DA. Seasonal height velocity variation in boys and girls age 8 to 18 years. *Am J Hum Biol* 1997;9:709-715.

88. Fehily AM, Coles RJ, Evans WD, et al. Factors affecting bone density in young adults. *Am J Clin Nutr* 1992;56:579-586.

89. Holbrook TL, Barrett-Connor E. Calcium intake: covariates and confounders. *Am J Clin Nutr* 1991;53:741-4.

90. Lukert BP, Johnson BE, Robinson RG. Estrogen and progesterone replacement therapy reduces glucocorticoid-induced bone loss. *J Bone Miner Res* 1992;

91. Reid IR. *Vitamin D and its metabolites in the management of osteoporosis*. In: Marcus R, Feldman D, Kelsey J, ed. Osteoporosis. San Diego, CA, USA: Academic Press, 1996: 1169-1190.

92. Stein MS, Scherer SC, Walton SL, et al. Risk factors for secondary hyperparathyroidism in a nursing home population. *Clin Endocrinol (Oxf)* 1996;44:375-383.

93. O'Dowd KJ, Clemens TL, Kelsey JL, et al. Exogenous calciferol (vitamin D) and vitamin D endocrine status among elderly nursing home residents in the New York City area. *J Am Geriatr Soc* 1993;41:414-21.

94. Nordin BEC, Baker MR, Horsman A, et al. A prospective trial of the effect of vitamin D supplementation on metacarpal bone loss in elderly women. *Am J Clin Nutr* 1985;42:470-474.

95. Buring K, Hulth AG, Nilsson BE, et al. Treatment of osteoporosis with Vitamin D. *Acta Med Scand* 1974;195:471-472.

96. Riggs BL, Seeman E, Hodgson SF, et al. Effect of the fluoride/calcium regimen on vertebral fracture occurrence in postmenopausal osteoporosis. Comparison with conventional therapy. *N Engl J Med* 1982;306:446-50.

97. Christiansen C, Christiansen MS, McNair P, et al. Prevention of early postmenopausal bone loss: controlled 2 year study in 315 normal females. *Eur J Clin Invest* 1980;10:273-279.

98. Chapuy MC, Arlot ME, Duboef F, et al. Vitamin D3 and calcium to prevent hip fracture in elderly women. *N Engl J Med* 1992;327:1637-1642.

99. Turner RT, Kidder LS, Kennedy A, et al. Moderate alcohol consumption suppresses bone turnover in adult female rats. *J Bone Miner Res* 2001;16:589-594.

100. Slovik DM, Adams JS, Neer RM, et al. Deficient production of 1,25 dihydroxyvitamin D in elderly osteoporotic patients. *N Engl J Med* 1981;305:372-374.

101. Ebeling PR, Sandgren ME, MiMagno EP, et al. Evidence of age-related decrease in intestinal responsiveness to vitamin D: Relationship between serum 1,25-(OH)2D and intestinal vitamin D receptor concentrations in normal women. *J Clin Endocrinol Metab* 1992;75:176-182.

102. Tilyard NW. Treatment of postmenopausal osteoporosis with calcitriol or calcium. *N Engl J Med* 1992;326:357-362.

103. Gallagher JC. Prevention of bone loss in postmenopausal and senile osteoporosis with vitamin D analogues. *Osteoporos Int* 1993;3:172-5.

104. Cumming RG, Nevitt MC. Calcium for prevention of osteoporotic fractures in postmenopausal women. *J Bone Miner Res* 1997;12:1321-1329.

105. Chevalley T, Rizzoli R, Nydegger V, et al. Effects of calcium supplementation on femoral bone mineral density and vertebral fracture rate in vitamin-D-replete elderly patients. *Osteoporos Int* 1994;4:245-252.

106. Recker R, Hinders S, Davies KM, et al. Correcting calcium nutritional deficiency prevents spine fractures in elderly women. *J Bone Miner Res* 1996;11:1961-1966.

107. Dawson-Hughes B, Harris SS, Krall EA, et al. Effect of calcium and vitamin D supplementation on bone density in men and women 65 years of age or older. *N Engl J Med* 1997;337:670-676.

108. Lips P. Vitamin D deficiency and osteoporosis: the role of vitamin D deficiency and treatment with vitamin D and analogues in the prevention of os-

teoporosis-related fractures. *Eur J Clin Invest* 1996;26:436-42.

109. Chapuy MC, Arlot ME, Delmas PD, et al. Effect of calcium and cholecalciferol treatment for three years on hip fractures in elderly women. *Br Med J* 1994;308:1081-2.
110. Lips P, Netelenbos JC, Jongen MJM, et al. Histomorphometric profile and vitamin D status in patients with femoral neck fracture. *Metab Bone Dis Relat Res* 1982;4:85-93.
111. Friedlander AL, Genant HK, Sadowsky S, et al. A two-year program of aerobics and weight training enhances bone mineral density of young women. *J Bone Miner Res* 1995;10:574-585.
112. Bikle DD, Genant HK, Cann C, et al. Bone disease in alcohol abuse. *Ann Int Med* 1985;103:42-48.
113. Wezeman F, Emanuele M, Emanuele N, et al. Chronic alcohol consumption during male rat adolescence impairs skeletal development through effects on osteoblast gene expression, bone mineral density, and bone strength. *Alcohol Clin Exp Res* 1999;23:1534-1542.
114. Felson DT, Zhang Y, Hannan MT, et al. Alcohol intake and bone mineral density in elderly men and women. *Am J Epidemiol* 1995;142:485-492.
115. Holbrook TL, Barrett-Connor E. A prospective study of alcohol consumption and bone mineral density. *Br Med J* 1993;306:1506-1509.
116. Laitinen K, Välimäki M, Keto P. Bone mineral density measured by dual-energy x-ray absorptiometry in healthy Finnish women. *Calcif Tissue Int* 1991;48:224-231.
117. Angus RM, Pocock NA, Eisman JA. Nutritional intake of pre- and postmenopausal Australian women with special reference to calcium. *Eur J Clin Nutr* 1988;42:617-625.
118. Hansen MA, Overgaard K, Riis BJ, et al. Potential risk factors for development of postmenopausal osteoporosis - examined over a 12 year period. *Osteoporos Int* 1991;1:95-102.
119. Hannan MT, Felson DT, Dawson-Hughes B, et al. Risk factors for longitudinal bone loss in elderly men and women: the Framingham Osteoporosis Study. *J Bone Miner Res* 2000;15:710-20.
120. Law MR, Hackshaw AK. A meta-analysis of cigarette smoking, bone mineral density and risk of hip fracture: recognition of a major effect. *Br Med J* 1997;315:841-6.
121. Krall EA, Dawson-Hughes B. Smoking and bone loss among postmenopausal women. *J Bone Miner Res* 1991;6:331-338.
122. Seeman E. *The effects of tobacco and alcohol use on bone.* In: Marcus R, Feldman D, Kelsey J, ed. Osteoporosis. San Diego, CA, USA: Academic Press, 1996: 577-597.
123. Hopper JL, Seeman E. The bone density of female twins discordant for tobacco use. *N Engl J Med* 1994;330:387-392.
124. Yuhara S, Kasagi S, Inoue A, et al. Effects of nicotine on cultured cells suggest that it can influence the formation and resorption of bone. *Eur J Pharmacol* 1999;383:387-93.
125. Johnston JD. Smokers have less dense bones and fewer teeth. *J Royal Soc Health* 1994;114:265-269.
126. Michaelsson K, Bergstrom R, Holmberg L, et al. A high dietary calcium intake is needed for a positive effect on bone density in Swedish postmenopausal women. *Osteoporos Int* 1997;7:155-161.
127. Dawson-Hughes B, Jacques P, Shipp C. Dietary calcium intake and bone loss from the spine in healthy postmenopausal women. *Am J Clin Nutr* 1987;46: 685-687.
128. Wark JD. Osteoporotic fractures: background and prevention strategies. *Maturitas* 1996;23:193-207.
129. Specker BL. Evidence for an interaction between calcium intake and physical activity on changes in bone mineral density. *J Bone Miner Res* 1996;11:1539-1544.
130. Lips P, Graafmans WC, Ooms ME, et al. Vitamin D supplementation and fracture incidence in elderly persons. A randomized, placebo-controlled clinical trial. *Ann Intern Med* 1996;124:400-6.
131. Wark, JD. *Osteoporotic fracture risk and the menopause. Prevention strategies.* In International Conference on Reproductive Health. WHO Publications 2001: (in press).

part III

Evidence and Prescription: A Life Span Approach

Having discussed the properties of bone (part I) and the important determinants of bone that are *not* related to exercise (part II), we are now ready to examine the evidence that shows that physical activity promotes bone health. In this part we also provide a variety of practical exercise programs for bone health for different stages of the life span.

In the book to date, we have used the term *physical activity* to refer to activity that might be related to bone health. We have not been specific about the kinds of physical activity needed to be effective, and so we did not define terms such as *exercise* or *targeted bone loading*. Now that we are launching into the "activity" chapters of this book, we need to distinguish among physical activity, exercise, and targeted bone loading, as these terms have differing implications for bone health. Chapter 9 begins with the definitions of these terms and then outlines the challenges that researchers encounter when assessing physical activity. It summarizes current approaches to measurement and makes practical suggestions for those currently working in the field of physical activity and bone health.

Chapters 10 through 13 examine the roles of various forms of physical activity in promoting bone health in children and adolescents (chapter 10), premenopausal women (chapter 11), postmenopausal women (chapter 12), and men (chapter 13). These chapters describe how the skeleton responds to activity and loading across the life span and between genders. It appears that the growing years are the time of maximal bone mineral accrual, the adult years are essentially a period of bone mineral maintenance, and the postmenopausal years are associated with bone loss.

Exercise is only one, albeit important, determinant of bone mineral (see part II), and so we must recognize that there is a limit to how much exercise intervention can influence bone mineral gain in, for example, elderly people. Fortunately, bone mineral gain is not the only way of maintaining bone health. Physical activity makes an important, and increasingly recognized, contribution to fall prevention and this is outlined in Chapter 14.

Despite even Herculean efforts at primary prevention, osteoporosis can and docs occur. This is not the time to give up on exercise. Chapter 15 outlines the important role of exercise in the treatment of osteoporosis and the prevention of further fractures, a topic that is generally given short shrift in traditional medical textbooks. This chapter provides important background information on the current clinical assessment methods and the pharmaceutical management of osteoporosis. It may be especially helpful to exercise scientists and exercise specialists.

Before focusing on the challenge of osteoporosis, let us begin with an equally daunting challenge—that of measuring physical activity accurately and reliably. Chapter 9 will help you decide whether you think this is something that can ever be achieved.

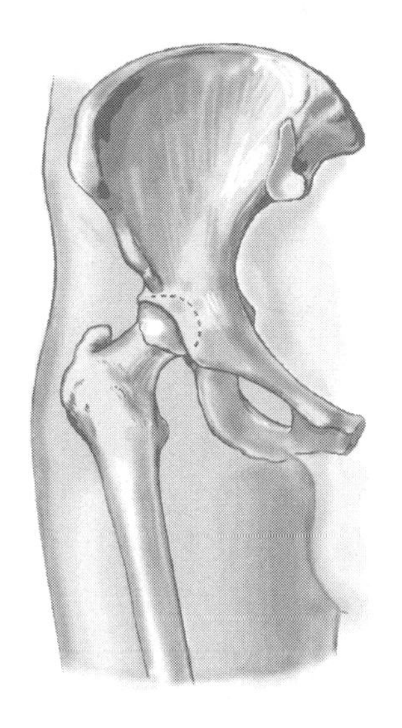

Chapter 9

Measurement of Physical Activity

In this chapter we define physical activity, exercise, and targeted bone loading and outline the challenge of measuring these variables. We then outline the different methods that are currently used to measure physical activity. We conclude by providing practical suggestions that may help with your own work in this field.

Physical activity is an umbrella term defined as any bodily movement produced by skeletal muscles that results in energy expenditure above the basal level [1]. The dimensions of physical activity—Frequency, Intensity, Time, and Type (FITT)—each result in different health outcomes. Elegant animal experiments suggest that the type of physical activity is by far the most important dimension affecting bone [2].

Exercise represents a subcategory of physical activity that is "planned, structured, repetitive and purposive in the sense that improvement or maintenance of one or more components of physical fitness is the objective" [3]. Thus, when physical activity is targeted specifically toward one of the dimensions of physical activity (F, I, T, or T) and has specific health-related or sport-specific objectives, it is considered to also be exercise.

We use the term *targeted bone loading* (or, where the context is clear, *targeted loading*) to refer to force-generating activity that provides stimulus to a specific bone or bone region *beyond that* provided by activities of daily living. It is a subcategory of both physical activity and exercise (figure 9.1). An activity may provide bone loading at one site, but not at another. Thus, jumping provides bone loading of the lower limb, but not the upper limb. Although all exercise commonly involves some form of loading through muscles and joints, some exercise (such as swimming) does not apply targeted bone loading to the skeleton. Thus, while virtually all physical activity reduces the risk of cardiovascular disease, not all physical activity will augment a person's BMD.

In this book, gardening, shopping, and child's play are considered physical activities but are not considered examples of targeted loading to a specific bone. Swimming is both physical activity and exercise but does not qualify as targeted

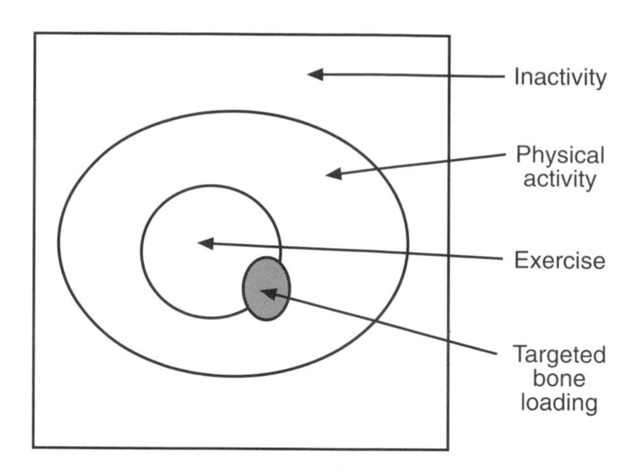

FIGURE 9.1 Venn diagram showing that life can be divided into physical activity and inactivity. Exercise is a subcategory of physical activity that is planned, structured, and repetitive and has the purpose of maintaining or enhancing a component of physical fitness. Targeted bone loading refers to physical activity that generates forces on bone that exceed those provided by activities of daily living. Although there are many examples of targeted loading in exercise, work-related activities can also generate targeted loading (e.g., a military parachutist landing hard on the earth as part of his work).

bone loading. The sport of triple jump provides physical activity, exercise, and targeted bone loading (figure 9.2). Research studies of targeted bone loading are evaluated separately from studies of physical activity.

Distinguishing targeted bone loading within physical activity should not be seen as downplaying the role of generalized physical activity in maintaining general health, or even bone health. Such activity provides health benefits in addition to its influence on bone mass. With respect to bone, the correct exercise prescription (as discussed in each of the following chapters in this part) can increase muscle mass and improve balance, thereby decreasing the risk of falling and, subsequently, fracture (see chapter 14).

Inherent Limitations

Physical activity is difficult to measure because it is a complex behavior [1]. The task is even more difficult for the bone researcher since not all physical activities influence bone in the same way or to the same extent. For example, riding a stationary bike at maximal effort is a demanding activity, yet it does very little to benefit bone. The fact is that a simple, accurate way of measuring the types of physical activity that influence bone still eludes us.

FIGURE 9.2 A 74-year-old woman illustrates that physical activity, exercise, and targeted bone loading can all be undertaken simultaneously in the sport of triple jump.

We have already stated that not all physical activities impose a load on the skeleton, and thus do not produce a measurable change in bone. In chapter 3 we described how bone responds to certain specific mechanical stimuli. In fact, the primary function of bone (and bone remodeling) is to maintain its load-bearing capacity [4]. At the risk of oversimplifying, we can say that activity that produces strains in the physiologic loading zone (200-2000 microstrain) maintains remodeling at a steady state, which maintains bone strength. Physical activities such as cycling and swimming that may load the skeleton *less* than what is necessary for the formation of new bone will result in bone maintenance, rather than gain. Activities such as walking and running, although recommended for all individuals, provide very little stimulus to augment bone mass in the healthy skeleton.

Questionnaires may have limited validity, reliability, and accuracy. Although some excellent questionnaires exist to assess physical activity itself, they do not necessarily assess the type of physical activity that is relevant for bone health (figure 9.3). Direct measures of physical activity are still in their infancy and suffer from some of the same limitations as questionnaires.

Biomechanists can measure the ground reaction forces that activities generate by using a force platform. However, ground reaction forces do not precisely reflect forces experienced by individual bones. Furthermore, since bone is not one uniform tissue, forces will vary in different anatomical areas and even within the same bone. The effect of loading will differ in cortical and trabecular bone as well. Thus, landing from a jump will load the cortical bone and the trabecular bone of the femoral neck differently, even though they are adjacent to each other. Clearly, a jump will also load the femur differently from the lumbar spine and upper limbs.

The difficulty of measuring physical activity generally is well accepted [1]. The simple examples described in the remainder of this chapter show how difficult it is to measure *targeted bone loading*. To help all those grappling with this problem, we first summarize the current methods of assessing all types of physical activity and then outline some suggestions as to how activity can best be measured for bone health studies.

FIGURE 9.3 Researcher Kerry MacKelvie (left) administering a physical activity questionnaire to a participant in a study of bone health in men.

Traditional Methods

Scientific interest in physical activity arose largely because of its association with the prevention of cardiovascular disease. For this reason, most of the methods of measuring physical activity are more relevant for cardiovascular health. Because the measurement of physical activity in the cardiovascular field is further advanced than that in the bone field, we must examine it and decide what to adopt, and what to disregard, for skeletal research.

The two main types of measurement of physical activity are self-reporting and direct measurement.

Self-Report in Adults

Self-report has been the most common method of measuring physical activity in studies that measured BMD [5-7]. One problem with this method has been the somewhat ad hoc approach to categorizing physical activity, which has resulted in wide-ranging categories reported in published studies. Many studies used hours per week of weight-bearing physical activity as a measure [8], but what constitutes such activity can be difficult to distinguish. Are in-line skating to work, working on the floor of the stock exchange, cleaning houses, or working as a fitness consultant at a desk all weight-bearing physical activities? This lack of a common method of quantifying self-reported physical activity has made it impossible to compare the outcome of different studies.

Self-report measures of physical activity in cardiovascular studies, on the other hand, often provide quite strict definitions of physical activity [9, 10]. Physical activity is then often further broken down into four categories: vigorous, moderate, low, or sedentary. These four categories of activity have been shown in cross-sectional studies to be correlated with biological indices of cardiovascular disease risk [11]. However, when scrutinized, even these rather more sophisticated measures of physical activity do not meet all the criteria necessary to be considered ideal measures [1]. Furthermore, they are unlikely to translate as appropriate measures of targeted bone loading.

Questionnaire Methods in Children

In chapter 10 we discuss the important association between childhood physical activity and bone mineral development as well as the relationship between child activity patterns and adult bone health. As adult measures of physical activity are not suitable for children, measurement of physical activity in children must be discussed separately.

Self-report physical activity instruments, including self-administered recall or an activity diary, are inexpensive and convenient to administer [1]. Of these methods, diaries may be the most accurate, but they require complete cooperation and are limited to use by children with good reading and writing skills.

Unlike self-reported questionnaires that can be administered to an entire classroom at one visit, interviewer-administered recalls are expensive and labor intensive. Most proxy reports of children's physical activity have not been shown to be valid, probably because neither parents nor teachers can observe children all day. In one study the mothers of children aged 5 to 14 years were asked to rate their children's activities relative to other children (on a five-point scale from much less to much more) and to estimate their children's hours of vigorous activity (table 9.1) [12].

The study found that the repeat measures reliability for the mothers' reports was no better than the children's. For comparison, Pearson's *r* correlations between mothers' and children's reports of hours spent watching television were approximately 0.60 for both weekends and school nights. Thus, the questionnaire was reproducible as a correlation coefficient of 0.60, which is considered good for a measure of time spent at a task [13].

Children are unable to self-report physical activity reliably before 9 or 10 years of age [1]. Even with children between the ages of 9 and 15 years self-report should be used cautiously. Adolescents aged 15 and above appear to report as accurately as adults. Short-term recall (one day) appears more reliable than long-term recall (e.g., seven days).

In studies of children below 9-10 years, direct observation may be the best method of quantifying physical activity [14]. The cost of doing this method, however, makes it impractical for most studies [1, 15].

Direct, Objective Measures

Devices such as the accelerometer and heart rate monitors have been used to measure physical activity more objectively. At present, doubly labeled water is considered the closest there is to a gold standard. However, each of these methods has limitations both for measurement of physical activity per se and for measuring activity that loads bone specifically.

Accelerometers (or motion sensors) (figure 9.4) detect body movement via a lever that is displaced and generates electrical current proportional to the energy of the acceleration [1, 16]. The current can be used as a raw number to indicate movement, which is ideal for bone studies. Many studies have used Caltrac accelerometers (Caltronics Division of Hemokinetics, Madison, Wisconsin), which have high inter-instrument reliability [16]. In a recent study, seven-day monitoring provided reliable estimates of usual physical activity behavior in children and adolescents and accounts for potentially important differences in weekend versus weekday activity behavior [17].

TABLE 9.1

Correlations between baseline and six-month questionnaires for children and their mothers

Children	r	Mothers	r
Biking	0.14	Relative to others	0.66
Swimming	0.39	Hours of vigorous activity	0.47
Baseball	0.55		
Weightlifting	0.40		
Basketball	0.41		
Soccer	0.77		
Tennis	0.58		
Total weight-bearing	0.62		

Because heart rate is highly correlated with oxygen uptake, especially during exercise, heart rate monitors (figure 9.5) have been used to measure physical activity. Because bone can respond to activity that does not increase heart rate some researchers combined accelerometers with heart rate monitors [18].

Doubly labeled water is a method that uses isotope (hydrogen and oxygen) tracers and the dilution principle to assess energy expenditure. Although this expensive method of measurement is a candidate for being the gold standard for energy expenditure [16], it is not a specific measure of physical activity and therefore does not lend itself to bone studies.

Suggestions for Measuring Physical Activity in Bone Research

An important first step in comparing the results of physical activity and bone research would be to use a standardized questionnaire to assess physical activity. This step, however, would not overcome the fundamental problem that the activity measures are only a surrogate measure of targeted bone loading. In addition, such a questionnaire is far easier to propose than to compose. The instrument would need to cater to subjects of different ages (the very young, the very old, etc.), address geographical and cultural differences, and overcome the difficulty of assessing the direct effect of any loading activity on bone.

At present, we suggest that researchers attempt to locate and use the instruments developed by the leading research groups in each specific age group [13, 19] or by specific interest groups (for example, glucocorticoid-induced bone loss [20]). We have found that researchers from these groups are willing to share their questionnaires.

A way of partially managing physical activity within research is to perform a randomized, controlled trial using physical activity as an intervention. Although the physical activity that subjects and controls undertake outside the intervention is a control variable that should be measured, randomization should minimize the

FIGURE 9.4 Tritrak accelerometer senses the frequency of movement with walking, providing a measure of physical activity.

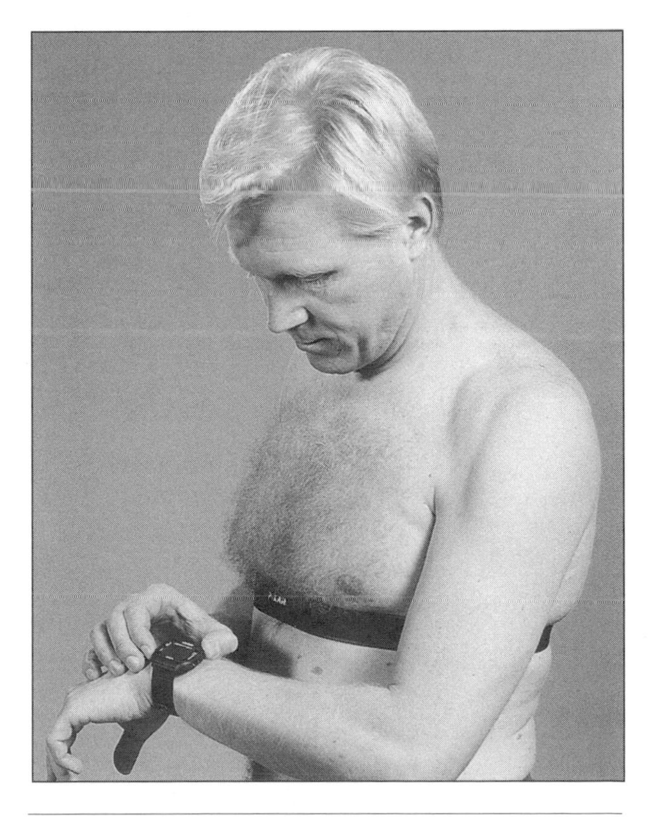

FIGURE 9.5 A heart rate monitor can be used to quantify physical activity.

degree of this potential confound. Furthermore, in a prospective intervention study, measuring change in bone mineral will allow the specific effect of the exercise intervention to be determined, unlike what happens in cross-sectional studies. Although very few exercise intervention studies measured bone as a primary outcome measure before 1990, this study design is now increasingly prevalent (chapters 10-13, 15). As a result, our knowledge of activity and bone has become much more sound.

Another approach to measuring activity and bone is to measure the ground reaction forces associated with the activity. This could be termed the "biomechanical approach." In chapter 3 we learned that the effect of physical activity was related not only to strain magnitude (intensity of the activity) but also to the pattern of distribution of the stimulus. This was termed the error strain distribution hypothesis [21]. In short, bone cells respond to unfamiliar patterns of loading. Thus, researchers with access to a biomechanics laboratory should attempt to characterize the loading pattern (including rates of force and landing strategy) associated with key physical activities (i.e., the intervention being used), rather than merely measure maximum ground reaction force. Eventually, kinematic measures (studies of the forces being transmitted to bones and joints) would also be measured.

Several researchers have combined the two previously discussed approaches to measure the biomechanical forces associated with their intervention. Heinonen and colleagues [22] assessed the ground reaction forces associated with the physical activity intervention in premenopausal women. A study of children in grades 4, 5, and 6 in Vancouver, Canada [23, 24], also assessed the ground reaction forces associated with diverse activities used in a circuit training intervention program administered within elementary school physical education [25]. The loading characteristics of these activities were also assessed and the outcome was used to develop and modify a progressive bone loading intervention [26].

There is an urgent need for valid and standardized questionnaires directed toward specific age groups and specific ethnicities. It remains to be seen whether a combination of measurement approaches, such as questionnaires used with motion sensors on a subset of subjects, measures targeted loading better than either instrument alone.

SUMMARY

- Physical activity is bodily movement produced by skeletal muscles that results in energy expenditure above the basal level.
- Exercise, a subcategory of physical activity, is planned, structured, repetitive, and purposive to improve fitness. Exercise does not necessarily augment bone mineral or bone strength, as bone responds to site-specific loading that has adequate strain rate and strain magnitude. Such exercise is called targeted bone loading.
- Although a number of ways to measure targeted bone loading exist, none are ideal. The measurement of targeted bone loading is a major unresolved challenge for bone health researchers.
- Until the ideal instrument is devised, researchers might consider using the instruments favored by experienced bone researchers. "Instruments" could be in the form of questionnaires, randomized, controlled trials of well-described interventions, or others.

References

1. Sallis JF, Owen N. *Physical activity and behavioural medicine*. Thousand Oaks, CA: Sage, 1998:210 pages.
2. Kannus P, Sievanen H, Vuori I. Physical loading and bone. *Bone* 1996;18:1S-3S.
3. Reid C, Dyck L, McKay H, et al., eds. *The benefits of physical activity for women and girls: A multidisciplinary perspective*. Vancouver: British Colum-

...hia Centers of Excellence in Women's Health,

ing and physical activity behaviours and associated events. *J App Behav Anal* 1991;24:141-151.

15. McKenzie TL. Observational measures of children's physical activity. *Journal of School Health* 1991;61:224-227.
16. Montoye HJ, Kemper HCG, Saris WHM, et al. *Measuring physical activity and energy expenditure.* Champaign, IL: Human Kinetics, 1996:191 pages.
17. Trost SG, Pate RR, Freedson PS, et al. Using objective physical activity measures with youth: how many days of monitoring are needed? *Med Sci Sports Exerc* 2000;32:426-31.
18. Haskell WL, Yee MC, Evans A, et al. Simultaneous measurement of heart rate and body motion to quantitate physical activity. *Med Sci Sports Exerc* 1993;25:109-115.
19. Bailey DA. The Saskatchewan bone mineral accrual study: Bone mineral acquisition during the growing years. *Int J Sports Med* 1997;18 (Suppl3):S191-194.
20. Lukert BP, Raisz LG. Glucocorticoid-induced osteoporosis. *Rheum Dis Clin North Am* 1994;20:629-650.
21. Lanyon LE, Goodship AE, Pye CJ, et al. Mechanically adaptive bone remodeling. *J Biomechanics* 1982;15:141-154.
22. Heinonen A, Kannus P, Sievänen H, et al. Randomised control trial of effect of high-impact exercise on selected risk factors of osteoporotic fractures. *Lancet* 1996;348:1343-1347.
23. McKay HA, Petit M, Schutz R, et al. Lifestyle determinants of bone mineral: A comparison between prepubertal Asian- and Caucasian-Canadian boys and girls. *Calcif Tissue Int* 2000;66:320-324.
24. McKay HA, Petit MA, Schutz RW, et al. Augmented trochanteric bone mineral density after modified physical education classes: A randomized school-based exercise intervention in prepubertal and early-pubertal children. *J Pediatr* 2000;136:156-162.
25. McKay H, Tsang G, Heinonen A, et al. Ground reaction forces in children jumping. *(submitted)* 2001.
26. MacKelvie KJ, McKay HA, Khan KM, et al. Lifestyle risk factors for osteoporosis Asian and Caucasian girls. *Med Sci Sports Exerc* 2001 (in press).

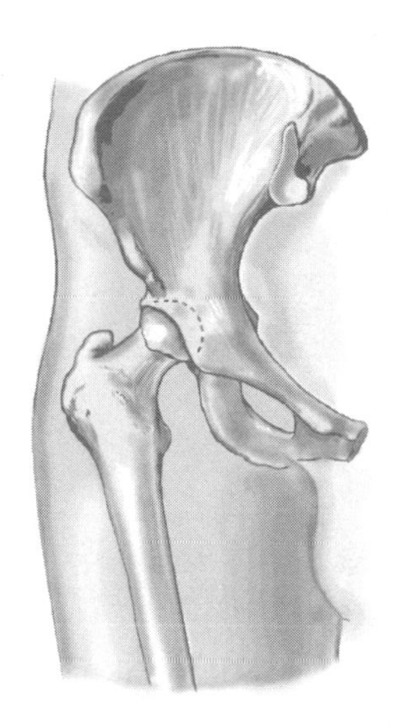

Chapter 10

Physical Activity and Bone in Childhood and Adolescence

with Kerry MacKelvie

Until very recently, a book on bone health would have paid scant attention to childhood and adolescence. However, in the late 1980s and early 1990s some researchers argued that osteoporosis—traditionally a geriatric condition—may have its antecedents in childhood [1-3]. They, and others, realized that peak bone mass is a determinant of adult bone mineral density and thus osteoporotic fracture risk [3]. Furthermore, since relatively little, if any, bone mineral can be added to the skeleton in adulthood with currently available safe interventions [4-6], it follows that the growing years are the key time to initiate osteoporosis prevention programs. Unfortunately, children are becoming progressively more inactive [7], and so increasing rates of osteoporosis may be inevitable.

peak bone mass—Highest bone mineral content during adulthood.

This chapter begins by summarizing the animal studies that provide unique information about how exercise affects bone. (If you skipped chapter 3, you may want to review it briefly before tackling this section.) We then share exciting recent data about the pattern of normal bone mineral accrual in childhood and adolescence. Next we summarize studies in both animals and humans that have addressed the question, Does childhood and adolescent exercise influence adult bone mineral density and thus the risk of osteoporosis? We then provide exercise prescription guidelines and examples to optimize bone health in children and adolescents.

This chapter will interest a wide range of readers—from bone researchers and pediatricians to those involved in primary school education and government officials who are responsible for framing public policy. Pediatric exercise and bone mineral accrual is one of the most important topics currently being studied in the applied sciences.

Exercise, Bone Mineral Response, and Age

Studies of growing, exercising animals provide information about the effect of mechanical loading on bone growth. Further, if animals are sacrificed, bone can be studied more fully than in in vivo studies.

In a number of experiments with young, rapidly growing animals, exercise groups gained substantially more bone mineral than their nonexercising contemporaries or older animals.

Researchers intervened with running for periods of 6-7 weeks in a rat model [8], 14 weeks in a horse model [9], and 12 months in a swine model [10]. The effect of sudden impact loading on mineral and mechanical properties was also assessed in slightly older animals who were subjected to 50 (week 1) to 200 (week 6) impacts 5 times per week for 9 weeks [11].

In these studies, the outcomes compared to control animals were consistent and conclusive. The young running rats showed 15-25% greater BMC at the mechanically loaded sites [8]. In the high-impact study, the impact animals demonstrated a 10% increase ($p = 0.02$) in cortical wall thickness and a significant enhancement (8.8%) in femoral breaking strength compared to sedentary animals [11] while BMC showed no difference. The exercising horses had increased weight, volume, moment of inertia, and BMC of their metacarpals [9]. Finally, the immature swine demonstrated a 17% increase in cortical thickness and 23% increase in cortical cross-sectional area compared with controls, but there was no change in the material properties of the bone itself [10]. (Material properties are discussed in chapter 3.) Thus, bone responded by becoming larger, rather than by changing its material properties. The magnitude of bone change found in each of these studies was much greater than has been reported in studies of mature animals [12].

Some of these and other studies also directly compared the effect of exercise on growing versus adult animals [12, 13]. Saville and Whyte [14] found that bone formation rate in the younger rats was 16 times higher than that in the older rats following a five-days-per-week, seven-week training program. Figure 10.1 illustrates the relative bone formation rates for immature rats and mature rats subjected to a four-point bending on the right tibia.

In the Steinberg and Trueta study [13] the young rats showed an increase in BMD and in cortical thickness at three weeks, whereas the mature animals showed no changes in the exercise group or in the controls. Raab and colleagues [12] found no difference in breaking strength after exercise in young and mature turkeys, which at first appears to contradict other studies. However, the mechanism by which equivalent breaking strength was achieved differed between the young and the old animals. The bone of young animals increased in cross-sectional area without a change in ultimate stress (strength per unit area). The bone of older animals increased in weight and ultimate stress without increasing in cross-sectional area [8, 12].

Finally, in a study using the isolated turkey ulna model, there was substantial change in bone cross-sectional area and periosteal mineralizing surface of the ulnae of one-year-old compared with three-year-old turkeys [16]. Although the geometric properties of the ulnae of the older animals did not change, they did show compensatory osteonal mean wall thickness and lengthened remodeling periods compared with the young animals. The bone of older animals increased in weight and ultimate stress without increasing in cross-sectional area [8, 12]. This provides clear evidence that growing bone has greater capacity to add new bone than does mature bone [8].

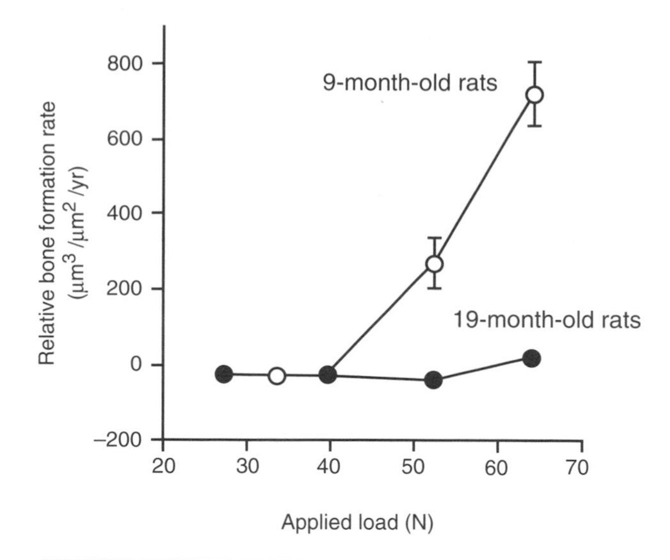

FIGURE 10.1 Relative bone formation rate (rBFR = BFR [right tibia] - BFR [left tibia]) on the endocortical surface for 9- and 19-month-old rats subjected to four-point bending of the right tibia (mean ± SD). For applied loads of 52 and 64 N, the mechanically induced bone formation rate was over 16-fold higher in younger rats compared with older rats [15].

Overall, very little reliable experimental data is available confirming the effects of similar exercise protocols on growing and adult animal bone. Also, the effect of detraining on exercise-induced bone gain in animals is largely unknown.

Normal Bone Mineral Accrual

The most convincing data regarding the normal pattern of bone mineral accrual around the ages 8 to 18 years come from the University of Saskatchewan Pediatric Bone Mineral Accrual study [17, 18]. The researchers measured the bone mineral content of approximately 200 children annually for 7 years. The findings from these unique year-to-year data are summarized for boys and girls at four anatomical regions (total body, total proximal femur, femoral neck, and lumbar spine) in figure 10.2 a-d.

The velocity curves provide an excellent illustration of the sex difference in the timing of peak

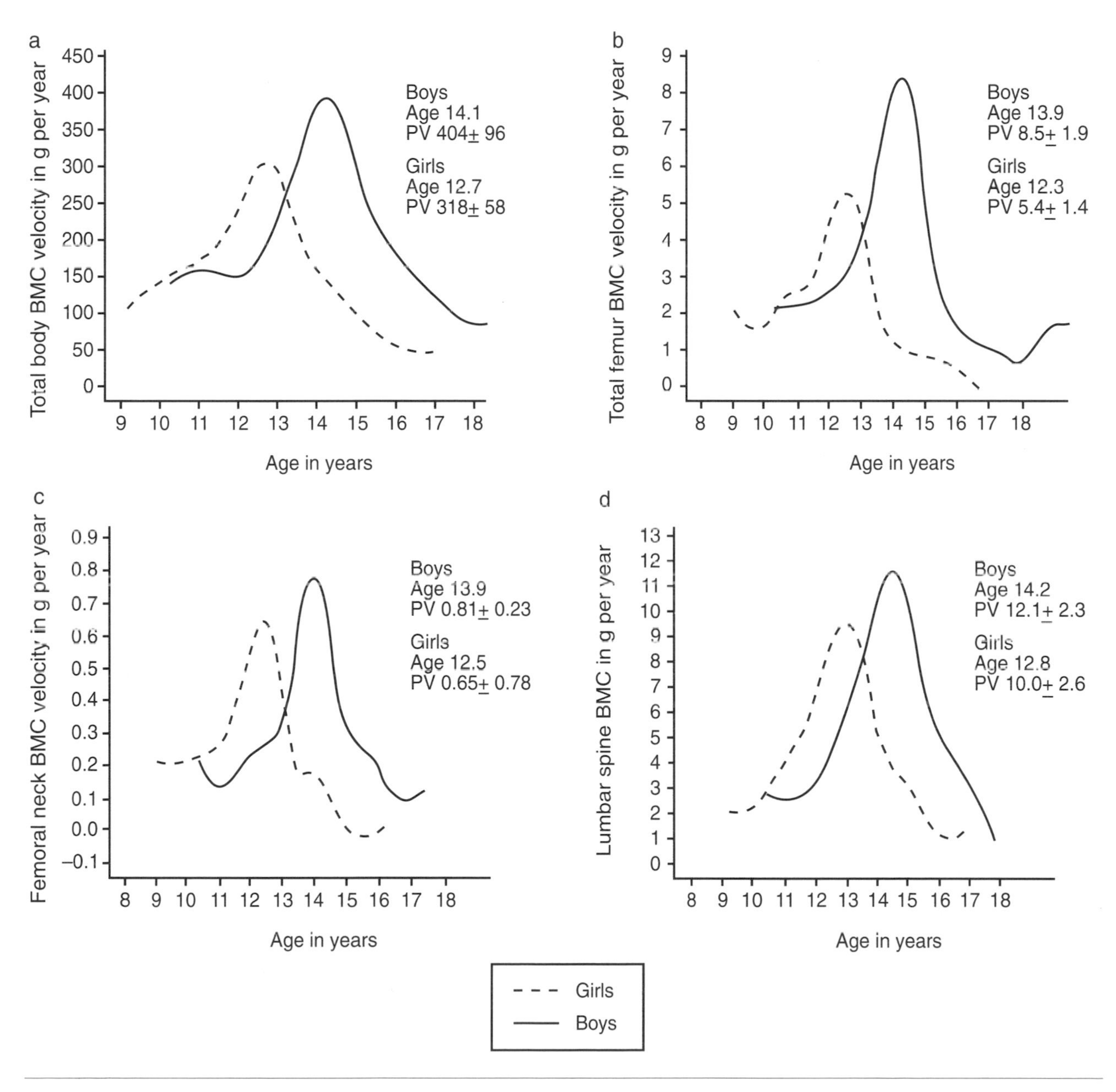

FIGURE 10.2 Velocity curves for bone mineral accrual for the total body, the total proximal femur, the femoral neck of the proximal femur, and the lumbar spine for boys and girls. Note the difference in the magnitude of the bone mineral gain at peak between boys and girls. Also note the timing of peak bone accrual velocities, with the girls always slightly in advance of boys.

bone mineral accrual, which, for the total body, occurs about 1.4 years earlier in girls than in boys and is of a 20% lesser magnitude (318 ± 58 g/y for girls versus 404 ± 96 grams per year for boys). When the researchers compared these overall accrual rates, controlling for size differences between girls and boys, they found that boys gained significantly ($p < 0.001$) more bone at peak and during the two years around peak as compared to girls [17]. The Saskatchewan group has also shown that for girls and boys the age of peak height velocity (PHV) preceded the age of peak bone mineral accrual by more than half a year [17], while the timing of menarche (first menstrual period) coincided with the age of peak accrual [19]. This is of clinical interest as the dissociation between accelerated linear growth (PHV) and peak bone mineral accrual may constitute a period of relative bone fragility that may partly explain the increased fracture rate observed during adolescence [20, 21].

menarche—The first menstrual period.

peak height velocity (PHV)—Maximum growth in height (measured in cm/yr or inches/yr).

peak bone mineral accrual—The period of most rapid gain in bone mineral, also called peak bone mineral content (BMC) velocity.

During the pubertal years a three- to four-year window, sometimes referred to as the "critical years," occurs, during which time a great proportion of bone mineral is accrued. In a study of healthy adolescent girls, maximal gain in lumbar spine and femoral neck BMC was observed at age 11-14 years with marked diminution of bone gain two years after [22]. A cross-sectional report from the University of Saskatchewan, which used 3020 DXA scans, determined the area under the curve showed that BMC accrued between developmental ages two years on either side of peak height velocity (PHV) [18]. About 35% of total body and lumbar spine BMC, and over 27% of femoral neck BMC, was laid down during the four years around PHV [18]. In a subsequent analysis of their seven-year longitudinal data these investigators found that, on average, 26% of adult total body bone mineral was accrued during the two years around the age of peak bone mineral accrual (figure 10.3) [17].

When designing this study the authors accounted for two factors that often confound studies of growing children—maturity and size differences. These authors aligned every subject on the same maturational land mark (PHV) and then controlled for body size in their analysis. Had this had not been done, critics could have suggested that the results merely reflected between-group differences in maturity or size at baseline or differences in rates of growth.

The amount of bone mineral gained was 30% greater than cross-sectional estimates [17], which demonstrates the "blunting" of values that occurs when cross-sectional data are used to represent longitudinal change. In another cross-sectional study of girls, the entire growth spurt

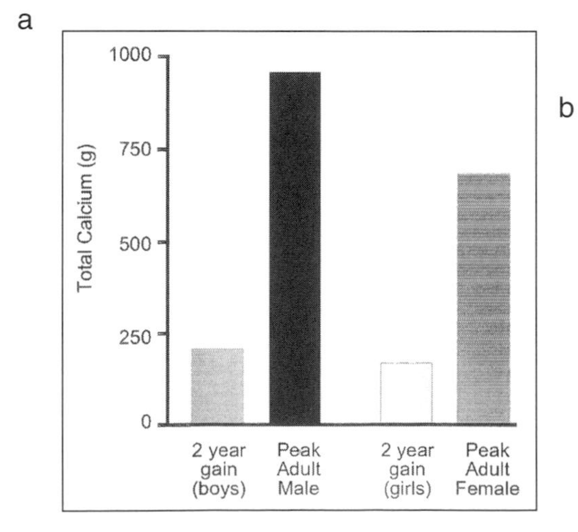

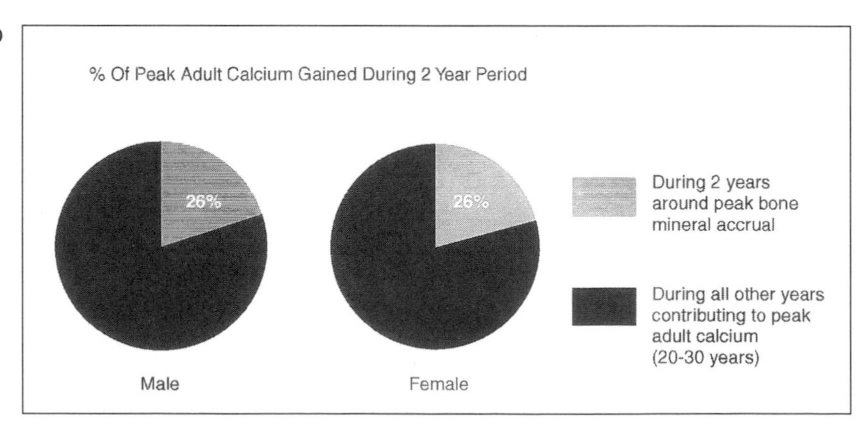

FIGURE 10.3 There is a substantial gain in total body calcium during the two-year period of fastest bone mineral accrual. This is depicted *(a)* as an absolute value of calcium gained (g) and *(b)* as a percentage of the peak adult value.

reportedly contributed 51% of peak bone mass [23]. These gains in BMC are more closely related to pubertal stage than to chronological age [24, 25].

A subsequent report from the University of Saskatchewan determined that calcium was being added to the skeleton at a rate of 359 ± 82 mg/day for boys and 284 ± 59 mg/day for girls [17]. This was calculated based on the assumption that 32% of total body bone mineral is calcium [26]. Based on these data, mean apparent calcium retention efficiencies at the age of peak bone mineral accrual were 36.5 ± 12.3 % for boys and 29.6 ± 8.5% for girls. However, there was considerable individual variability. At this rate of retention, a mean dietary calcium intake of approximately 1077 mg/day for boys (aged 14.0 years) and 852 mg/day for girls (aged 12.5 years) would be required to meet the demands of the skeleton during this rapid growth period. These values can be compared to the current recommended dietary intakes of 800 mg/day for both boys and girls aged 4-8 years and 1300 mg/day for children 9-18 years in Canada and the United States.

Having reviewed these key data on normal bone mineral accrual, we now ask, What role does physical activity play in modulating the normal pattern of bone mineral accrual? We review the effect of targeted bone loading, such as in a monitored intervention, separately from everyday physical activity that may not be focused on bone health.

Targeted Bone Loading and Bone Mineral

The effect of targeted bone loading during childhood and adolescence is discussed in three subsections according to the experimental study design used. The first examines the small number of intervention studies that have been undertaken in the pediatric age group. The second reviews cross-sectional studies and observational studies of activity and bone in childhood. Finally, we summarize those retrospective studies in which bone density was measured in adult populations with subjects asked to recall their childhood activity.

Intervention Studies: Targeted Loading

Only a few researchers have undertaken controlled exercise intervention studies in girls [27-30], boys [31], or both [32, 33] and evaluated change in bone mineral as the primary outcome measure. Self-selection into intervention and control groups was permitted in some of the studies in girls [28, 34], whereas in others, selection was done by the school for practical and compliance purposes [28, 29]. In the latter studies the schools were matched as closely as possible for ethnicity and socioeconomic status. In two studies children were randomized into intervention and control groups by school [31-33].

The outcome of the studies varied depending on the maturity level of the cohort at intervention. Studies that targeted pre- or early pubescent children [28, 29, 31, 33] reported significant increases in BMD. However, studies with postpubertal girls [27, 30] reported no significant difference in bone mineral between control and intervention groups at the conclusion of the studies. Comparing the results of intervention studies is fraught with danger as methodological differences among the studies might account for these differences. Some of these differences are outlined in the following discussion.

An eight-month prospective study of normally active prepubertal 9- and 10-year-old schoolgirls added three weight-bearing activity classes per week. (Baseline activity level was 31 METS average.) Gains in bone mineral content or density for girls in the intervention group, compared with controls, were on average 5.5% greater for the total body, 3.6% greater at the lumbar spine, and 10.3% greater at the femoral neck [28]. A similar study in prepubescent boys elicited a much smaller response for legs (3%), total body (1%), and lumbar spine (2.5%) [31].

A recent prospective study randomized 10 grade 3 and 4 classes (mean age 8.9 ± 0.7 years) into exercise and control groups. After an 8-month intervention with tuck jumps and hopping and skipping games for 10 to 20 minutes within school physical education classes (described at the end of this chapter), the exercise group showed significantly greater change in femoral trochanteric BMD (4.4% exercise versus 3.2% controls; $p = 0.03$) than the classes that continued with regular physical education [33]. The results from these studies with pre- and early pubertal children for the total body and the proximal femur are compared in figure 10.4.

Heinonen and colleagues [29] compared the effects of a nine-month step aerobics intervention on BMC in premenarcheal and postmenarcheal girls. Bone gain was significant in the premenarcheal group (exercisers vs. controls, 9.3% vs. 4.8% for lumbar spine BMC and 9.6% vs. 5.1% for femoral neck BMC), while no difference was reported between exercisers and controls in the postmenarcheal group. Findings of a positive bone response to loading are consistent with Frost's mechanostat model for regulation of bone strength as discussed in chapter 3 [35-37].

A recent study of postpubescent girls intervened with a progressive program of plyometric jumps over nine months [30]. Although the exercising group showed improved knee extensor strength, there was no difference between groups for BMC at the hip or spine or for total body. Similarly, a 26-week resistance training intervention in postmenarcheal girls showed a 20% increase in upper arm strength, a 53% increase in knee flexion, and a 22% increase in squat press strength, but no greater increase in total body or lumbar spine BMC in the training group.

In the younger groups, lean body mass was the most powerful predictor of bone mass [28, 31, 33]. Change in lean body mass was also an independent predictor of the change in vertebral and hip BMD that accounted for 10-58% of the variance on bone mineral accrual [28]. While young children have shown a capacity to undertake resistance training [38, 39], the position statement on strength training from the American Academy of Pediatrics states that such programs for prepubescent, pubescent, and postpubescent children should be permitted only if designed for the specific maturity level of the child and only when conducted by well-trained and qualified adults [83].

A few intervention studies have been undertaken in clinical groups of children with low bone mineral density [40, 41] whose movement ability is compromised. In one eight-month study a small number of children with cerebral palsy were given simple weight-bearing activities by a physiotherapist during a 60-minute program three times per week. Outcomes demonstrated a 9.6% increase in bone mineral at the hip in the intervention group, whereas those children who were not loading lost 5.8% of their bone mineral (figure 10.5) [42]. Studies such as this demonstrate the need for even a modest amount of loading for otherwise nonambulatory children.

The intervention studies presented here have begun the important task of elucidating the role of exercise in bone mineral accrual. However, the cellular and molecular basis for the differences in the response of the mature versus the immature skeleton are not known [21]. Further, the optimal timing for an exercise intervention to elicit the greatest increase in bone mineral and strength is also unknown. Factors that must be considered are the variable rates at which children mature, the different temporal patterns of growth and bone accrual between and within skeletal sites, and the hormonal regulation of the growth process. Given the complexity of these events, it will be some time before the mechanisms behind the skeleton's response to loading are completely understood.

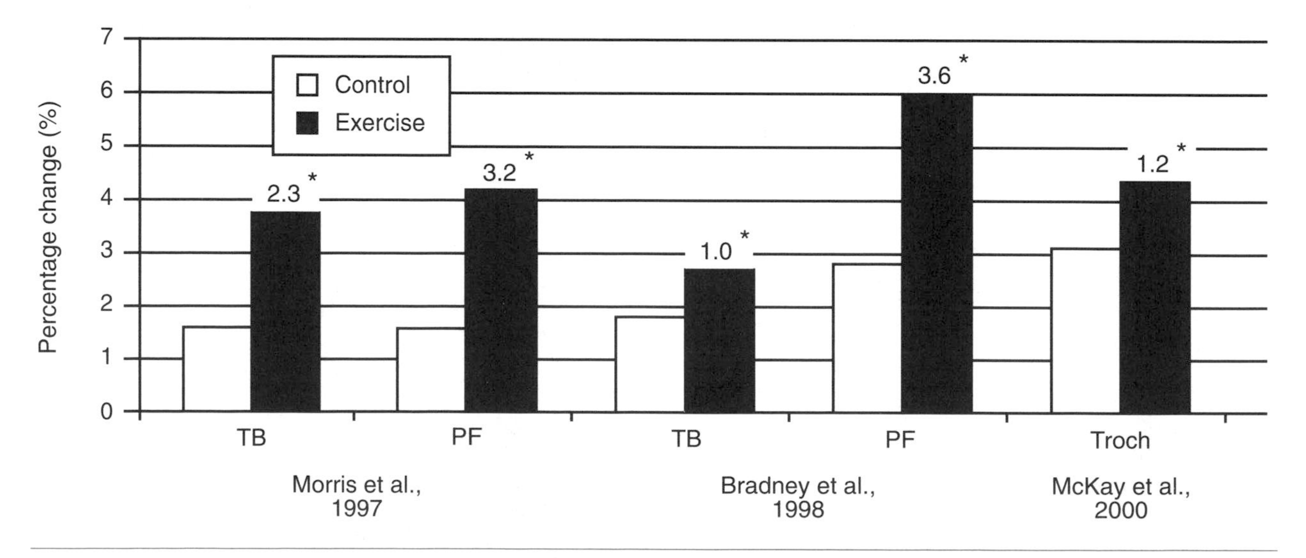

FIGURE 10.4 Results of prospective exercise intervention studies in pre- and early pubertal girls [28], boys [31], and both [33] for the total body (TB), total proximal femur (PF), and the trochanteric region of the proximal femur (Troch). Numbers over exercise bars represent difference in percent change between exercise and control groups.

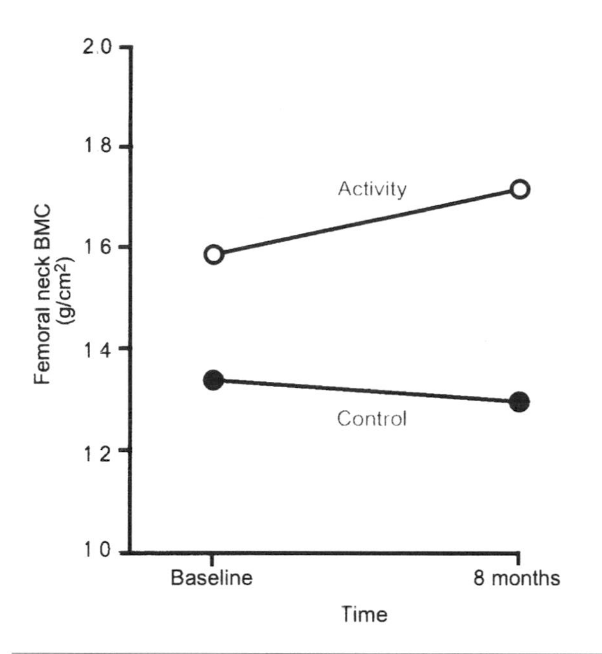

FIGURE 10.5 Results of an intervention study in a group of 18 children with cerebral palsy who showed a significant gain in bone mineral at the proximal femur and the femoral neck following a modest eight-month weight-bearing intervention. The control group of children with cerebral palsy lost 5.8% of their proximal femur bone mass during the same time frame [42].

Observational Studies in Sports

Sports including tennis, gymnastics, and Olympic weightlifting have been used to study the association between exercise and bone mineral in children. Studies of sporting populations are not easily extrapolated to the general population as they can potentially be confounded by selection bias and dropout bias. Also, the overall numbers of children involved in elite sporting activities is relatively small, and thus these finding are relevant only to a small percentage of the population. Despite these inherent limitations, the studies can still provide insight into the effect of high levels of mechanical loading on growing bone.

Tennis. The prolific Bone Research Group of the UKK Institute for Health Promotion Research in Tampere, Finland, published results of a cross-sectional study that evaluated the side-to-side difference in upper limb BMD in pre- and peripubertal female tennis players and matched controls [43]. In this study a researcher evaluated Tanner maturity stage using standard procedures that evaluate breast and pubic hair development with a series of photographs [83]. This study by Heidi Haapasalo and colleagues was the first to examine the pubertal stage at which the effect of physical activity on bone became evident. The investigators compared relative side-to-side arm differences in BMD between tennis players and controls (i.e., examination of the training effect) and revealed no BMD gain in Tanner stage I players even with very strenuous training. The obvious and significant BMD difference became apparent in the girls at each of Tanner stages III, IV, and V [43, 44]. Tanner stage III corresponds with the time of the adolescent

Finer Point: Study Design

The abovementioned Finnish researchers had previously reported a two to four times greater side-to-side difference in the playing arm as compared to the nonplaying arm of racket sport players who had started their playing careers before or at menarche than in those who started playing more than 15 years after menarche [45]. This study by Pekka Kannus and colleagues is discussed in detail on page 119. Although the outcomes of Haapasalo's and Kannus' studies may appear different at first glance, they are difficult to compare, as different study designs were used. The Kannus et al. study [45] was a retrospective recall study of 27-year-old women (± 9 years) in which menarche was the sole maturity indicator. In the Haapasalo et al. [43] cross-sectional study of 7- to 17-year-old girls, more specific Tanner staging was used. In fact, the Haapasalo et al. study was designed to further explore this group's earlier observations. Tanner stage IV-V approximates menarche, although considerable individual variation exists [84]. Therefore the majority of the Tanner IV-V girls in the Haapasalo study would likely have been near, at, or just past menarche.

growth spurt (mean chronological age was 12.6 years in these subjects). The authors proposed that "before puberty bone's responsiveness to loading is, in general, rather poor and that beyond the mechanical loading itself there must be many, still largely unknown (genetic, hormonal) factors that modulate the response of a growing bone to loading" [44].

Tanner maturity stage rating—A five-stage rating system of breast and pubic hair development.

Gymnastics. The pediatric bone and physical activity literature boasts a relatively large number of gymnast studies. This athletic population is ideal to study, given the extremely high landing ground reaction forces (15 times body weight) associated with the sport.

In a 12-month prospective study of 45 10-year-old active prepubertal elite female gymnasts and 48 controls, gymnasts had a 30-85% greater increase in BMD than controls at the total body, spine, and legs [46, 47]. At baseline, gymnasts' mean BMD values were already 0.4 to 2.1 standard deviations higher than the predicted mean at the arm, spine, and hip sites. Volumetric bone density was estimated using geometric formulae and BMD and was 12% higher at the lumbar spine and 16% higher at the femoral midshaft. The z scores (standardized scores, see chapter 4) increased with increasing duration of training.

A similar study with Tanner stage I and II boys demonstrated significantly higher ultrasound velocity at the calcaneus, distal radius, and phalanx at baseline compared to normally active controls [48]. Over 18 months ultrasound attenuation increased more than three times in the gymnasts but did not change significantly in controls. Sound wave attenuation reflects bone microstructure and certain qualities of the nonmineralized portions of bone. Higher-quality bone architecture, and therefore improved mechanical bone strength, would theoretically attenuate a greater amount of the ultrasound signal.

In a shorter (eight-month) prospective study, bone mineral accrual in 26 collegiate women gymnasts was compared with that in 36 runners and 11 nonathletic women. The percent change in lumbar spine BMD was significantly greater ($p = 0.0001$) for the gymnasts ($2.8 \pm 2.4\%$) than in the runners ($-0.2 \pm 2.0\%$) or the controls ($0.7 \pm 1.3\%$). The $1.6 \pm 3.6\%$ increase in femoral neck BMD for gymnasts was also significantly greater than that in runners ($-1.22 \pm 3.0\%$; $p < 0.05$) and approached significance for controls ($-0.9 \pm 2.2\%$; $p = 0.06$) [49].

In a smaller cross-sectional study of 7- to 11-year-old female gymnasts ($n = 16$) and controls, gymnasts had significantly greater femoral neck (8%) and trochanteric (16%) areal BMD as well as whole body (7%), femoral neck (20%), and lumbar spine (8%) BMAD compared with controls [50]. (BMAD represents bone mineral areal density and is a measure of the three-dimensional nature of bone. See chapter 4.) At the radius, total bone, trabecular, and cortical BMD as measured by pQCT were all significantly higher in gymnasts than controls. There was a trend toward greater radial cross-sectional area and more trabecular area in the gymnasts. The authors concluded that the primary response to the targeted bone loading is predominantly qualitative, reflecting proportional increases in the density of cortical and trabecular bone, with the possibility of slight increases in bone size, or hypertrophy.

hypertrophy—An increase in the volume of a tissue or organ produced by cell enlargement.

Weightlifting. In a cross-sectional study of 15- to 20-year-old male Olympic weightlifters, the mean distal and proximal forearm BMC of subjects was 51% and 41% above the age-matched controls, respectively [51]. These boys started training on average at age 14 years and had trained for an average of four years at the time of the study [21]. The differences are illustrated in figure 10.6.

Ballet. A study of 8- to 10-year-old ballet dancers who had been dancing 4.3 ± 1.4 years on average found that even young dancers have a site-specific bone mineral advantage compared with age- and sex-matched controls (figure 10.7) [52]. Interestingly, the relatively greater BMD in the lower limb and relatively poorer BMD in the upper limb in the young dancers mirrored the pattern in professional retired elite ballet dancers [53]. Self-selection bias is one potential explanation of these findings; however, in this instance, the young novice dancers had not undergone a selection process to gain admittance to the ballet schools.

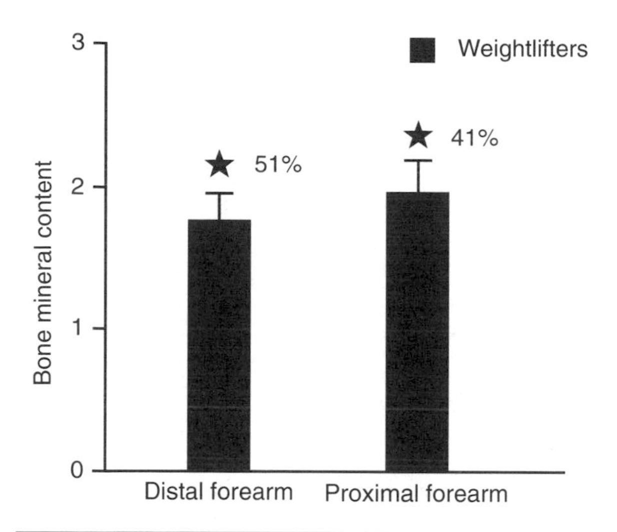

FIGURE 10.6 Individual forearm BMC values in 59 junior male weightlifters (aged 15-20 years) were 40-50% (or more than 2 standard deviations) higher than age-matched controls [51].

FIGURE 10.7 Subjects in the 8- to 10-year-old ballet and BMD study [52]. Although these girls are only novice dancers, they already display a site-specific advantage in bone mineral compared with controls.

With permission from The Herald and Weekly Times [86].

Adult Bone Mineral and Childhood Physical Activity Recall

Some researchers have suggested that augmented peak bone mass during childhood may afford a degree of protection against osteoporosis in later life [54, 55]. Figure 10.8 is a schematic representation of the effect of increasing adolescent peak bone mass by 1 standard deviation on risk of fracture after the menopause. If this illustration held true, and if bone loss rates are held constant, there would be a dramatic effect on the age of incidence of osteoporotic fracture. However, childhood bone gain could not be expected to provide absolute immunity against the rapid bone loss associated with such factors as immobilization, corticosteroid use, and early menopause—although it could theoretically temper the severity of the effect.

To examine possible long-term benefits of childhood and adolescent bone gain, we review retrospective studies in older individuals who were high-level athletes during childhood. Despite this study method's limitations (recall bias, etc.) it helps to shed some light on whether bone mineral gains achieved during childhood persist into adult life independently of adult levels of physical activity. We return to studies from the sports of tennis and gymnastics that suggest that childhood training may indeed influence adult bone mass until at least the mid-20s.

Tennis. The precursor to the aforementioned study by Haapasalo and colleagues examined bone mineral content of playing and nonplaying arms in 105 nationally ranked female racket sport players and 50 healthy women of similar age (mean 27 ± 9), height, and weight who served as controls [45]. The sportswomen had been playing on average 10 ± 6 years; the controls were comprised of a group of recreationally active students, none of whom were involved in activities that affected the dominant or nondominant arm only. Although controls had approximately 4% higher bone mineral content in the dominant than in the nondominant humerus (attributed to activities of daily living), racket sport players had on average approximately a 13% side-to-side difference in

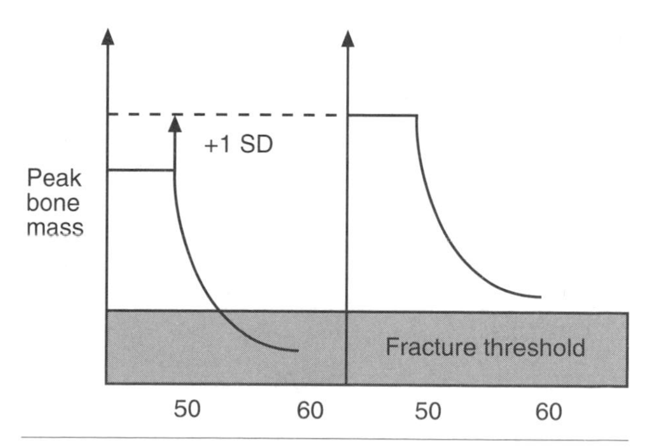

FIGURE 10.8 Schematic representation of the effect of increasing adolescent peak bone mass by 1 standard deviation on risk of fracture after the menopause.

humeral BMC, reflecting the strains experienced during tennis and squash (figure 10.9a). A key finding of this study was that when training was begun before or at menarche, difference in side-to-side bone mineral content of the arms ranged from 11% to 24% compared to 2% to 10% when training began more than 15 years after menarche (figure 10.9b) [45].

A four-year follow-up of male tennis players demonstrated that bone gained in adolescence appeared to persist further into adulthood, despite reduced playing activity [56]. This prospective study of "deconditioning" seems to support the residual bone benefits of childhood physical activity.

Gymnastics and Ballet. Gymnasts with a mean age of 25 ± 1 years who had retired, on average, 8 years previously had actual BMD z scores 0.5 to 1.5 standard deviation higher than predicted at the hip, arm, and lumbar spine sites [46, 47].

BMD z scores—Standardized scores for bone mineral density.

Australian researchers examined the relationship between reported hours of ballet classes per week undertaken as a child and adult bone mineral density (BMD) at the spine, hip, and forearm in 99 female dancers (mean age 51 years; SD = 14 years) who had been retired 26 years on average, and 99 normal controls [57]. A strength of this study was the careful documentation of balletic activity dating back many years. Self-reported hours of ballet class undertaken per week at each age between 10 and 12 were positively associated with differences in BMD between dancers and controls at both the femoral neck site (beta coefficient = 0.73, $p < 0.001$) and the total hip site (0.55, $p < 0.01$). These associations were unaffected by adjustment for covariates including measures of adult activity (current physical activity, years of full-time ballet).

The association between the starting age of training and bone mass was observed at only the load-bearing sites, whereas at the nonloaded sites (such as the upper limb), no association was noted. This suggests that the findings were most likely the result of ballet training at a young age rather than a genetic predisposition to high bone mass. Thus, classical ballet classes undertaken between the ages of 10 and 12 years were independently and positively associated

a

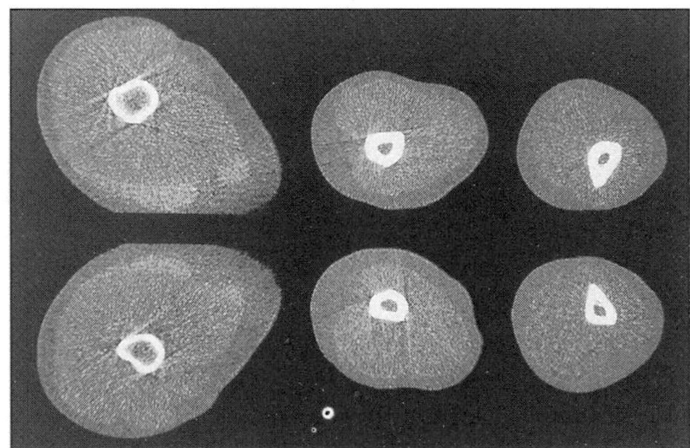

b

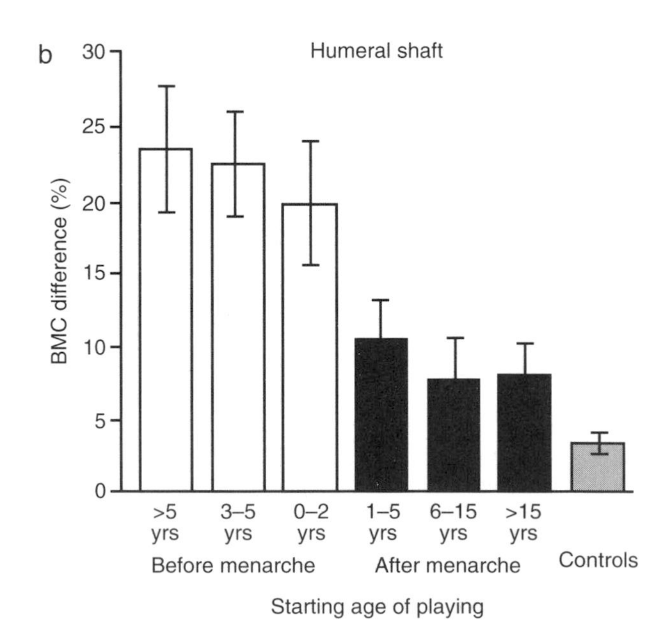

FIGURE 10.9 *(a)* pQCT images of the playing arm (top row) and the nonplaying arm (bottom row) of elite tennis players. The images show greater cortical bone dimensions in the playing arm [87]. *(b)* The side-to-side differences in bone mass at the humerus was two to four times higher in players who had started training before or at menarche (average difference 10.5-23.5%) than those who had started more than 15 years after menarche (2.4-9.6%) [45]. Bars represent 95% confidence intervals.

with a difference in hip BMD between dancers and controls, suggesting that the proximal femur is particularly responsive to weight-bearing exercise at a particular stage of development.

Generalized Physical Activity and Bone Mineral

We now examine the association between everyday activity, not necessarily targeted bone loading sporting activities, and bone mineral. The University of Saskatchewan six-year longitudinal study investigated the relationship between everyday physical activity and peak bone mineral accrual in a group of healthy Canadian children (53 girls and 60 boys) passing through adolescence [58]. Physical activity was assessed by a previously validated questionnaire [84] administered two to three times a year for the six-year duration of the study. The researchers derived a composite activity score to the age of peak height velocity for every child and compared children by quartiles of physical activity. With height and weight at peak height velocity controlled for, the study demonstrated a 9% and 17% greater total body BMC for active boys and girls, respectively, over their inactive peers one year after the age of peak bone mineral increase. Similarly, the most active group accrued more bone than the least active group at peak velocity and in the two years around the bone mineral spurt for the total body and the femoral neck (but not the lumbar spine) (figure 10.10).

peak bone mineral content velocity—The maximum annual gain in total body bone mineral as measured by dual energy X-ray absorptiometry (DXA; g/y).

Several cross-sectional studies found an association between childhood weight-bearing activity and bone mineral at certain sites. In American children aged 5 to 14 BMD of both the spine and the proximal femur was positively associated with total hours of weight-bearing activity including sports and play [59]. Similarly, the weekly duration of sports activity was positively associated with BMD at both the vertebral and the femoral sites in pre- and peripubertal French girls [60]. Finnish children with a mean age of 12 years who were more physically active than their counterparts had greater femoral neck, but not lumbar spine, BMD [61].

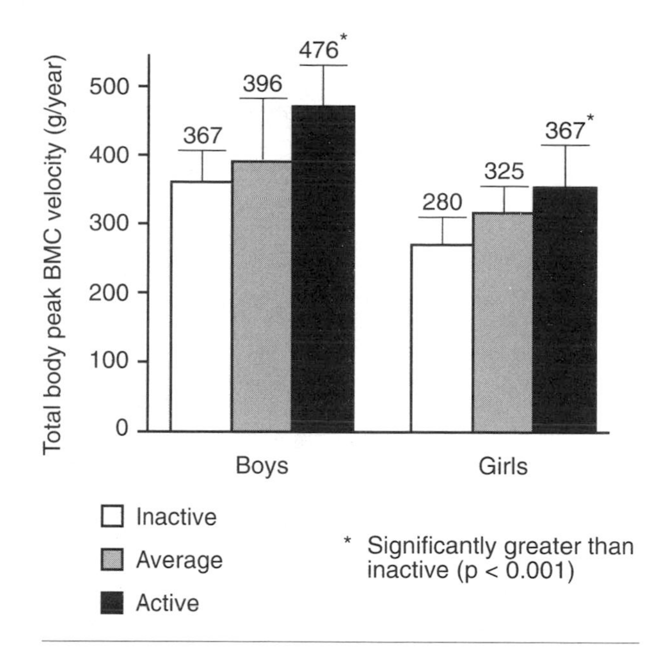

FIGURE 10.10 In the two years around the growth spurt, active boys and girls gain more total body BMC per year than their less active counterparts.

In a prospective study, self-reported physical activity (including sports, but not restricted to them) was significantly and positively associated with subsequent changes in femoral BMD over three years in 45 prepubertal monozygotic twin pairs (32 girls, 15 boys; mean age 7.4 ± 1.5 years) [62]. Prepubertal children with activity in the highest 25% of physical activity increased femoral neck BMD by 0.14 g/cm^2 (over three years) compared with 0.05 g/cm^2 for the children in the lowest 25% of physical activity [85]. The positive associations between physical activity and femoral neck BMD in both peripubertal and postpubertal children were smaller and not statistically significant [62]. This study once again directs us to a "critical time" when exercise appears to have an optimum effect on bone.

Associations have been reported between recreational levels of childhood activity and adult BMD. The length of recall demanded of the subjects varied from less than 10 years in subjects in their early 20s to a lifetime in studies such as the Rancho Bernardo Epidemiology Study. Since determinants of adult BMD can differ at various ages, these studies are reviewed according to the age of subjects at the time BMD was measured.

Researchers found a positive association between time spent playing sports at age 12 and radial BMD as measured by single photon absorptiometry (the forerunner of DXA scanning) in women aged 20 to 23 years [63]. That association was stronger than the relationship between current sports involvement and BMD. In a longitudinal study of the relationship among childhood growth, lifestyle, and peak bone mass in women (mean age 21 years), physical activity (outdoor walking) was the major lifestyle determinant of BMD after body build, which itself may be partially lifestyle determined [64]. The relationship was more marked at the proximal femur (where BMD was up to 12% greater in the highest activity group than the lowest) than at the spine. The playing of outdoor school sports also corresponded with proximal femur BMD [64]. A separate study showed that women who had played sports in childhood had greater BMD in their mid-20s than women who had not [65].

One study reported that high school, rather than current, activity predicted total body BMC and regional BMC [66]. This finding does not rule out the possibility of primary school sports influencing BMD since levels of physical activity are known to remain relatively constant over long periods of time [67-69]. The more active high school children are likely to have also been the more active primary school children [70].

Valimaki found that weight, adolescent exercise, and age together predicted 38% of the variance in BMD in women who were studied at a mean age of 27 years [71]. The effect size was 7.6-10.5% higher in those who exercised three times a week than in those who did not exercise. (Effect size is the annualized percentage difference between the treatment and the control group [21, 72, 73]). In women aged in their late 20s, calcaneal BMD measured by QCT was associated with self-reported physical activity in childhood [74] independent of adult activity.

In a slightly older cohort ranging in age from 20 to 50 years, lifetime physical activity was associated with radial BMD [76]. Kriska [75] found that self-reported physical activity between the ages of 14 and 21 years, but not during later years, was associated with increased adult bone area [75] (mean age 57 years). On the other hand, in the Rancho Bernardo Study (men and women, mean age 73 years), lifetime and current physical activity were both associated with higher hip BMD [77]. Past exercise was not associated with BMD at any site for either sex. It may be that childhood activity is no longer a determinant of BMD during the advanced years. However, the instrument for measuring physical activity in that study was designed to capture the cardioprotective effect of physical activity rather than the bone loading character.

It is important to recognize the limitations of retrospective studies that suggest an association between childhood activity and BMD at various adult ages. Such studies can only suggest; they cannot establish a causative relationship between activity and bone density. Subjects may be inaccurate in self-reporting. In addition, athleticism may be linked with genetically greater bone mineral density. Finally, ascertainment and dropout bias can also strongly influence outcomes of observational studies.

Exercise Prescriptions

This section focuses on children from eight years old through to adolescence and young adulthood.

Data from studies with children in these age groups have been used to develop the following recommendations [78].

- Make a lifelong commitment to physical activity and exercise.
- For developing bone, choose weight-bearing activities such as basketball and gymnastics over weight-supported activities such as swimming and cycling. However, it is important to remember that all of these activities are important for improving muscular strength and cardiovascular health. Short, intense daily activity is more effective than prolonged activity done infrequently.
- Perform activities that will increase muscle strength, such as running, hopping, or skipping games.
- Choose activities that work all muscle groups. Gymnastics is an excellent example of this.
- Avoid immobilization and periods of immobility; when this is not possible (as in bed rest during sickness), even brief daily weight-bearing movements can help to reduce bone loss.

Some fundamentals for exercise prescription to enhance bone health include the following:

- Exercise programs do not necessarily benefit the whole skeleton. Bone strength is specifically gained at sites that are stressed.
- The skeletal response to exercise is greatest at the site of maximum stress.

- To effect an adaptive response, the training stimulus must be greater than that habitually encountered.
- If physical activity ceases, bone mineral diminishes back to baseline levels.

The FITT principle can be adapted for exercise programs that benefit bone health in children and adolescents.

Frequency: Distributed bouts of targeted bone loading are more effective than one long bout.

Intensity: High-intensity exercises and high strain rates promote increased bone mineral more than low-intensity endurance-type activity does.

Time: The number of strain cycles can be small, and so the duration of each exercise need not be long.

Type: Exercises that include a variety of loading patterns and a novel and unusual strain distribution promote increased bone mineral more than do exercises that involve normal or regular loading patterns. Static loads applied continuously do not promote increased bone mineral.

Furthermore, exercise must be combined with a healthy lifestyle to optimize bone gain. Therefore:

- Girls should avoid lifestyle habits known to delay menarche or contribute to chronic menstrual disturbance (see chapter 16) such as inadequate energy intake combined with excess endurance activity.
- Athletes and coaches should be instructed on the potential skeletal hazards of menstrual dysfunction.
- Children should eat a well-balanced diet that meets the recommended nutrient intake for calcium. They should especially avoid substituting soft drinks for milk.
- Children should avoid cigarettes, as these are antiestrogenic and may interfere with the normal attainment of peak bone mass.
- Children, adolescents, parents, coaches, and teachers should all be alerted to the dangers of disordered eating patterns, which often begin in adolescence and are destructive to the skeleton.

Sample Osteogenic Intervention Program

Specific exercise intervention programs proved effective for increasing bone mineral. One such program from Melbourne, Australia, demonstrated significant increases in bone mineral at both the lumbar spine and all regions of the proximal femur in 9- to 10-year-old girls [28, 31]. Note that this program incorporates extra classes for the children, in addition to school physical education.

Frequency: Three times per week

Intensity: High impact

Time: 30 minutes of physical activity after school in addition to normal activities and physical education

Type:

1. Aerobic workouts: aerobics, modified football (soccer and Australian Rules), step aerobics, bush dance, skipping, ball games, modern dance, and weight training
2. Circuit training:
   - 20-minute weight-bearing, strength-building circuit that incorporated 10 exercises designed to load the biceps, triceps, pectoralis major, latissimus dorsi, trunk, deltoids, rectus abdominis, quadriceps femoris, hamstrings, gastrocnemius, soleus
   - Approximately one minute per station
   - One set of 10 repetitions progressing to 3 sets of 10 over time

Modified School Physical Education

The Healthy Bones (HB) I and II studies were conducted in the Richmond School District, a suburb of Vancouver, Canada. These programs incorporated moderate-impact (HB I) and high-impact (HB II) loading programs into the existing physical education curriculum. Teachers were provided with a resource manual [79] and instructed in how to teach the modified physical education class. Because the classroom teachers were able to teach the program, it was cost effective. The HB I program is outlined here.

Frequency: Three times per week

Intensity: Moderate impact

Time: 10-25 minutes of each physical education class

Type: The resource book included activities with interesting names such as Jump the Shot, Leapin' Lizards, Kangaroos and Frogs, Funny Farm, and Frog in the Sea. It also has creative skipping routines such as Capture the Sea, Canadian Hopscotch, and Obstacle Course Relays. We have provided one example of these jumping activities.

Leapin' Lizards

Divide students into five or six relay teams. Line each team up behind a start line with cones lined up in front of them at 8-ft (2.43-m) intervals. There should be one to two cones in each lane per team member. At the start signal, the first individual in each team leaps over each of the cones picking up the last one (farthest from the team). Turning, they run back to the start line, leaping over the cones in reverse order before tagging the next player in line. Participants in turn run and leap over the cones, picking up the farthest one and returning to the back of their line. The first team to return to the start line with all their cones picked up wins.

Activity time: 15 minutes

Equipment: 30-36 cones (or substitute)

Impact emphasis: Change the skills that students are required to use to run through and over the cones (different types of hopping, galloping, etc.).

SUMMARY

- Strong evidence suggests that weight-bearing physical activity plays a key role in the normal growth and development of a healthy skeleton.
- We recommend diverse kinds of physical activity for all children and adolescents, but high-intensity exercise of short duration appears to elicit the greatest bone accrual response in the growing skeleton.
- Although every age is a good age to exercise, there appears to be a critical time during childhood when the skeleton is most responsive to bone loading. Until better evidence is available, common sense should prevail when providing guidelines for children's exercise and sport. These instructions should

always take into account the child or adolescent as an entity—not just her bones or cartilage.

• The exercise guidelines we have provided are important information for children and their caregivers, educators, and coaches. They are also relevant to public health authorities who are responsible for school curricula.

• Children who are growing up today suffer more from the repercussions of an excessively sedentary lifestyle [80] than an excessively active one. For example, in our society, children and adolescents are at much greater risk of becoming obese than of undertaking the type of excessive physical activity required to damage growing bone and cartilage.

References

1. Dent CE. *Keynote address: Problems in metabolic bone disease*. In: B F, M PA, Duncan H, ed. Proceedings of the international symposium on clinical aspects of metabolic bone disease. Detroit: Henry Ford Hospital, 1972: 1-7.
2. Bailey DA, McCulloch R. Osteoporosis: Are there childhood antecedents for an adult health problem? *Can J Paediatr* 1992;4:130-134.
3. Chesnut III CH. Is osteoporosis a paediatric disease? Peak bone mass attainment in the adolescent female. *Public Health Report* 1989;S1:50-58.
4. Heinonen A, Kannus P, Sievanen H, et al. Randomised controlled trial of effect of high-impact exercise on selected risk factors for osteoporotic fractures. *Lancet* 1996;348:1343-1347.
5. Heinonen A, Oja P, Sievänen H, et al. Effect of two training regimes on bone mineral density in healthy perimenopausal women: a randomised control trial. *J Bone Miner Res* 1998;13:483-490.
6. National Osteoporosis Foundation. Osteoporosis: Review of the evidence for prevention, diagnosis, treatment and cost-effective analysis. *Osteoporos Int* 1998;8:Suppl 4: S3-S6.
7. Andersen RE. Relationship of physical activity and television watching with body weight and level of fatness among children: results from the Third National Health and Nutrition Examination Survey. *JAMA* 1998;279:938-942, 959-960.
8. Forwood MR, Burr DB. Physical activity and bone mass: exercises in futility? *Bone Miner* 1993;21:89-112.
9. McCarthy RN, Jeffcott LB. Effects of treadmill exercises on cortical bone in the third metacarpus of young horses. *Res Vet Sci* 1992;52:28-37.
10. Woo SL-Y, Kuei SC, Amiel D, et al. The effect of prolonged physical exercise on the properties of long bone: a study of Wolff's law. *J Bone Joint Surg* 1981;63A:780-787.
11. Jarvinen TLN, Kannus P, Sievanen H, et al. Randomized controlled study of effects of sudden impact loading on rat femur. *J Bone Miner Res* 1998;1475-1482.
12. Raab DM, Smith EL, Crenshaw TD, et al. Bone mechanical properties after exercise training in young and old rats. *J Appl Physiol* 1990;68:130-134.
13. Steinberg ME, Trueta J. Effects of activity on bone growth and development in the rat. *Clinical Orthopedics and Related Research* 1981;156:52-60.
14. Saville PD, Whyte MP. Muscle and bone hypertrophy. Positive effect of running exercise in the rat. *Clin Orthop* 1969;65:81-88.
15. Turner CH, Takano Y, Owan I. Aging changes mechanical loading thresholds for bone formation in rats. *J Bone Miner Res* 1995;10:1544-1549.
16. Rubin CT, Bain SD, McLeod KJ. Suppression of the osteogenic response in the aging skeleton. *Calcif Tissue Int* 1992;50:306-313.
17. Bailey DA, Martin AD, McKay HA, et al. Calcium accretion in girls and boys during puberty: a longitudinal analysis. *J Bone Miner Res* 2000;15:2245-2250.
18. Bailey DA. The Saskatchewan bone mineral accrual study: Bone mineral acquisition during the growing years. *Int J Sports Med* 1997;18 (Suppl3): S191-194.
19. McKay HA, Bailey DA, Mirwald RL, et al. Peak bone mineral accrual and age of menarche in adolescent girls: A 6-yr longitudinal study. *J Pediatr* 1998;133:682-687.
20. Bailey DA, Wedge JH, McCulloch RG, et al. Epidemiology of fractures at the distal end of the radius in children associated with growth. *J Bone Joint Surg* 1989;71-A:1225-1231.
21. Parfitt AM. The two faces of growth: benefits and risks to bone integrity. *Osteoporos Int* 1994;4:382-398.
22. Theintz G, Buchs B, Rizzoli R, et al. Longitudinal monitoring of bone mass accumulation in healthy adolescents: Evidence for a marked reduction after 16 years of age at the levels of lumbar spine and femoral neck in female subjects. *J Clin Endocrinol Metab* 1992;75:1060-1065.
23. Gordon CL, et al. The contributions of growth and puberty to peak bone mass. *Growth Dev Aging* 1991;55:257-262.
24. Kröger H, Kotaniemi A, Kröger L, et al. Development of bone mass and bone density of the spine

and femoral neck — a prospective study of 65 children and adolescents. *Bone Miner* 1993;23:171-182.

25. Grimston SK, Morrison K, Harder JA, et al. Bone mineral density during puberty in Western Canadian children. *Bone Miner* 1992;19:85-96.
26. Ellis K, Shypailo R, Kergengroeder A, et al. Total body calcium and bone mineral content: a comparison of dual-energy x-ray absorbtiometry with neutron activation analysis. *J Bone Miner Res* 1996;11:843-848.
27. Blimkie CJR, Rice S, Webber CE, et al. Effect of resistance training on bone mineral content and density in adolescent females. *Can J Physiol Pharmacol* 1996;74:1025-1033.
28. Morris FL, Naughton GA, Gibbs JL, et al. Prospective 10-month exercise intervention in premenarcheal girls: positive effects on bone and lean mass. *J Bone Miner Res* 1997;12:1453-1462.
29. Heinonen A, Sievänen H, Kannus P, et al. High-impact exercise and bones of growing girls: A 9-month controlled trial. *Osteoporos Int* 2000;11:1010-1017.
30. Witzke KA, Snow CM. Effects of plyometric jump training on bone mass in adolescent girls. *Med Sci Sports Exerc* 2000;32:1051-1057.
31. Bradney M, Pearce G, Naughton G, et al. Moderate exercise during growth in prepubertal boys: changes in bone mass, size, volumetric density and bone strength. A controlled study. *J Bone Miner Res* 1998;13:1814-1821.
32. McKay HA, Petit M, Schutz R, et al. Lifestyle determinants of bone mineral: A comparison between prepubertal Asian- and Caucasian-Canadian boys and girls. *Calcif Tissue Int* 2000;66:320-324.
33. McKay HA, Petit MA, Schutz RW, et al. Augmented trochanteric bone mineral density after modified physical education classes: A randomized school-based exercise intervention in prepubertal and early-pubertal children. *J Pediatr* 2000;136:156-162.
34. Heinonen A, Kannus P, Oja P, et al. Good maintenance of high-impact activity-induced bone gain by voluntary, unsupervised exercises: An 8-month followup of a randomised control trial. *J Bone Miner Res* 1999;14:125-128.
35. Frost HM. The mechanostat: A proposed pathogenetic mechanism of osteoporoses and bone mass effects of mechanical and nonmechanical agents. *Bone Miner* 1987;2:73-85.
36. Kimmel DB. A paradigm for skeletal strength homeostasis. *J Bone Miner Res* 1993;8:S515-S522.
37. Lanyon LE. Control of bone architecture by functional load bearing. *J Bone Miner Res* 1992;7:S369-S375.
38. Lohman T. Exercise training and body composition in childhood. *Can J Spt Sci* 1992;17:284-287.
39. Payne VG, Morrow JR, Johnson L, et al. Resistance training in children and youth: a meta-analysis. *Res Q Exerc Sport* 1997;68:80-88.
40. Henderson RC, Lin PP, Greene WB. Bone-mineral density in children and adolescents who have spastic cerebral palsy. *J Bone Joint Surg Am* 1995;77:1671-81.
41. Chad KE, McKay HA, Zello GA, et al. Body composition in nutritionally adequate ambulatory and non-ambulatory children with cerebral palsy and a healthy reference group. *Developmental Medicine and Child Neurology* 2000;42:334-339.
42. Chad KE, Bailey DA, McKay HA, et al. The effect of a weight-bearing physical activity program on bone mineral content and estimated volumetric density in children with spastic cerebral palsy. *J Pediatr* 1999;135:115-117.
43. Haapasalo H, Kannus P, Sievänen H, et al. Effect of long-term unilateral activity on bone mineral density of female junior tennis players. *J Bone Miner Res* 1998;13:310-319.
44. Haapasalo H. Physical activity and growing bone [PhD thesis]. University of Tampere, Finland, 1998.
45. Kannus P, Haapasalo H, Sankelo M, et al. Effect of starting age of physical activity on bone mass in the dominant arm of tennis and squash players. *Ann Int Med* 1995;123:27-31.
46. Bass S, Pearce G, Bradney M, et al. Exercise before puberty may confer residual benefits in bone density in adulthood: studies in active prepubertal and retired female gymnasts. *J Bone Miner Res* 1998;13:500-507.
47. Bass S. Heterogeneity in growth of the axial and appendicular skeleton: implications for the pathogenesis of osteoporosis [PhD]. The University of Melbourne, 1996.
48. Daly RM, Rich PA, Klein R, et al. Effects of high impact exercise on ultrasonic and biochemical indices of skeletal status: A prospective study in young male gymnasts. *J Bone Miner Res* 1999;14:1222-1230.
49. Taaffe DR, Robinson TL, Snow C, et al. High-impact exercise promotes bone gain in well-trained female athletes. *J Bone Miner Res* 1997;12:255-260.
50. Dyson K, Blimkie CJR, Davison KS, et al. Gymnastic training and bone density in pre-adolescent females. *Med Sci Sports Exerc* 1997;29:443-450.
51. Virvidakis K, Georgiou E, Korkotsidis A, et al. Bone mineral content of junior competitive weightlifters. *Int J Sports Med* 1990;11:244-246.
52. Bennell KL, Khan KM, Matthews B, et al. Activity-associated differences in bone mineral are evident before puberty: a cross-sectional study of 130 novice dancers and controls. *Pediatr Ex Sci* 2000;12:91-122.
53. Khan KM, Green R, Saul A, et al. Retired elite female ballet dancers have similar bone mineral density at weightbearing sites to nonathletic controls. *J Bone Miner Res* 1996;11:1566-1574.

54. Kelly PJ, Eisman J, Sambrook PN. Interaction of genetic and environmental influences on peak bone density. *Osteoporos Int* 1990;1:56-60.

55. Gutin B, Kaspar MJ. Can vigorous exercise play a role in osteoporosis intervention? A review. *Osteoporos Int* 1992;2:55-69.

56. Kontulainen S, Kannus P, Haapasalo H, et al. Changes in bone mineral content with decreased training in competitive young adult tennis players and controls: a prospective 4-year follow-up. *Med Sci Sports Exerc* 1999;31:646-652.

57. Khan KM, Bennell KL, Hopper JL, et al. Self-reported ballet classes undertaken at age 10-12 years and hip bone mineral density in later life. *Osteoporos Int* 1998;8:165-173.

58. Bailey DA, McKay HA, Mirwald RA, et al. The University of Saskatchewan Bone Mineral Accrual Study: A six-year longitudinal study of the relationship of physical activity to bone mineral accrual in growing children. *J Bone Miner Res* 1999;14:1672-1679.

59. Slemenda CW, Miller JZ, Hui SL, et al. Role of physical activity in the development of skeletal mass in children. *J Bone Miner Res* 1991;6:1227-1233.

60. Ruiz JC, Mandel C, Garabedian M. Influence of spontaneous calcium intake and physical exercise on vertebral and femoral bone mineral density of children and adolescents. *J Bone Miner Res* 1995;10:675-682.

61. Kröger H, Kotaniemi A, Vainio P, et al. Bone densitometry of the spine and femur in children by dual energy X-ray absorptiometry. *Bone Miner* 1992;17:75-85.

62. Slemenda CW, Reister TK, Hui SL, et al. Influences on skeletal mineralization in children and adolescents: evidence for varying effects of sexual maturation and physical activity. *J Pediatr* 1994;125:201-207.

63. Fehily AM, Coles RJ, Evans WD, et al. Factors affecting bone density in young adults. *Am J Clin Nutr* 1992;56:579-586.

64. Cooper C, Cawley M, Bhalla A, et al. Childhood growth, physical activity and peak bone mass in women. *J Bone Miner Res* 1995;10:940-947.

65. Talmage RV, Anderson JJB. Bone density loss in women: effects of childhood activity, exercise, calcium intake and estrogen therapy (abstr). *Calcif Tissue Int* 1984;36(suppl 2):S52.

66. Teegarden D, Proulx WR, Kern M, et al. Previous physical activity relates to bone mineral measures in young women. *Med Sci Sports Exerc* 1996;28:105-113.

67. Kuh DL, Cooper C. Physical activity at 36 years: patterns and predictors in a longitudinal study. *J Epidemiol Comm Health* 1992;46:114-119.

68. Janz KF, Mahoney LT. Three-year follow-up of changes in aerobic fitness during puberty: the Muscatine Study. *Res Q Exerc Sport* 1997;68:1-9.

69. Pate RR, Baranowski T, Dowda M, et al. Tracking of physical activity in young children. *Med Sci Sports Exerc* 1996;28:92-96.

70. Malina RM. Tracking of physical activity and fitness across the lifespan. *Res Q Exerc Sport* 1996;67 (3 Suppl):S48-S57.

71. Valimaki MJ, Karkkainen M, Lamberg-Allardt C, et al. Exercise, smoking, and calcium intake during adolescence and early adulthood as determinants of peak bone mass. *Br Med J* 1994;309:230-235.

72. Cumming RG. Calcium intake and bone mass: A quantitative review of the evidence. *Calcif Tissue Int* 1990;47:194-201.

73. Specker BL. Evidence for an interaction between calcium intake and physical activity on changes in bone mineral density. *J Bone Miner Res* 1996;11:1539-1544.

74. McCulloch RG, Bailey DA, Houston CS, et al. Effects of physical activity, dietary calcium intake and selected lifestyle factors on bone density in young women. *Can Med Assoc J* 1990;142:221-227.

75. Kriska AM, Sandler RB, Cauley J, et al. The assessment of historical physical activity and its relation to adult bone parameters. *Am J Epidemiol* 1988;127:1053-1061.

76. Halioua L, Anderson JJB. Lifetime calcium intake and physical activity habits: independent and combined effects on the radial bone of healthy premenopausal women. *Am J Clin Nutr* 1989;49:534-541.

77. Greendale GA, Barrett-Connor E, Edelstein S, et al. Lifetime leisure exercise and osteoporosis. The Rancho Bernardo Study. *Am J Epidemiol* 1995;141;951-959.

78. Bailey DA, Faulkner RA, McKay HA. *Growth, physical activity, and bone mineral acquisition*. In: Holloszy JO, ed. Exercise and Sport Sciences Reviews. Baltimore: Williams & Wilkins, 1996; 24:233-266.

79. Wilson G, McKay H, Waddell L, et al. The health benefits of a "Healthy Bones" physical education curriculum. A Richmond, BC Schools Study. *Physical & Health Education Journal* 2000;66:22-28.

80. Gortmaker SL, Must A, Sobol AM, et al. Television viewing as a cause of increasing obesity among children in the United States, 1986-1990. *Arch Pediatr Adolesc Med* 1996;150:356-362.

81. Witzke, K.A., and C.M. Snow. (2000). Effects of plyometric jump training on bone mass in adolescent girls. *Med Sci Sports Exerc* 32: 1051-1057.

82. American Academy of Pediatrics. Strength training, weight and power lifting and body building by children and adolescents. *Pediatrics* 1990;86:801-803.

83. Tanner J. *Growth at Adolescence*. (2nd ed.) Oxford: Oxford Blackwell Scientific Publication, 1962.

84. Crocker P, Bailey D, Faulkner P, et al. Measuring general levels of physical activity: preliminary evidence for the Physical Activity Questionnaire for

Older Children. *Med Sci Sports Exerc* 1997;29:1344-1349.

85. Slemenda CW, Reister TK, Hui SL, et al. Influences on skeletal mineralization in children and adolescents: evidence for varying effects of sexual maturation and physical activity. *J Pediatr* 1994;125:201-207.

86. Alexander C. Girls exercise judgement on osteoporosis. *Herald Sun*, Dec 8, 1997.

87. Haapasalo H, Kontulainen S, Sievänen H, et al. Exercise-induced bone gain is due to enlargement in bone size without a change in volumetric bone density: A peripheral quantitative computed tomography study of the upper arms of male tennis players. *Bone* 2000;27:351-357.

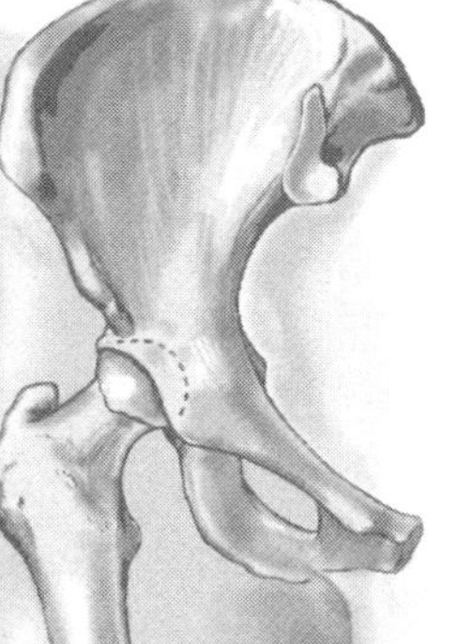

Chapter 11

Physical Activity, Targeted Bone Loading, and Bone Mineral in Premenopausal Women

with Ari Heinonen

This chapter reviews studies of the effect of physical activity on the skeleton in premenopausal women over 18 years of age. This group of women is becoming increasingly aware that osteoporosis prevention is far preferable to attempts at cure. They often ask exercise specialists and health care providers about lifestyle factors that can decrease their risk of osteoporosis and fracture.

The aim of this chapter is to summarize the many studies that have examined exercise and bone mineral density (BMD). We reemphasize that BMD is not synonymous with bone health or bone strength (see chapter 3). Nevertheless, as most studies have used BMD as an outcome measure, it is a logical common denominator with which to evaluate, at least partly, the role of exercise [1].

Because longitudinal studies of exercise intervention provide us with the best quality information about exercise and BMD, we begin the chapter with a review of those studies. We consider four of the randomized studies in some detail. We then summarize the cross-sectional studies that used the athlete-control model to determine the role of exercise, a model with some limitations.

This chapter also outlines studies that used the method of regression analysis to see if reported physical activity was a determinant of bone mineral in normally active women (i.e., not elite athletes). Regression is a statistical method that is used commonly to determine how much one aspect contributes to a multifactorial outcome (see chapter 5). Our discussion then moves to the mechanism that underpins bone change in this age group, and we speculate whether bone mineral is augmented by increased new bone formation or by an antiresorptive effect. The chapter concludes by proposing an evidence-based exercise prescription for premenopausal women.

Longitudinal Studies: Exercise Intervention

In longitudinal studies of premenopausal women, bone loading intervention has generally augmented BMD by about 1-3% at the loaded sites compared with controls (tables 11.1 and 11.2). In some studies, the difference in bone mineral between groups was achieved essentially via bone mineral gain in the exercise group [2, 3]. In one study, the difference was achieved mainly via reduction of loss in the exercise group [4], and in others, both bone mineral gain in exercisers and loss in controls combined to create a significant difference between groups [4]. Theories regarding the potential mechanism whereby bone responds to mechanical loading are discussed later in this section.

Studies by Friedlander, Lohman, Heinonen, Bassey, and their respective colleagues [2, 4-6] provide the best evidence to date that exercise intervention can augment bone mass in premenopausal women. These studies were performed with between 14 and 49 exercising normal women, subject groups were randomized, and studies were a minimum of five months' duration.

Friedlander and colleagues at UCLA in California [4] found a difference in excess of 2% between subject and control groups in hip BMD following a two-year aerobics exercise intervention that emphasized jumping activities. Spine BMD gains were of the order of 1% when measured by DXA (cortical and trabecular bone), and 2% when measured by QCT (trabecular bone only) (figure 11.1). Given that vertebrae consist of about equal parts trabecular and cortical bone, these data suggest that cortical bone BMD did not change.

Lohman and colleagues in Arizona, USA [6], directed an 18-month exercise intervention that consisted of a 60-minute resistance training program using free weights three times a week. Although the intensity of the program was increased from 70% of 1-RM at baseline (1-RM is the maximum mass of a free weight or other resistance that can be moved by a muscle group through the full range of motion with good form one time only) to 75% and then 80% of 1-RM at the 6-month and 12-month stages, most of the gain occurred in the first 6 months. This pattern of change differed from that of the California study, in which BMD continued to change throughout the second year [4]. The resistance training program proved to be particularly effective in increasing BMD at the lumbar spine [6] (see table 11.1).

Heinonen and colleagues from the UKK Institute in Tampere, Finland, tested the ground reaction forces generated by various jumping exercises and then incorporated those with highest impact into a 20-minute aerobics and stepping program (figure 11.2) that was particularly effective in increasing lower limb BMD [2]. Hip, femur, tibia, and calcaneal BMD increased by between 1% and 4% depending on the site, with BMD response greatest toward the periphery of the lower limb. This pattern is consistent with the theory that bone responds most to greatest mechanical loading (highest impact at the heel, with impact progressively attenuated by the ankle, knee, and hip joints). This pattern has been reported in cross-sectional studies comparing the BMD of athletes and controls at various skeletal sites [7-9].

In a smaller randomized prospective study (n = 14) using 50 small jumps daily (ground reaction force of each jump = 5-6 times body weight) the intervention group BMD increased 3% at the hip (figure 11.3) [5]. When the researchers applied an identical intervention to a group of postmenopausal women, they found no effect (see chapter 12) [5]. The authors [11] speculated that inherent differences between pre- and postmenopausal women may influence the response to the intervention. For example, the endocrine milieu is clearly different between these groups. The practical implication is that premenopausal women might have a greater

TABLE 11.1

Randomized controlled trials of exercise intervention in premenopausal women using BMD as the outcome variable

Authors	No of subjects and mean (SD) age in yr	Description of program (duration of sessions × no. of sessions per week)	Sites loaded	Effect size (% BMD difference between subjects and controls)
Snow-Harter et al. 1991 [21]	22 subjects aged 20 (1) yrs	Jogging or weightlifting × minute × times per week for 8 months	Hip Spine	Spine BMD increased 1.3 ± 1.6% in runners, 1.2 ± 1.8% in weight trainers ($p < 0.05$). No change in controls.
Lohman et al. 1995 [6]	22 subjects aged 34 yrs (3)	Strength training at 70-80% of 1-RM, 12 exercises, about 1 hr, 3 times weekly for 18 months	Hip Spine	Lumbar spine BMD increased 2.3% and femoral trochanter 1.8% in exercisers (both, $p < 0.05$), no change in controls.
Sinaki et al. 1996 [14]	50 subjects aged 36 yrs (3)	Nonstrenuous 30 minute weight-lifting exercise program, supervised once a week, done twice more per week. (3 yr interventiuon)	Hip Spine	No effect at hip, spine, or midradius
Friedlander et al. 1995 [4]	32 subjects aged 28 yrs (7)	Aerobics and weight training; 3 supervised 1-hr classes per week for 2 years	Hip Spine	Hip: exercise group gained 0.3 ± 2.6%, control group lost 1.9 ± 5.7% ($p < 0.05$) Spine: controls lost 3.0 ± 3.1% using QCT to measure trabecular bone; exercise group lost 0.5 ± 4.6% ($p < 0.05$)
Bassey et al. 1994 [5]	14 subjects aged 32 yrs (3)	High-impact jumps—daily at home and once weekly under supervision. Jumping component was 6-10 minutes of the 1-hr program. (6 month intervention)	Hip	Hip: Exercise group increased BMD by 3.4% at the trochanter ($p < 0.01$) and this differed significantly from controls ($p < 0.05$)
Bassey et al. 1994 [5]	7 subjects aged 30 yrs (4)	High-impact jumps—daily at home and once weekly under supervision. Jumping component was 6-10 minutes of a 1-hr program. (6 month intervention)	Hip	Hip: Exercise group increased BMD by 4% at the trochanter ($p < 0.01$)
Heinonen et al. 1996 [2]	49 subjects aged 39 yrs (3)	Jumping exercises with high ground reaction forces, aerobics, stepping, 1-hr, three times a week. Jumping/stepping component took 20 minutes of each session. Total program duration = 18 months	Spine, hip, femur, leg	Lumbar spine: exercise group gained 1.8% in BMD; controls exercisers gained 1.6%; controls 0.6%. Femoral neck: exercise group gained 1.6%; controls 0.6% Significant gains at distal femur, tibia, and calcaneum

TABLE 11.2

Nonrandomized controlled trials of exercise intervention in premenopausal women

Authors (no. in exercise group)	Mean (SD) age in yr	Description of program (duration of sessions × no. of sessions per week)	Sites loaded	Effect size (% difference between subjects and controls)
Gleeson et al. 1990 [53] (n = 34)	33 (6)	Moderate weightlifting	Spine	Spine 1.3%
Rockwell et al. 1990 [16] (n = 10)	36 (3)	8-station resistance training including chest press, leg press, overhead press, leg curl, rowing, leg extension, and abdominal and back exercises	Hip Spine Arm	Spine: –4% in 9 months
Smith et al. 1989 [54] (n = 80 pre- and post; pre not specified)	39 (3)	Forearm exercises (4 yr)	Forearm	Radius: 1.3% (mean of left and right radius)

Note: The study of Recker et al. 1992 is not included because it was observational. Also, the study of Peterson et al. 1991 was not included because the paper did not separate the data for pre- and postmenopausal women.

propensity to obtain small bone gains than their postmenopausal counterparts.

A recent meta-analysis evaluated the effect of exercise training programs on bone mass [12]. Randomized, controlled trials showed very consistently that the exercise training programs prevented or reversed about 1% of bone loss per year at both the lumbar spine and femoral neck in both pre- and postmenopausal women (see chapter 12). These findings were echoed in another systematic analysis [13].

Intervention Studies Resulting in Bone Gain

The common feature of the successful exercise intervention studies (i.e., those in which exercise intervention resulted in significant bone gain at loaded sites) appeared to be the nature of the intervention, since subjects' ages and the study durations varied substantially (see table 11.1). The type of exercise prescribed in each study was designed to mechanically load bone. Heinonen and colleagues even quantified the mechanical loading generated by their intervention using a force platform [2]. Bassey, Friedlander, and Heinonen's protocols all involved jumping exercises that can create ground reaction forces up to six times body weight [2, 4, 5]. Lohman's protocol involved no jumping but included high-intensity strength training [6]. By contrast, Sinaki and colleagues performed a three-year low-impact intervention study in premenopausal women that increased muscle mass but did not augment BMD (figure 11.4) [14]. That intervention may not have met the threshold required to promote osteogenesis. The lack of gain in bone mineral despite the length of the study suggests that in premenopausal women exercise duration does not compensate for lack of intensity.

How long the exercise prescription need be remains unclear. One study found that maximal change in bone mineral occurred after 18 months or longer [4], but others found changes occurred within six to nine months [3, 5, 6]. Some exercise intervention studies have been undertaken for less than six months [15]. We believe that these data, taken together with bone physiology such as the bone modeling transient (see chapter 2), suggest that an intervention aimed at augmenting bone mineral in premenopausal women should last for a minimum of six to nine months, but preferably longer.

Friedlander and colleagues [4] postulated that the skeletons of 20- to 30-year-old subjects may respond better to loading than would those of older premenopausal women, but Heinonen showed that BMD can be augmented even in older premenopausal women aged up to 45 years [2].

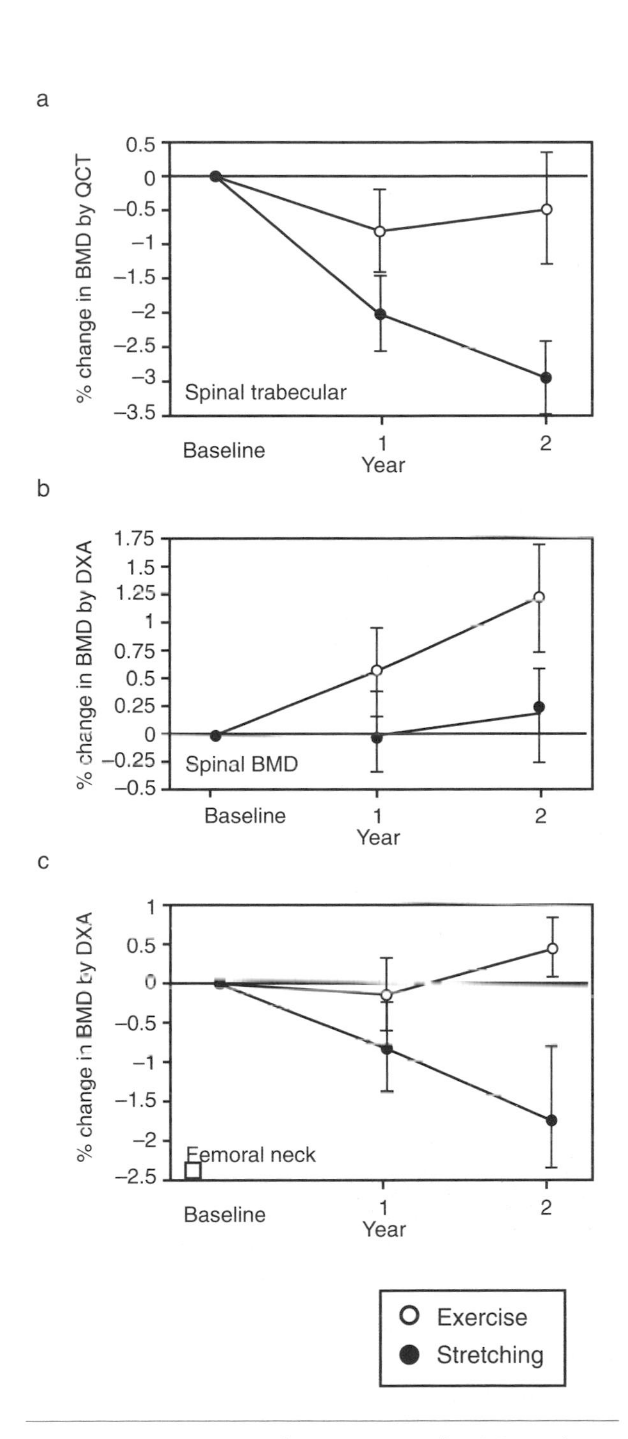

FIGURE 11.1 Percent change in BMD by QCT and DXA in a two-year exercise intervention study. *(a)* Change in spinal trabecular BMD by QCT. *(b)* Change in spinal BMD by DXA. *(c)* Change in femoral neck BMD by DXA.

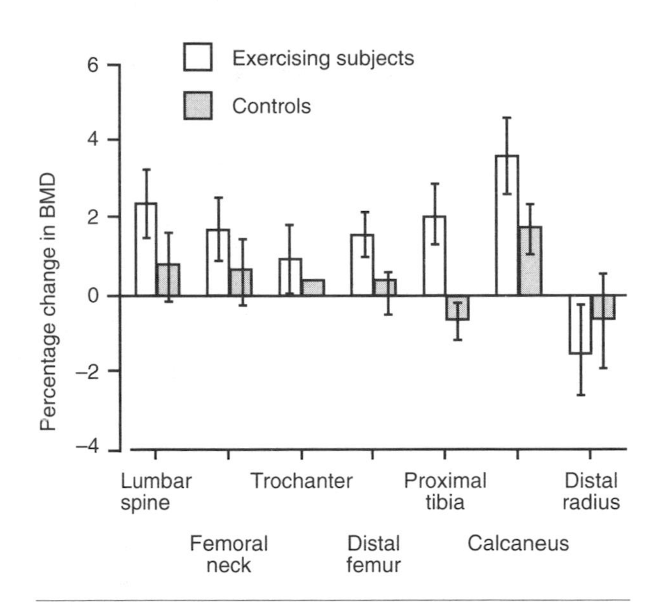

FIGURE 11.2 Premenopausal women undertaking high-impact exercises at the UKK Institute in Finland. Lower limb BMD increased by 1-4% compared with controls [2].

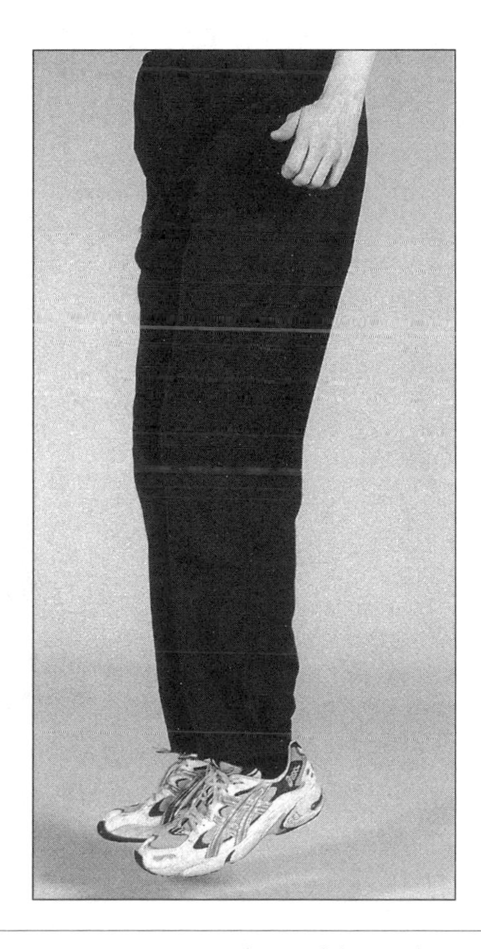

FIGURE 11.3 In Bassey and Ramsdale's study [5], the premenopausal women undertook very small jumps, as illustrated, yet gained 3% in BMD at the hip.

Finer Point: Resistance Training and Bone Loss

In general, the majority of the interventions have not seen bone loss in the premenopausal exercises. However, a study that requires mention is that of Rockwell and colleagues [16], in which a 4% loss of vertebral bone mineral was reported in ten 36-year-old women who undertook a mild weight-training intervention. The lumbar BMD of the seven controls did not change. Why would healthy young women lose 4% of their lumbar BMD in nine months?

Rockwell and colleagues' nonrandomized study design permitted the subjects to choose whether to exercise or be controls, which may have confounded the results, particularly since there were so few subjects. None of the controls and only four of the intervention subjects (40%) were regular exercisers before entering the study. Baseline serum osteocalcin concentration in the exercise group was five times that of the control group, consistent with greater levels of activity [16]. Thus, the exercising group may have been inherently more athletic than the controls and therefore the given interventions may have added relatively little to the exercisers' weekly loading.

Another important point to note in Rockwell and colleagues' study is that the baseline measurements were obtained in October (after summer), the midpoint measurements were obtained in February (the last winter month), and the final measurements were obtained in May (after winter). At baseline, the BMD of the exercising group (1.25 g/cm^2) was higher than that of the controls (1.19g/cm^2) and dropped to 1.22 g/cm^2 and 1.20 g/cm^2 in February and May, respectively. At both the midpoint and final measurements, the mean lumbar spine BMDs of controls were 1.18 g/cm^2. Thus, the mechanical loading activity of the exercising subjects may have decreased in the winter months, despite the twice-weekly intervention. Controls, on the other hand, may not have substantially altered their bone-loading physical activity in the winter if they were habitually rather inactive.

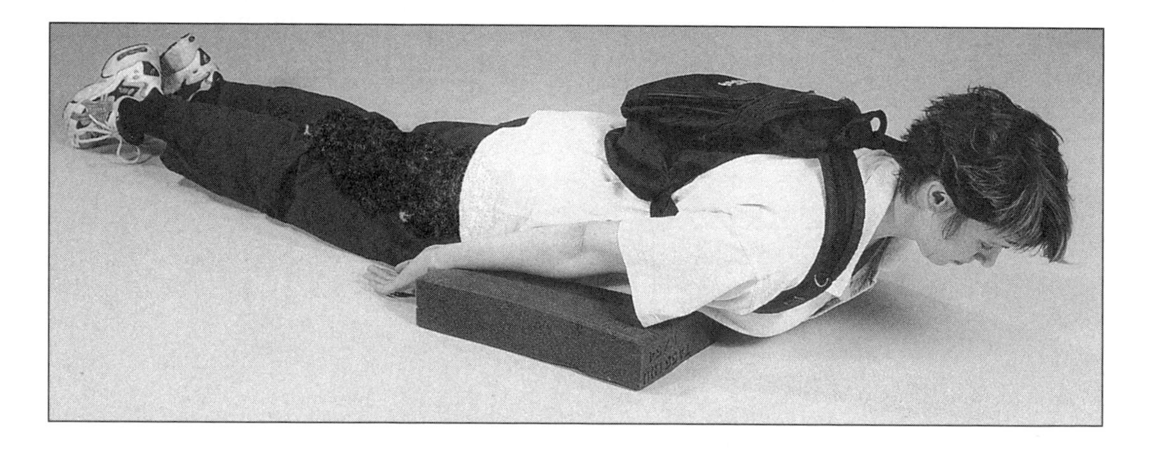

FIGURE 11.4 Back strengthening exercises. In a study of back strengthening there was no increase in BMD over three years [14].

Cross-Sectional Studies: Athletes Versus Controls

The multitude of cross-sectional studies that have reported that premenopausal female athletes have higher bone mineral than controls have been thoroughly reviewed [18-26]. The conclusion researchers have reached is that sportswomen have about 10% higher bone density than controls [19, 21, 25, 27, 28].

The limitations of these cross-sectional athlete–control studies are well recognized [19, 24, 25, 28]. Cross-sectional athlete studies are normally not controlled for either the amount of bone that was present before the athletes started the sporting activity (self-selection) or the amount of physical activity that was performed during childhood or adolescence. In addition, many athletes may adopt lifestyle habits (healthy nutrition, not smoking, etc.) that are not always controlled for. Also, sampling techniques are often not rigorous or not reported, which can lead to a volunteer effect. Thus, these studies do not answer the question, Can exercise augment bone mineral in the mature skeleton? Despite these limitations, cross-sectional studies with careful attention to research design can be useful "if they are related to the age of onset of activity, to the adolescent growth spurt and to previous physique" [24].

A much better study design to see whether exercise can augment bone mineral in the mature skeleton is the side-to-side comparison of the upper arm BMD or BMC between racket sport players who had started their playing careers at adulthood and controls. Using this study design, in which selection bias and confounding factors can be controlled for, Kannus and colleagues [29] showed that tennis players that started in adulthood had significant increases (about 5%) in the BMC of the playing arm humerus (see chapter 10).

Longitudinal Studies: Athletes Versus Controls

To our knowledge, there has been only one longitudinal study of BMD change over time in athletes versus controls when the skeleton was essentially mature. In both male and female subjects aged 17-27 years BMD changes over a 12-month period were independent of exercise status except at the lumbar spine [30]. Bone biomarker studies showed that there was no difference in bone formation or turnover between athletes and controls at this age.

Cross-Sectional Studies: Generalized Physical Activity

Another method of studying the association between physical activity and premenopausal bone mineral density is by using linear regression analysis of data from observational studies. In this instance the question asked is, What proportion of the bone mineral density can be attributed to physical activity? A drawback of this method is that physical activity can be a difficult variable to both define and measure, as discussed in chapter 9.

A longitudinal observational study found that generalized physical activity was associated with bone mineral accrual during the early premenopausal years [31]. In this well-controlled four-year study, 156 young women wore physical activity monitors for four days prior to each six-month visit for data collection. The physical activity monitor results correlated well with activity questionnaires and energy intake data. Physical activity positively and significantly correlated with change in vertebral BMD [31]. Multiple linear regression including age and a measure of nutritional intake (calcium/protein ratio) with physical activity provided an r value of 0.31. Since none of the women in the study were exercising heavily, nor had calcium intakes that were extremely high, the authors concluded that "modest increases in physical activity and calcium intake in young adult women under the age of 30" could lead to "highly significant reductions of fracture risk in later life" [31]. Also noteworthy was the fact that the women in this study experienced very minimal bone gain despite being only in their 20s (figure 11.5).

Another cross-sectional study also used physical activity monitors for data acquisition and found that after controlling for appropriate confounds, activity was positively associated with bone mineral at the lumbar spine, radius, and total body [32].

Recently, two large population studies in the United States and Finland have documented a positive association between physical activity, as assessed by questionnaire, and BMD in premenopausal women. In Pennsylvania, USA, ex-

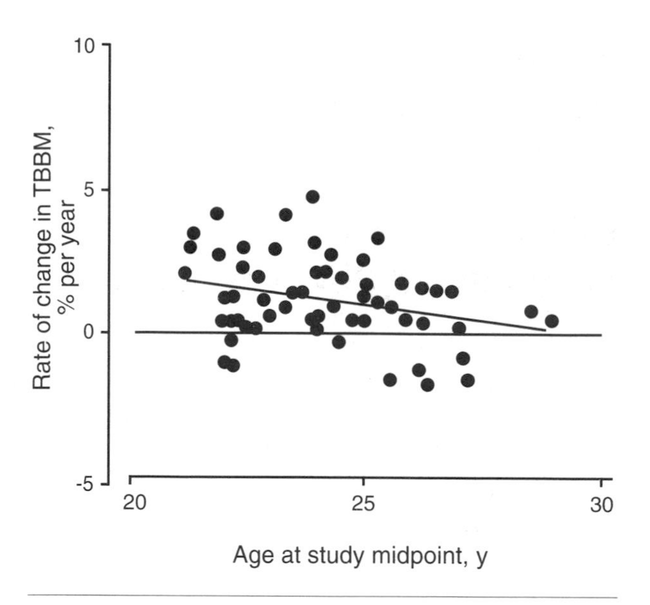

FIGURE 11.5 Bone loss in the third decade. Annual percentage change in total body bone mineral content plotted as a function of age shows a trend from small amounts of gain in the early 20s to no change around 30 years.

ercise was associated with total body BMD measured by DXA in 470 healthy premenopausal women [33]. Similarly, a population study of 1600 women in Kuopio, Finland, found that physical activity predicted femoral but not spinal BMD [34]. However, even in studies in which physical activity is measured, rarely is mechanical loading quantified in any way.

Although some publications have reported an association between maximal oxygen uptake and BMD [35-37] in women over a range of ages, most studies in both premenopausal and postmenopausal women have not found such an association [38-42]. Nelson and colleagues found no difference between endurance-trained and sedentary groups in BMD at either the spine or the hip [43]. It appears that any relationship between fitness and BMD is likely to be mediated through the weight-bearing stimulation that the activity provides the skeleton [23].

Mechanism of Bone Augmentation

The mechanism underlying the effect of exercise on bone in hormonally replete women is still not clear. In animal studies, exercise promotes increased bone formation in sexually mature rats [44, 45]. In Friedlander and colleagues' study the difference in BMD between the intervention group and control group continued to increase in the second year, suggesting that intervention stimulated new bone formation [4]. In Lohman and colleagues' study, osteocalcin, a marker of bone formation, increased in the exercising group that gained bone mineral [6].

On the other hand, Lohman's study also reported a fairly rapid increase in BMD with exercise followed by a period in which BMD remained constant [6]. An initial BMD increase followed by a plateau is characteristic of any treatment that acts mainly to reduce the rate of resorption (decreasing activation frequency), thereby reducing the remodeling space [46-48]. This question requires further study.

Exercise Prescriptions

Because exercise produces site-specific effects, not all bone responds to a particular training program. Under the subheadings that follow, we describe and illustrate exercise programs that have been shown to be effective in certain populations. While we do not guarantee that the exercise protocols will be equally effective in other settings, we believe they provide a sound starting point for fitness leaders and others who are prescribing exercise to adults.

Increasing Bone Mass in the Femoral Neck

The following protocol (table 11.3) was designed by Ari Heinonen and colleagues at the UKK Institute in Tampere, Finland. It proved successful in increasing femoral neck BMD in 35- to 45-year-old Finnish women [2].

TABLE 11.3

Exercise prescription that increased femoral neck BMD in 35 to 45-year-old women

Activity	Duration
Warm-up	15 min
High-impact jumps	20 min
Stretching and nonimpact activities	15 min
Cool-down	10 min

The high-impact jump training consisted of either an aerobic jump program or a step program (figure 11.6). This alternated every two weeks with jumps from a step or over a foam fence increasing from a height of 4 to 10 in. (10-25 cm). At the start the jumps were low, but there were more of them (200); as the height of the jumps increased, the number of jumps was progressively reduced to 150, 120, and 100.

Bassey and Ramsdale [5] found that a 10-minute program of 50 3-in. (8-cm) jumps per day was also effective in increasing femoral neck BMD. The jump technique required subjects to land on their toes and then follow through to a heel strike (figure 11.3). Note that this technique was not effective in maintaining BMD in postmenopausal women [11, 49] (see chapter 12).

In women aged 52 and 53 years (a mix of pre-, peri-, and postmenopausal women) a high-intensity endurance exercise protocol maintained BMD at the femoral neck but no other sites [50]. The protocol consisted of a 10-minute warm-up, 30 minutes of effective exercise time, and a 10-minute cool-down four times per week. The exercise prescription requires participants to maintain 55-75% $\dot{V}O_2max$ while either walking, jogging, stair-climbing, or moving on a graded treadmill. While this exercise prescription is safe, inexpensive, and easy to undertake without trained supervision, it requires a high intensity of training and a substantial time commitment.

Multi-exercise endurance training maintains BMD at the clinically important femoral neck site (only) in women of this age. It must be performed four times per week for 18 months and includes walking, jogging, ergometry, cycling, stair-climbing, and walking on a graded treadmill at 55-75% $\dot{V}O_2max$ [4].

FIGURE 11.6 High-impact jumps. Step program illustrating *(a)* side-side jumps and *(b)* jumping jacks.

Maintenance and Acquisition of Bone Mass in the Femoral Trochanteric Region and Lumbar Spine

Two resistance training protocols proved to be highly successful in augmenting BMD at the femoral trochanter and the lumbar spine [4, 6]. In a trial undertaken in women aged 20-35 years, aerobic training was alternated with strength training. The women participated in 3 one-hour classes per week that included three types of exercise. The first and last 10 minutes of each class were for warming up and cooling down (table 11.4).

TABLE 11.4

Resistance training protocol that used three different classes per week to augment BMD at the femoral trochanter and lumbar spine

Class 1—Participants alternate quickly (every 12 min) between exercise stations and high-impact aerobic activities. Exercise stations consisted of push-ups, sit-ups (figure 11.7), arm curls with dumbbells, and military presses with barbells (figure 11.8).
Class 2—Special emphasis is put on the gluteus maximus, erector spinae, and shoulder girdle muscles. Lifts for this purpose were bent knee dead lifts, bent rows (figure 11.9 a and b), shoulder presses , and cleans (lifts to shoulder height) (figure 11.10a and b). Lifts were started from ground level, which further exercised the erector spinae muscles and corresponding stabilizing muscle groups. In this study only moderate weights were used: dumbbells (3, 6, 12 lb), barbells (16-36 lb), and ankle/wrist weights (0.5 to 2.5 lb).
Class 3—This vigorous, high-impact aerobic workout aims to keep the heart rate between 70% and 85% of maximum, which was estimated as being 220 minus age.

From [4].

The protocol that used resistance training only [51] consisted of exercises similar to those described in Class 2 in table 11.4. Specifically, the program consisted of three sets of 8-12 repetitions for 12 weightlifting exercises (free weights) designed to load all major muscle groups. Intensity began at 70% of 1-RM and was reassessed every 6 to 8 weeks. After 6 months, intensity was increased to 75% 1-RM, and after 12 months it was increased again to 80% of 1-RM. Women trained for about 1 hour, 3 days a week for 18 months.

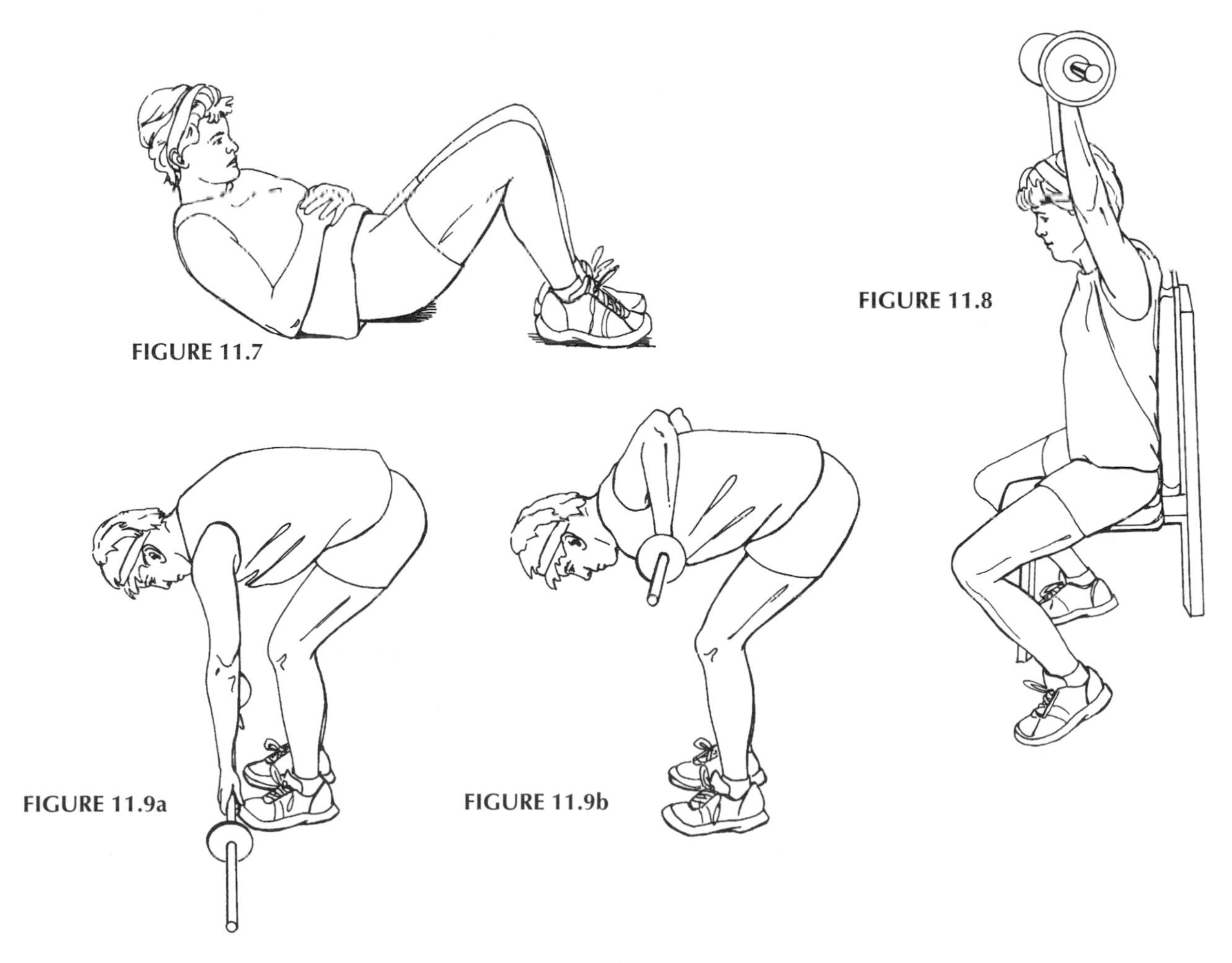

FIGURE 11.7

FIGURE 11.8

FIGURE 11.9a

FIGURE 11.9b

SUMMARY

- Bone mineral is not easily augmented once the skeleton has reached maturity. The main role of exercise in premenopausal women appears to be conservation, not acquisition, although targeted, high-intensity exercise can lead to modest bone accrual.
- Nevertheless, the bone-preserving action of premenopausal adult exercise may still maintain bone strength and prevent age-related fractures since even small increases in bone mineral can significantly reduce the risk of fracture [10, 52].

References

1. Jarvinen TLN, Kannus P, Sievanen H. Have the DXA-based exercise studies seriously under-estimated the effects of mechanical loading on bone? (Letter). *J Bone Miner Res* 1999;14:1634-1635.
2. Heinonen A, Kannus P, Sievänen H, et al. Randomised control trial of effect of high-impact exercise on selected risk factors of osteoporotic fractures. *Lancet* 1996;348:1343-1347.
3. Snow-Harter C, Bouxsein ML, Lewis BT, et al. Effects of resistance and endurance exercise on bone mineral status of young women: a randomized exercise intervention trial. *J Bone Miner Res* 1992;7:761-769.
4. Friedlander AL, Genant HK, Sadowsky S, et al. A two-year program of aerobics and weight training enhances bone mineral density of young women. *J Bone Miner Res* 1995;10:574-585.
5. Bassey EJ, Ramsdale SJ. Increase in femoral bone density in young women following high-impact exercise. *Osteoporos Int* 1994;4:72-75.
6. Lohman T, Going S, Pamenter R, et al. Effects of resistance training on regional and total bone mineral density in premenopausal women: a randomized prospective study. *J Bone Miner Res* 1995;10:7.
7. Bennell KL, Malcolm SA, Thomas SA, et al. Risk factors for stress fractures in female track-and-field athletes: a retrospective analysis. *Clin J Sport Med* 1995;5:229-235.
8. Fehling PC, Alekel L, Clasey J, et al. A comparison of bone mineral densities among female athletes in impact loading and active loading sports. *Bone* 1995;17:205-210.
9. Heinonen A, Oja P, Kannus P, et al. Bone mineral density in female athletes representing sports with different loading characteristics of the skeleton. *Bone* 1995;17:197-203.
10. Kannus P. Preventing osteoporosis, falls, and fractures among elderly people. *Br Med J* 1999;318:205-206.
11. Bassey EJ, Rothwell MC, Littlewood JJ, et al. Pre- and postmenopausal women have different bone mineral density responses to the same high-impact exercise. *J Bone Miner Res* 1998;13:1805-1813.
12. Wolff I, van Croonenborg, Kemper HCG, et al. The effect of exercise training programs on bone mass: a meta-analysis of published controlled trials in pre- and postmenopausal women. *Osteoporos Int* 1999;9:1-12.
13. Ernst E. Exercise for female osteoporosis. A systematic review of randomised clinical trials. *Sports Med* 1998;25:359-68.
14. Sinaki M, Wahner HW, Bergstralh EJ, et al. Three-year controlled, randomized trial of the effect of dose-specified loading and strengthening exercises on bone mineral density of spine and femur in non-athletic, physically active women. *Bone* 1996; 19:233-244.
15. Beverly M, Rider T, Evans M, et al. Local bone mineral response to brief exercise that stresses the skeleton. *Br Med J* 1989;299:233-235.
16. Rockwell JC, Sorensen AM, Baker S, et al. Weight training decreases vertebral bone density in premenopausal women: A prospective study. *J Clin Endocrinol Metab* 1990;71:988-993.
17. Nishiyama S, Tomoeda S, Ohta T, et al. Differences in basal and postexercise osteocalcin levels in athletic and nonathletic humans. *Calcif Tissue Int* 1988;43:150-154.
18. Suominen H. Physical activity and bone. *Ann Chir Gynaecol* 1988;77:184-188.
19. Suominen H. Bone mineral density and long term exercise. An overview of cross-sectional athlete studies. *Sports Med* 1993;16:316-30.
20. Smith EL, Gilligan C. Physical activity effects on bone metabolism. *Calcif Tissue Int* 1991;49:S50-S54.
21. Snow-Harter C, Marcus R. Exercise, bone mineral density, and osteoporosis. *Exerc Sport Sci Rev* 1991;19:351-358.
22. Myburgh KH, Micklesfield L. Exercise and bone mass in mature premenopausal women. *S A J Sports Med* 1995;2:15-21.
23. Marcus R, Drinkwater B, Dalsky G, et al. Osteoporosis and exercise in women. *Med Sci Sports Exerc* 1992;24:S301-S307.

24. Forwood MR, Burr DB. Physical activity and bone mass: exercises in futility? *Bone Miner* 1993;21:89-112.

25. Chilibeck PD, Sale DG, Webber CE. Exercise and bone mineral density. *Sports Med* 1995;19:103-122.

26. Gutin B, Kaspar MJ. Can vigorous exercise play a role in osteoporosis intervention? A review. *Osteoporos Int* 1992;2:55-69.

27. Drinkwater BL. C H McLoy Research Lecture: Does physical activity play a role in preventing osteoporosis. *Res Q Exerc Sport* 1994;65:197-206.

28. Bailey DA, Faulkner RA, McKay HA. *Growth, physical activity, and bone mineral acquisition*. In: Holloszy JO, ed. Exercise and sport sciences reviews. Baltimore: Williams & Wilkins, 1996;24:233-266.

29. Kannus P, Haapasalo H, Sankelo M, et al. Effect of starting age of physical activity on bone mass in the dominant arm of tennis and squash players. *Ann Int Med* 1995;123:27-31.

30. Bennell KL, Malcolm SA, Khan KM, et al. Bone mass and bone turnover in power athletes, endurance athletes and controls: A 12 month longitudinal study. *Bone* 1997;20:477-484.

31. Recker RR, Davies KM, Hinders SM, et al. Bone gain in young adult women. *JAMA* 1992;268:2403-2408.

32. Zhang J, Feldblum PJ, Fortney JA. Moderate physical activity and bone density among perimenopausal women. *Am J Public Health* 1992;82:736-738.

33. Salamone LM, Glynn N, Black D, et al. Determinants of premenopausal bone mineral density: the interplay of genetic and lifestyle factors. *J Bone Miner Res* 1996;11:1557-1565.

34. Kroger H, Tuppurainen M, Honkanen R, et al. Bone mineral density and risk factors for osteoporosis - a population-based study of 1600 perimenopausal women. *Calcif Tissue Int* 1994;55:1-7.

35. Pocock NA, Eisman JA, Yeates MG, et al. Physical fitness is a major determinant of femoral neck and lumbar spine bone mineral density. *J Clin Invest* 1986;78:618-21.

36. Pocock NA, Eisman JA, Gwinn T, et al. Muscle strength, physical fitness and weight but not age, predict femoral neck bone mass. *J Bone Miner Res* 1989;4:441-447.

37. Henderson NK, Price RI, Cole JH, et al. Bone density in young women is associated with body weight and muscle strength but not dietary intakes. *J Bone Miner Res* 1995;10:384-393.

38. Dalsky GP, Stocke KS, Ehsani AA, et al. Weight-bearing exercise training and lumbar bone mineral content in postmenopausal women. *Ann Int Med* 1988;108:824-828.

39. Bevier WC, Wiswall RA, Pyka G, et al. Relationship of body composition, muscle strength and aerobic capacity to bone mineral density in older men and women. *J Bone Miner Res* 1989;4:421-432.

40. Cheng S, Suominen H, Rantanen T, et al. Bone mineral density and physical activity in 50-60 year old women. *Bone Miner* 1991;12:123-32.

41. Jonsson B, Ringsberg K, Josefsson PO, et al. Effects of physical activity on bone mineral content and muscle strength in women: a cross-sectional study. *Bone* 1992;13:191-195.

42. Welten DC, Kemper HCG, Post GB, et al. Weight-bearing activity during youth is a more important factor for peak bone mass than calcium intake. *J Bone Miner Res* 1994;9:1089-1096.

43. Nelson ME, Meredith CN, Dawson-Hughes B, et al. Hormone and bone mineral status in endurance-trained and sedentary postmenopausal women. *J Clin Endocrinol Metab* 1988;66:927-933.

44. Chen MM, Yeh JK, Aloia JF, et al. Effect of treadmill exercise on tibial cortical bone in aged female rats: a histopmorphometric and dual energy x-ray absorptiometry study. *Bone* 1994;15:313-319.

45. Yeh JK, Aloia JF, Chen MM, et al. Influence of exercise on cancellous bone of the aged female rat. *J Bone Miner Res* 1993;8:1117-1125.

46. Parfitt AM. Morphological basis of bone mineral measurements: transient and steady state effects of treatment in osteoporosis. *Miner Elect Metab* 1980;4:273-287.

47. Parfitt AM. The two faces of growth: benefits and risks to bone integrity. *Osteoporos Int* 1994;4:382-398.

48. Kanis JA. Calcium nutrition and its implications for osteoporosis. *Eur J Clin Nutr* 1994;48:757-767, 833-841.

49. Bassey EJ, Ramsdale SJ. Weight-bearing exercise and ground reaction force: a 12-month randomized controlled trial of effects on bone mineral density in healthy postmenopausal women. *Bone* 1995;16:469-476.

50. Heinonen A, Oja P, Sievänen H, et al. Effect of two training regimes on bone mineral density in healthy perimenopausal women: a randomised control trial. *J Bone Miner Res* 1998;13:483-490.

51. Lohman T. Exercise training and body composition in childhood. *Can J Spt Sci* 1992;17:284-287.

52. Hui SL, Slemenda CW, Johnston Jr CC. Age and bone mass as predictors of fracture in a prospective study. *J Clin Invest* 1988;81:1804-1809.

53. Gleeson PB, Profas EJ, LeBlanc AD, et al. Effects of weight lifting on bone mineral density in premenopausal women. *J Bone Miner Res* 1990;153-158.

54. Smith EL, Gilligan C, McAdam M, et al. Deterring bone loss by exercise intervention in premenopausal and postmenopausal women. *Calcif Tissue Int* 1989;44:312-321.

55. Alekel L, Clasey JL, Fehling PC, et al. Contributions of exercise, body composition, and age to bone mineral density in premenopausal women. *Med Sci Sports Exerc* 1995;27:1477-1485.

56. Aloia JF, Vaswani AN, Yeh JK, et al. Premenopausal bone mass is related to physical activity. *Arch Intern Med* 1988;148:121-3.

57. Davee AM, Rosen CJ, Adler RA. Exercise patterns and trabecular bone density in college women. *J. Bone Miner Res* 1990;5:245-50.

58. Halioua L, Anderson JJB. Lifetime calcium intake and physical activity habits: independent and combined effects on the radial bone of healthy premenopausal women. *Am J Clin Nutr* 1989;49:534-541.

59. Kanders B, Dempster DW, Lindsay R. Interaction of calcium nutrition and physical activity on bone mass in young women. *J Bone Miner Res* 1988; 3:145-149.

60. Kirk S, Sharp CF, Elbaum N, et al. Effect of long-distance running on bone mass in women. *J Bone Miner Res* 1989;4:515-522.

61. Mazess RB, Barden HS. Bone density in premenopausal women: Effects of age, dietary intake, physical activity, smoking and birth-control pills. *Am J Clin Nutr* 1991;53:132-142.

62. Sowers M, Wallace RB, Lemke JH. Correlates of forearm bone mass among women during maximal bone mineralization. *Preventive Medi* 1985; 14:585-96.

63. Stevenson JC, Lees B, Devonport M, et al. Determinants of bone density in normal women: Risk factors for future osteoporosis? *Br Med J* 1989; 298:924-928.

64. Uusi-Rasi K, Nygard CH, Oja P, et al. Walking at work and bone mineral density of premenopausal women. *Osteoporos Int* 1994;4:336-340.

65. Young D, Hopper JL, Nowson CA, et al. Determinants of bone mass in 10- to 26-year-old females: A twin study. *J Bone Miner Res* 1995; 10:558-567.

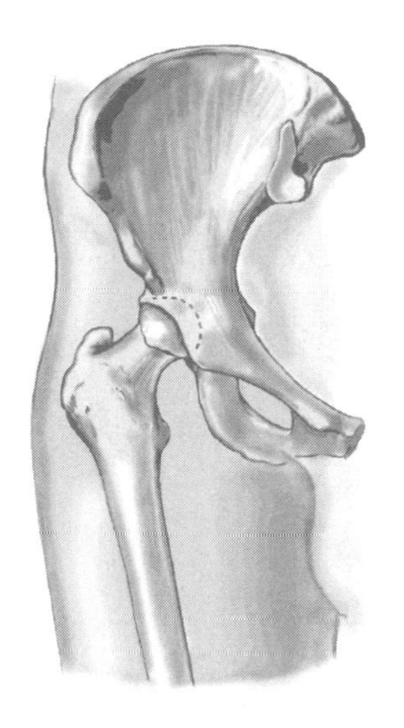

Chapter 12

Physical Activity, Targeted Bone Loading, and Bone Mineral in Postmenopausal Women

with Moira Petit

Postmenopausal women make up a large, and increasing, proportion of the population. As a group, they express the desire to increase their bone mass to prevent osteoporotic fracture. Whether this is feasible is the subject of this chapter. Because many of these women prefer not to take medication to achieve this, the question often arises, Can a postmenopausal woman augment or maintain her bone mineral by undertaking physical activity? In an attempt to answer this question, we summarize the findings of those studies that have used a prospective, controlled design. Several recent meta-analyses have addressed the same subject [1-5].

Because targeted bone loading has a site-specific effect, we have divided the chapter into separate sections to examine the effect of exercise on vertebral and proximal femoral bone mineral density. As in previous chapters, we distinguish between exercise interventions that appear to have provided substantial targeted bone loading and those that consisted essentially of general physical activity.

Most of the studies that have examined the effect of exercise on bone used densitometry (DXA, DPA, QCT) to measure bone mineral (BMC or BMD). Earlier studies that used other methods [6-8] are not reviewed here, as the techniques used to measure bone in those studies were much less precise than DXA.

Before continuing, it is necessary to provide an accepted definition of menopause. Although definitions in various studies differ, a World Health Organization (WHO) scientific group has defined natural menopause as the permanent cessation of menstruation resulting from the loss of ovarian follicular activity [9]. Because menopause occurs after the final menstrual period, it can only be known with certainty in retrospect a year or more after the event [9].

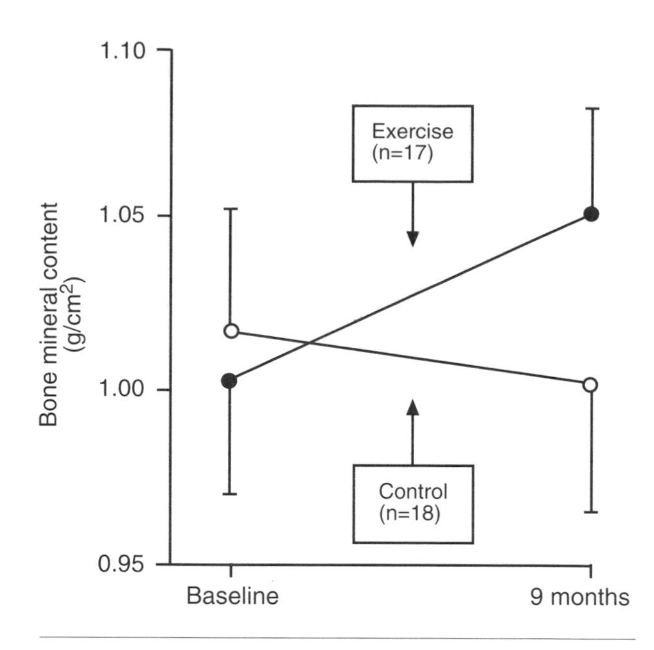

FIGURE 12.1 Relationship between high-impact exercise and bone mineral in postmenopausal women. Women who performed high-intensity training gained 5.2% more lumbar spine BMC over nine months than did controls.

Reprinted, by permission, from Dalsky et al., 1998, "Weight bearing exercise training and lumbar bone mineral content in postmenopausal women," *Annals of Internal Medicine*, (108):826.

Vertebral BMD and Targeted Bone Loading

Numerous workers have evaluated the effect of high-impact loading or strength training exercises or a combination of these on vertebral BMD of healthy postmenopausal women [3, 48]. The most often cited "high-impact" loading intervention study is that of Gail Dalsky and colleagues [10] who found that the lumbar spine BMC of exercising subjects increased a mean of 5.2%, which, when combined with a 1.4% decrease in the BMC of controls, represented a 6.6% greater BMC after the nine-month intervention (figure 12.1). Factors that may have contributed to such a substantial increase in BMC include the high-intensity, high-impact exercises, which included subjects running up and down concrete stairwells; the self-selection of highly motivated women into the subject group; and the fact that 4 of the 17 exercise subjects were taking ovarian hormone therapy and calcium supplementation to 1500 mg/day. Three of the controls were also on hormone therapy. It should be noted that hormone therapy combined with exercise has been shown to produce greater gain in bone mass in postmenopausal women than hormone therapy alone [11-13], although that has not been the case in every study [14].

How does high-impact activity compare with strength training as a stimulus for positive bone remodeling in postmenopausal women? A randomized, controlled study compared the effects of impact loading (jogging, running, stair-climbing) and strength training (weightlifting using both free weights and resistance equipment and using a rowing machine) on BMD in previously nonexercising postmenopausal women (figure 12.2) [15]. Impact loading exercises were estimated to cause ground reaction forces of five to six times body weight at the hip, which translates into a slightly lower force at the spine [16]. Intervention was performed for 45 minutes, three to four times a week for 11 months. Exercise prescription was individualized and updated weekly, and calcium was supplemented to 1500 mg/day. Both interventions achieved a 1-2% positive effect on lumbar spine BMD (figure 12.3).

Some interventions that consisted essentially of strength training have led to improvements in bone mineral compared with controls, while some have not. In a randomized, controlled trial, Nelson and colleagues [17] prescribed high-intensity dynamic exercises using free weights and resistance machines. The regimen was adjusted each month so that intensity was maintained at 80% of 1-RM (one-repetition maximum). Compared with controls, exercising

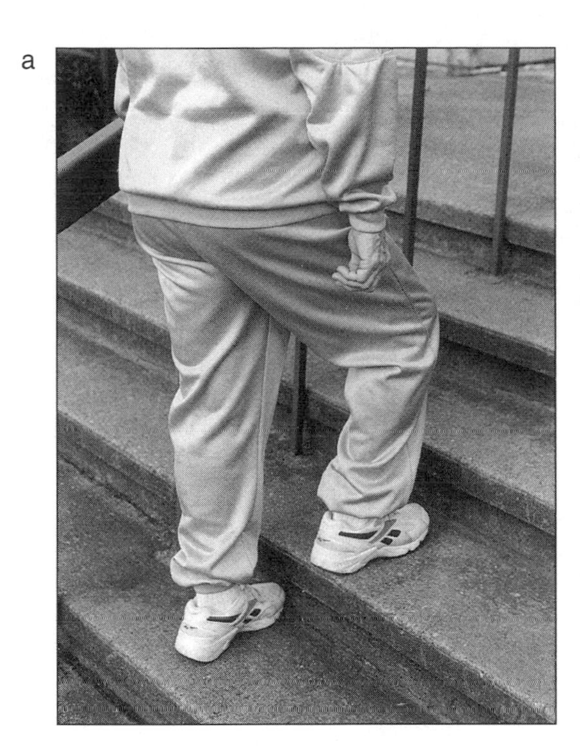

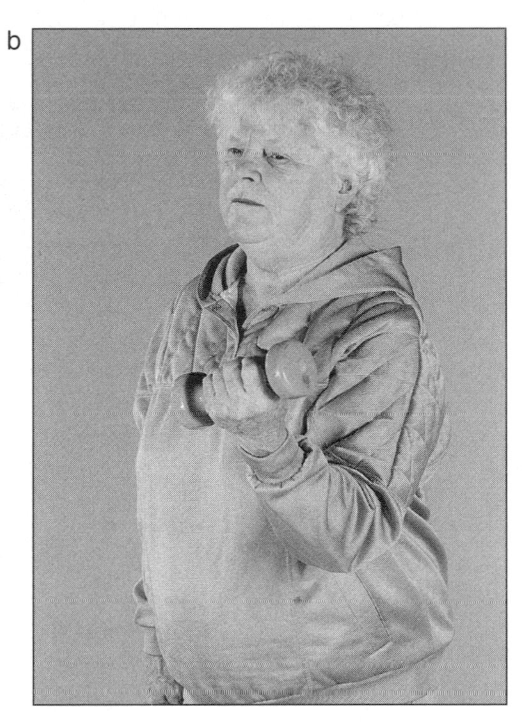

FIGURE 12.2 Kohrt and colleagues [15] compared the effect of *(a)* impact loading (e.g., stair climbing) versus *(b)* strength training (e.g., free weights) in postmenopausal women.

FIGURE 12.3 Percent change in lumbar bone mineral density in response to impact loading and resistance training exercise programs and in control subjects. Different from control group, †$p < 0.05$, •$p < 0.01$.

Reproduced from *J Bone Miner Res* 1997; 12: 1253-1261 with permission of the American Society for Bone Mineral Research.

subjects gained strength as well as total body BMC and lumbar spine and femoral BMD (tables 12.1 and 12.2). These results are comparable with those of Pruitt and colleagues [18] who used a similar protocol in a nonrandomized trial. On the other hand, Nichols and colleagues [19], also using a similar intervention but in subjects who were already physically active, found no change in lumbar BMD at 12 months. These latter subjects, although regular runners, had never performed weight training [19].

Thus, the main trend in intervention studies in which the spine was loaded was that vertebral bone mineral was increased 1-2% at the end of intervention relative to controls [12, 15, 17-21, 22]. These data suggest that exercise intervention in this age group can maintain bone mineral, but rarely serves to add substantial amounts of bone de novo.

Conserving bone mineral is of significant clinical benefit, but whether this benefit persists after exercise intervention ceases is not clear. In the study by Dalsky [23], subjects who had gained BMC with 9 months of training were remeasured after 13 months of detraining. A significant decrease in lumbar spine BMC had occurred in the detraining period. After both the training and the detraining period (22 months after entering the study), lumbar spine BMC was 1.1% higher than baseline, an insignificant difference.

TABLE 12.1

Prospective studies of vertebral BMD in postmenopausal women: Effect of targeted bone loading

Authors	Specific parts of intervention program likely to have loaded the lumbar vertebrae	Effect size (%) at the lumbar spine
Dalsky et al. 1988 [10]	Running up and down concrete stairwells	Exercisers: +5.4 Controls: –1.2
Grove & Londree 1992 [20] (high-impact group)	Jumping jack exercise (ground reaction force = 3.3 times bodyweight); knee to elbow with jump (ground reaction force = 2.8 times bodyweight)	Exercisers: +1.7 Controls: –6.1
Heinonen et al. 1998 [32]	Either calisthenics or endurance exercise 3 times per week for 30-50 minutes for 18 months	Calisthenics: No training effect Endurance: Maintained BMD at the femoral neck Control: No training effect
Kohrt et al. 1995 [12]	Vigorous walking, jogging, stair-climbing/descending	Exercisers: +2.3 Controls: –0.0
Kohrt et al. 1997 [15] (impact loading group)	Stair-climbing/descending	Exercisers: +1.8 Controls: +0.1
Kohrt et al. 1997 [15] (resistance training group)	Standing while doing resistance training, including squats	Exercisers: +1.5 Controls: +0.1
Nelson et al. 1991 [22]	Walking rapidly wearing an 8 lb (3.1 kg) belt	DPA spine: No change in exercise or control group QCT spine: Exercise: +0.5 and controls –7.0
Nelson et al. 1994 [17]	Dynamic strength training exercises	Exercisers: +1.0 Controls: –1.8
Nichols et al. 1995 [19]	Dynamic strength training exercises	Exercisers: –0.9 Controls: +1.4
Pruitt et al. 1992 [18]	Weight training using universal equipment. Trunk extension and trunk lateral flexion were designed to load the trunk.	Exercisers: +1.6 Controls: –3.6
Welsh & Rutherford 1996 [2]	High-impact step and jumping exercises, including abdominal flexion and lumbar extension. Light weights 2.5 to 10 lbs (1-4 kg) were also used to target most muscles.	Exercisers: +0.2 (NS) Controls: –0.7

Vertebral BMD and Minimal Bone Loading

Aerobic exercise that provides the lumbar spine with minimal loading would not be predicted to stimulate gain in bone mineral in postmenopausal women. For example, slow walking and fast walking have peak ground reaction forces only 1.2 and 1.5 times body weight, respectively [20]. These forces are likely to be attenuated before reaching the spine [16]. Despite this, numerous authors found that exercise that was not predicted to generate substantial strain at the spine nevertheless generated increases in lumbar BMD in postmenopausal women (table 12.3, figure 12.4, Appendix A, table A.3). In some cases the magnitude of BMD increase was comparable with that reported in studies that specifically targeted the lumbar spine [24-26].

Increased Vertebral Bone Mass

What could explain increases in vertebral BMD with minimal bone loading? Several factors could lead to an artifactual increase in BMD, while several

TABLE 12.2

Results of meta-analyses of the effect of exercise on BMD in postmenopausal women

Author	Skeletal site	Summary of material included in analysis	Years (inclusive)	Outcome
Bérard 1997 [1]	All sites	18 eligible studies	1966–1996	Effect size = 0.88 at the lumbar spine in studies published after 1991
Kelley 1998a [3]	Lumbar spine	17 effect sizes from 10 studies, 330 subjects (192 exercise, 138 nonexercise)	1975–1994	2.8% change in BMD in exercisers, consisting of 0.3% gain in exercisers compared with base-line and 2.5% loss in controls
Kelley 1998b [4]	Femur	13 treatment effects	1975–1995	0.9% gain (significantly different from zero)
Wolff et al. 1999 [5]	All sites	25 studies	1966–1996	1% effect in randomized controlled trials, 2% in uncontrolled trials

TABLE 12.3

Prospective studies of bone mineral in postmenopausal women at the hip: Effect of targeted bone loading

Authors	Exercises likely to have specifically loaded the hip	Effect size at the femoral neck (%) ($p < 0.05$ unless noted)
Bassey & Ramsdale 1995 [46]	50 heel drops (raising body weight onto toes and then letting it drop to the floor with the knees and hips extended)	Exercise: +0.1 Controls: –0.8 (NS)
Bloomfield et al. 1993 [24]	Stationary cycling against resistance at 60-80% maximum heart rate	Exercise: +0.1 Controls: –0.8 (NS)
Heikkinen et al. 1997 [14]	Exercise designed to load the hip—details not provided in paper	(Trochanter, assuming mean value of 0.820 g/cm^2) Exercise: +0.2 Controls: –2.1
Kerr et al. 1996 [31]	Weight training	(Trochanter, trained vs. untrained side) Exercise: +1.7 Controls: –0.6
Kerr et al. 2001 [48]	Weight training	(Intertrochanter, trained vs. untrained side) Exercise: +0.7 Controls: –1.2
Kohrt et al. 1995 [12]	Walking, jogging, stair-climbing	Exercise: +3.3 Controls: –0.5
Kohrt et al. 1997 [15]	Ground reaction force group: walking, jogging, stair-climbing	Exercise: +3.5 Controls: –1.5
Kohrt et al. 1997 [15]	Joint reaction force group: strength training with free weight and machines; program included squats and overhead press	Exercise: –0.2 Controls: –1.5 (NS)
Nelson et al. 1991 [22]	Walking with a weighted belt (moderate calcium group, i.e., no calcium intervention)	Exercise: –1.2 Controls: –1.0 (NS)
Nelson et al. 1991 [22]	Walking with a weighted belt (high calcium group, i.e., calcium intervention)	Exercise: +3.0 Controls: –1.2

(continued)

TABLE 12.3 *(continued)*

Authors	Exercises likely to have specifically loaded the hip	Effect size at the femoral neck (%) ($p < 0.05$ unless noted)
Nelson et al. 1994 [17]	Exercises using hydraulic resistance machines, including back extension and abdominal flexion	Exercise: +0.1 Controls: –0.8
Nichols et al. 1995 [19]	Isotonic training exercises including back extension, trunk flexion, bench press, and seated row	Exercise: –1.3 Controls: +0.8 (NS)
Welsh & Rutherford 1996 [21]	High-impact step and jumping exercises to load femur and spine; light weights of 2 to 10 lbs (1-4 kg) were also used	Exercise: +1.6 Controls: –1.9

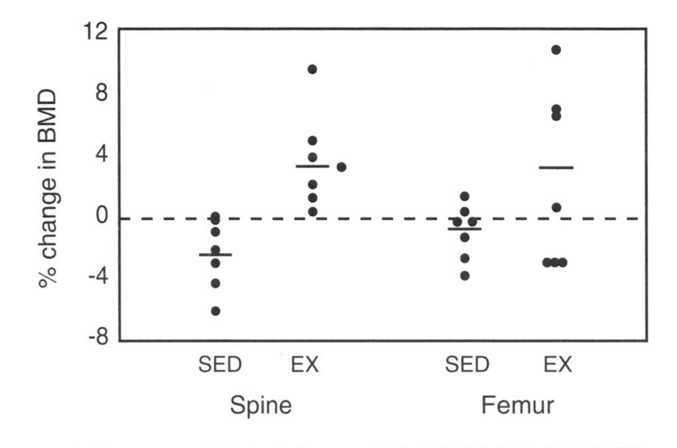

FIGURE 12.4 Trivial exercise and increased BMD.

mechanisms may explain a real increase in BMD. Technical aspects of measurement that could lead to an artifactual increase in bone mineral include (1) the use of neutron activation analysis [7], a technique much less precise than DXA; (2) the small numbers of subjects in some DXA studies so that error in subject positioning or scan analysis could substantially influence results; (3) accelerated osteoarthritis (biologically plausible) or vascular calcification (biologically paradoxical) in exercising subjects, resulting in elevated BMD; (4) changes in soft-tissue composition (e.g., increases in lean mass) (see chapter 6); and (5) subjects increasing their aerobic fitness during the study period with resultant increases in erythropoiesis and bone marrow change that could be interpreted as increased vertebral BMD by DXA [27].

Factors that could contribute to a real increase in vertebral BMD in this group of exercising subjects include (1) a tendency for people who undertake an exercise intervention study to increase their general physical activity (and perhaps their bone loading) outside the study [17]; (2) exercise-related endocrine changes such as increased growth hormone, adrenal androgens (providing estrogen via peripheral aromatization—a less likely scenario), and plasma androstenedione, which contributes slightly to BMD in postmenopausal women [28]; and (3) exercise-associated changes in nutrient intake.

A technical factor that would tend to mitigate *against* finding positive changes in BMD with intervention is that posteroanterior spine densitometry underestimates the extent of change (both positive and negative) in lumbar spine trabecular BMD. This occurs because DXA taken in the posteroanterior plane (the most common view, in contrast to the lateral view) includes cortical as well as trabecular bone when measuring BMD. In one study that measured lumbar spine BMD by both QCT and DPA, QCT detected a 7% change in trabecular bone while DPA detected none [22].

Meta-Analyses Findings: Lumbar Spine BMD

A recent meta-analysis of "the effectiveness of physical activity for the prevention of bone loss in postmenopausal women" [1] found that "overall, physical activity had no statistically significant effect on any of the sites measured" (see table 12.2). The authors of this meta-analysis did not distinguish between activities that differed in their extent of bone loading or the site of loading. They did, however, note "a significant effect of physical activity on BMD of the lumbar column in studies published after 1991." One could argue that around 1990 it became evident that "the form of exercise may determine whether the weightbearing and/or nonweightbearing segments of the skeleton are stimulated" [29]. In other words, recent interventions may have better targeted the lumbar spine. Also, the introduction of DXA around that time has permitted more precise measurement. However, the post-1991 data examined by Bérard and colleagues included both studies in which interventions consisted of targeted loading and those that appeared to lack this factor [1].

Kelley's meta-analysis [3] showed that exercisers enjoyed, on average, a 3% advantage in lumbar spine BMD than controls and this was largely the result of decreased BMD in the control groups. His data did not provide conclusive evidence as to what type of exercise prescription was most likely to optimize bone mass, as no relationship was demonstrated between training program characteristics and BMD [3]. Wolff's meta-analysis found a 1% per year increase in lumbar spine BMD [5].

Proximal Femoral BMD and Targeted Bone Loading

Proximal femoral BMD predicts femoral neck fractures [30]. Several groups of researchers have reported a positive effect of exercise intervention at the hip [12, 15, 17], while another found no change [18] (see table 12.3).

Kerr and colleagues [31] performed a one-year progressive resistance intervention study in 56 early postmenopausal women who performed exercises that stressed the ipsilateral forearm and hip region while the contralateral side acted as the control. Subjects were also randomized to either a strength-trained group (3 × 8 repetition maximum) or an endurance-trained group (3 × 20 repetition maximum). The bone mineral increase with the strength regimen was significantly greater at the trochanteric and intertrochanteric hip sites than with endurance training. However, no significant increase in proximal femoral BMD was noted with the endurance regimen. Muscle strength, measured by a one-repetition maximum (1-RM) test, increased significantly and to a similar degree in both the resistance- and endurance-trained groups. This study emphasizes that the femur (and radius, which was similarly trained with similar results) responds in a site-specific manner to progressive resistance training and that peak load is more important than the number of loading cycles [31, 48].

As discussed earlier, Kohrt and colleagues [15] trialed a strength training and a high-impact training intervention, both of which caused increased lumbar spine BMD. However, only the impact loading program augmented femoral neck BMD (figure 12.5). To prevent osteoporotic fractures, it may be "prudent to recommend exercises that generate ground reaction forces over those that generate joint reaction forces" (resistance

Finer Point: Inclusion Criteria for Bérard's Meta-Analysis

1. Full-length articles reporting prospective, controlled (but not necessarily randomized) intervention studies testing exercise of all forms and durations.
2. Studies in which experimental and control groups were not contaminated by treatments such as estrogen, vitamin D, or calcium (to avoid the possibility of a synergistic effect).
3. Studies that had BMC or BMD as an outcome in healthy, sedentary postmenopausal women with an average age of 50 years or more (i.e., no symptomatic osteoporosis at the time of entry into the study).

Because these inclusion criteria do not differentiate among different types or durations of exercise, a particularly effective type or duration would not stand out from the rest. By analogy with the treatment of bacterial pneumonia, if the effect of all antibiotics were studied together, the conclusion might be that antibiotics (as a group) do not cure pneumonia, when, in fact, *certain* antibiotics are particularly effective for treating bacterial pneumonia. Advocates of physical activity therefore contend that the Bérard meta-analysis does not provide the final word on this issue and illustrates a limitation of meta-analysis [1].

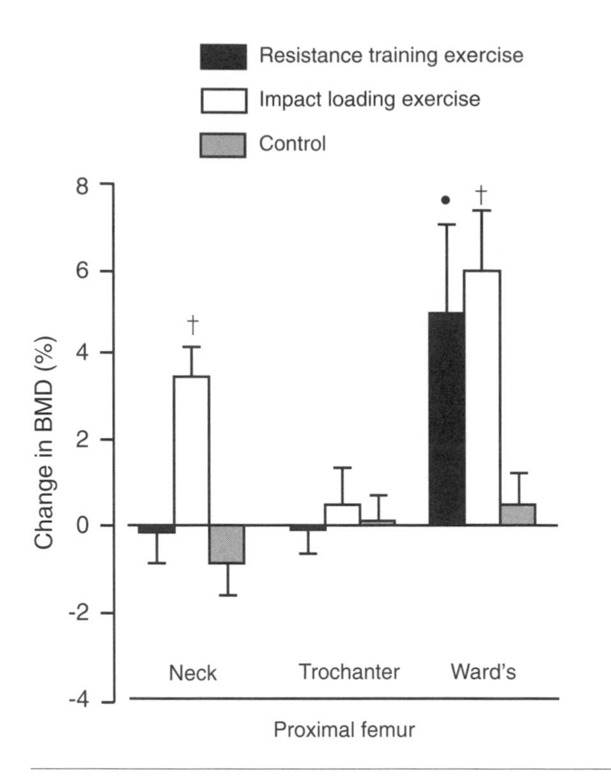

FIGURE 12.5 Percent change in femoral neck bone mineral density in response to ground reaction force and joint reaction force exercise training programs and in control subjects. * and ** represent significant difference from control group, $+p < 0.05$, $\bullet p < 0.01$.

Reproduced from *J Bone Miner Res* 1997; 12: 1253-1261 with permission of the American Society for Bone and Mineral Research.

training) [15]. The authors added that the increased lean mass and strength acquired via strength training also benefit the elderly, and therefore a program combining elements of both types of training may be optimal [15]. In an exercise program for women just entering the menopause, Heinonen found exercise interventions to be safe and effective for increasing lumbar spine BMD [32].

Mechanism of Bone Changes in Postmenopausal Women

Osteogenic response is maximized by just a few loading cycles [33, 34] when loading forces are high [35], applied at a fast rate [36, 37], and result in nonconventional strain distribution [38]. The principle that peak strain is more important than repetitions appears to apply in postmenopausal women [31].

Several researchers measured biochemical markers of bone metabolism (see also chapter 4) in exercising postmenopausal subjects. Pyridinium crosslinks, a measure of bone resorption, decreased in exercising subjects during the first 6 months of a 12-month intervention study; at 12 months differences had disappeared [21]. In three studies, subjects decreased their osteocalcin levels (an index of bone formation and mineralization) relative to controls despite increasing their BMD [10, 15, 39]. Other studies, however, have failed to find these changes in biochemical markers of bone turnover [12]. This may be due to the low reliability of bone markers, particularly those used in the early 1990s.

Whether targeted loading influences bone by increasing formation or decreasing resorption remains unclear. As mechanical stimuli are locally mediated via mechanotransduction (chapter 2), any factors that alter the hormonally controlled environment could subsequently alter the influence of the mechanical stimulus on bone coupling. For example, estrogen directly inhibits the resorptive phase of bone remodeling. When estrogen levels decrease (e.g., during menopause), the resorptive phase of remodeling increases. Although bone formation also increases, it cannot keep pace with excessive resorption. As a result, the process is "uncoupled" and net bone loss occurs. If estrogen again becomes available, resorption is slowed and bone mass may once again reach a homeostasis determined by mechanical loading. Although some have argued that estrogen may affect bone mineral by affecting the set-point for response to mechanical loading [40], Kohrt [12] argues that her data illustrating the additive effect of exercise and estrogen suggest that these two influences on the skeleton act via different mechanisms. Estrogen augments bone mineral by reducing the rate of resorption and, to a lesser degree, the rate of formation, whereas weight-bearing exercise has no effect on formation and reduces resorption by a different mechanism [12]. The details of this mechanism still need elucidation (see chapter 20).

Exercise Prescriptions

Exercise programs of proven benefit are outlined in tables 12.4 and 12.5. While we do not guarantee that the exercise protocol will be effective in every setting, it provides a rational basis for exercise prescription.

Improving Strength and Optimizing BMD

In postmenopausal women, a combination of resistance training (table 12.4) and rowing significantly increased proximal femoral and lumbar spine BMD, as well as augmenting lean mass and strength [15]. Women devoted about equal amounts of time to resistance training and rowing in each training session.

TABLE 12.4

Exercise program using a combination of free weights and strength training equipment to optimize BMD and quadriceps strength in postmenopausal women

Exercise	No. of reps to fatigue	No. of sets	No. of times/wk
Using free weights (performed standing whenever possible):			
Overhead press	8-12	2-3	2
Biceps curl	8-12	2-3	2
Triceps extension	8-12	2-3	2
Using equipment			
Leg press	8-12	2-3	2
Leg extension	8-12	2-3	2
Leg flexion (also known as hamstring curl)	8-12	2-3	2
Bench press	8-12	2-3	2
Using the Smith machine (counterbalance bar)			
Squats	8-12	2-3	2

Subjects had dietary calcium supplemented, where necessary, to 1500 mg/day.

The resistance training protocol included a one-week familiarization period. Subjects were prescribed two to three sets of each exercise at an intensity that resulted in fatigue after 8-12 repetitions (also known as 8-12 RM max). Resistance was increased whenever a participant was able to complete the prescribed number of sets and repetitions.

The rowing component consisted of an initial phase in which subjects rowed for 15 to 20 minutes at a moderate intensity (60-70% of maximal heart rate). Thereafter, the intensity was gradually increased so that participants performed two or three 10-minute bouts at an intensity that was 80-85% of maximal heart rate.

Table 12.5 shows the resistance training exercise protocol used by Deborah Kerr and colleagues [31, 48] to obtain a significant increase in proximal femoral and upper limb BMD. It also resulted in strength gains. Subjects underwent a three-week familiarization period and then started training at upper limb weights that represented 60% of 1-RM (maximum weight that can be lifted on one occasion maintaining good form) and lower limb weights of 40% 1-RM. Subjects rested for two to three minutes between sets. The exercise protocol was preceded and followed by stretching exercises. This study had a novel methodology—subjects exercised one side only. This permitted the authors to convincingly demonstrate the site specificity of exercise in an intervention [31].

TABLE 12.5

Exercise program using free weights only to optimize BMD and quadriceps strength in postmenopausal women

Exercise	No. of reps	No. of sets	No. of times/wk
Using free weights (performed standing):			
Biceps curl (figure 12.6 a and b)	8-RM	3	3
Wrist curl	8-RM	3	3
Reverse wrist curl	8-RM	3	3
Triceps extension (using pulley) (figure 12.7)	8-RM	3	3
Forearm pronation/supination (dumbbell)	8-RM	3	3
Leg press (figure 12.8)	8-RM	3	3
Hip abduction/adduction (figure 12.9 a and b)	8-RM	3	3
Hamstring curl (figure 12.10)	8-RM	3	3
Hip flexion	8-RM	3	3
Hip extension	8-RM	3	3

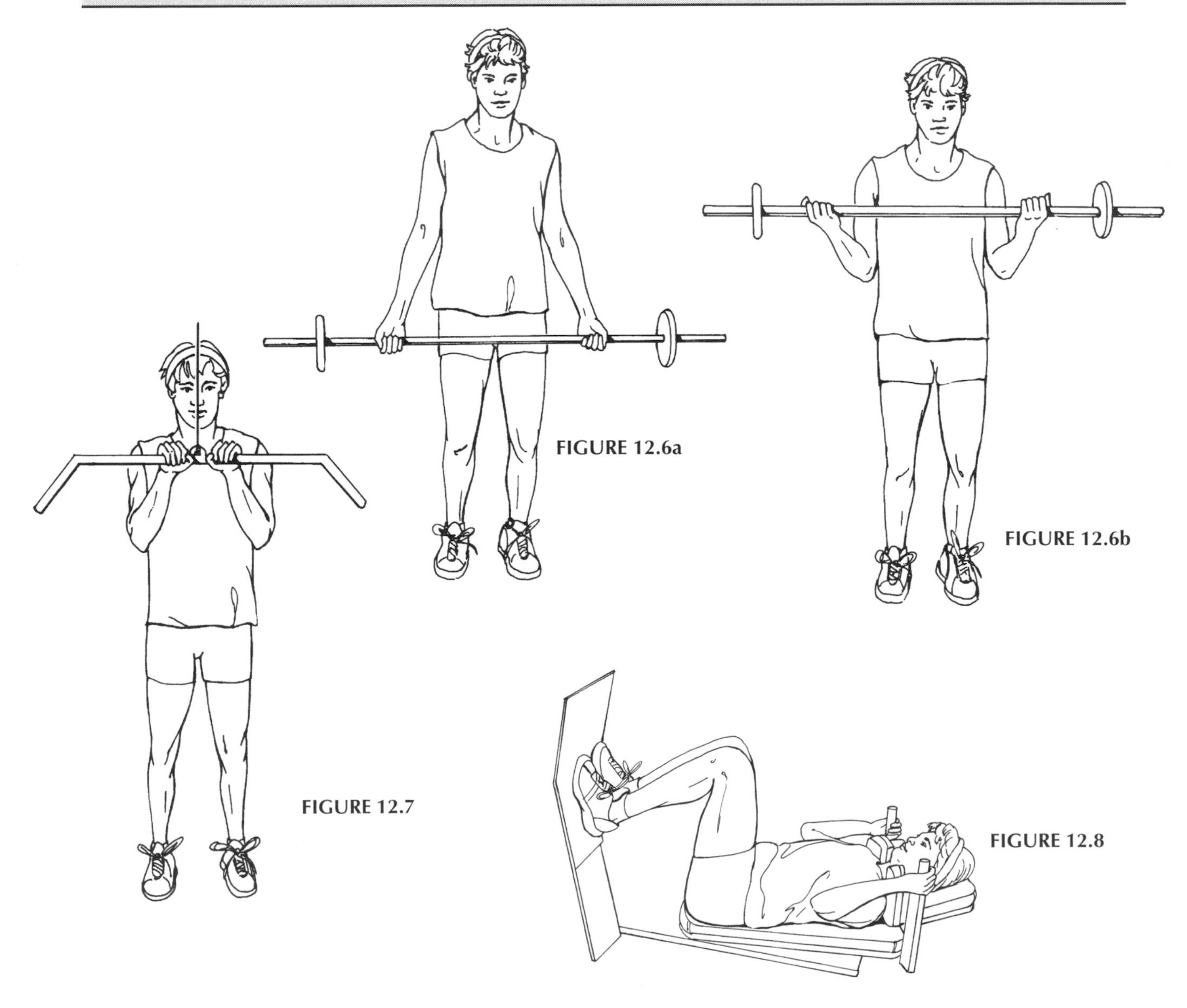

FIGURE 12.6a

FIGURE 12.6b

FIGURE 12.7

FIGURE 12.8

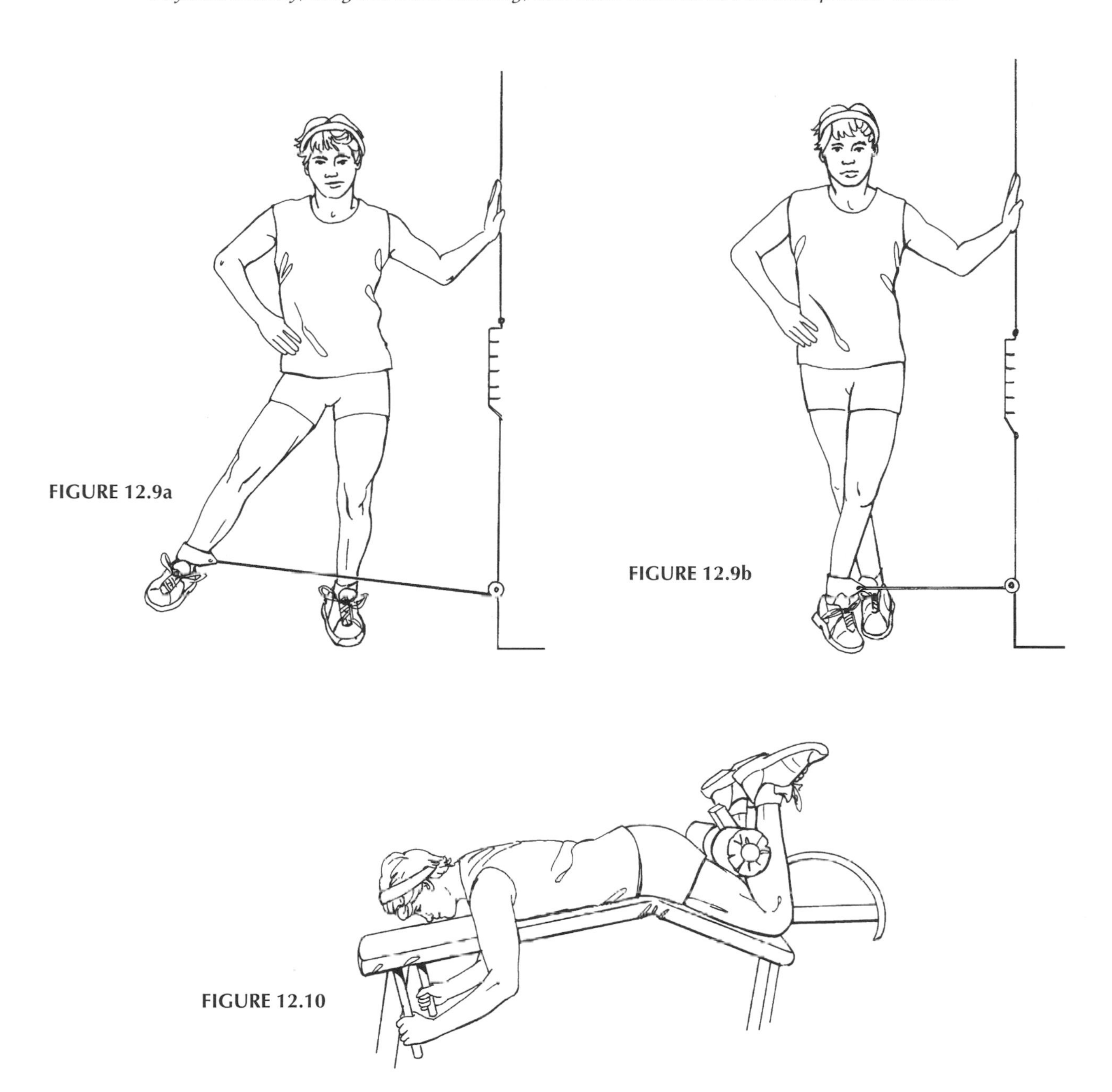

FIGURE 12.9a

FIGURE 12.9b

FIGURE 12.10

A walking/jogging/stair-climbing protocol [15] provided almost all of the benefits of the strength training protocol outlined in table 12.4. In this program, participants walked for 30 minutes (not including the warm-up or cool-down) at a moderate intensity (60-70% of maximal heart rate). The rate and duration of exercise were increased progressively until subjects exercised at least 45 minutes a day at an average intensity of 80-85% of maximal heart rate.

The only difference in outcome between this protocol and the resistance training protocol was that the walking/jogging/stair-climbing protocol failed to augment femoral neck BMD. Although this is a clinically important site, we note that BMD is only one determinant of hip fracture; quadriceps weakness is also a risk factor for osteoporotic fracture [41]. It is conceivable that strength gains obtained with the walking/jogging/rowing protocol may still lower the risk of fracture without increasing BMD, but prospective studies are needed to test this.

Exercise Prescription That Did Not Augment BMD

As well as highlighting exercise protocols that successfully augmented BMD, we provide examples of those that did not. Researchers at the Mayo Clinic [42] provided postmenopausal women with a backpack that contained weights equivalent to 30% of maximal isometric back muscle strength in pounds. In the prone position, subjects lifted the backpack 10 times. The maximal weight was 50 lb (22.7 kg). This was done once a day, five days per week. Although subjects gained more strength (which may prove protective against fracture) lumbar spine BMD did not improve relative to controls.

Other interventions that failed to improve BMD are shown in table 12.6. This illustrates that not all exercise serves to increase BMD, although some programs may provide other benefits, such as improved posture and eventual reduction of fracture.

TABLE 12.6

Interventions that failed to increase BMD

Exercise and author	No. of reps	No. of sets	No. of times/wk, duration
Small jumps Bassey and Ramsdale 1995 [46]	50	1	6 days/wk, for 12 mo
Endurance resistance training Kerr et al. 1996 [31]	Free-weight program (as in table 12.5)	3 sets of 20-RM	3 days/wk, for 12 mo
Walking below the anaerobic threshold Hatori et al. 1993 [39]	30 min of walking		3 days/wk, for 7 mo
Step-ups and 15-min upper body exercises while standing Lau et al. 1992 [47]	100 steps up and down 23 cm (9-in.) block	1	4 days/wk, for 10 mo

SUMMARY

- Postmenopausal women make up a large proportion of the community. Physicians, physiotherapists, athletic trainers, and gym instructors are often asked if exercise can help to augment BMD in this population.
- Several excellent prospective, controlled studies of high-impact intervention programs in postmenopausal women found that BMD increased at the targeted site, on average, from 1% to 3%. The magnitude of these changes is far less than is seen in growing children (5-10+%) (see chapter 10).
- The force that the femur needs to withstand to avoid fracturing during a fall is far in excess of the strength of the postmenopausal femur, even with 1-3% augmentations in BMD. On the other hand, the DXA-based exercise studies may have seriously underestimated the effect of targeted bone loading on bone strength, the bottom line (see chapters 3 and 4) [43, 44].
- If complete detraining follows exercise intervention, bone mineral tends to return to original (pretraining) levels. This issue requires further study before we can draw extended conclusions.

- While the evidence for exercise and bone health has, to date, mainly used BMD as an outcome measure, potential benefits of exercise in preventing osteoporotic fracture include increasing bone geometric properties.
- Exercise therapy is the only single therapy that can simultaneously augment muscle mass, muscle strength, balance, and bone strength [17] and decrease the risk of falling [45]. Exercise, therefore, has the potential to diminish fracture risk without necessarily altering BMD. Studies are needed to prove this potential.

References

1. Bérard A, Bravo G, Gauthier P. Meta-analysis of the effectiveness of physical activity for the prevention of bone loss in postmenopausal women. *Osteoporos Int* 1997;7:331-337.
2. Lewis RD, Modlesky CM. Nutrition, physical actitivity and bone health in women. *International Journal of Sport Nutrition* 1998;8:250-284.
3. Kelley DA. Aerobic exercise and lumbar spine bone mineral density in postmenopausal women: a meta-analysis. *Journal of the American Geriatrics Society* 1998a;46:143-152.
4. Kelley DA. Exercise and regional bone mineral density in postmenopausal women. *American Journal of Physical Medicine & Rehabilitation* 1998b;77:76-87.
5. Wolff I, van Croonenborg, Kemper HCG, et al. The effect of exercise training programs on bone mass: a meta-analysis of published controlled trials in pre- and postmenopausal women. *Osteoporos Int* 1999;9:1-12.
6. Aloia JF, Cohn SH, Ostuni JA, et al. Prevention of involutional bone loss by exercise. *Ann Int Med* 1978;89:356-8.
7. Chow RK, Harrison JE, Notarius C. Effect of two randomised exercise programmes on bone mass of healthy postmenopausal women. *Br Med J* 1987;295:1441-1444.
8. Simkin A, Ayalon J, Leichter I. Increased trabecular bone density due to bone loading exercises in postmenopausal osteoporotic women. *Calcif Tissue Int* 1987;40:59-63.
9. Report of a WHO Scientific Group. *Research on menopause*. Geneva: World Health Organization, 1981 WHO Technical Report Series; vol #670.
10. Dalsky GP, Stocke KS, Ehsani AA, et al. Weight-bearing exercise training and lumbar bone mineral content in postmenopausal women. *Ann Int Med* 1988;108:824-828.
11. Prince RL, Smith M, Dick IM, et al. Prevention of postmenopausal osteoporosis: a comparative study of exercise, calcium supplementation and hormone-replacement therapy. *N Engl J Med* 1991;325:1189-1195.
12. Kohrt WM, Snead DB, Slatopolsky E, et al. Additive effects of weight-bearing exercise and estrogen on bone mineral density in older women. *J Bone Miner Res* 1995;10:1303-1311.
13. Notelovitz M, Martin D, Tesar R, et al. Estrogen therapy and variable-resistance weight training increase bone mineral in surgically menopausal women. *J Bone Miner Res* 1991;6:583-590.
14. Heikkinen J, Kyllonen E, Kurttila-Matero E, et al. HRT and exercise: effects on bone density, muscle strength and lipid metabolism. *Maturitas* 1997; 26:139-149.
15. Kohrt WM, Ehsani AA, Birge SJ. Effects of exercise involving predominantly either joint-reaction or ground-reaction forces on bone mineral density in older women. *J Bone Miner Res* 1997;12:1253-1261.
16. Cappozzo A. Compressive loads in the lumbar vertebral column during normal level walking. *J Orthop Res* 1984;1:292-301.
17. Nelson ME, Fiatarone MA, Morganti CM, et al. Effects of high-intensity strength training on multiple risk factors for osteoporotic fractures. A randomized control trial. *JAMA* 1994;272:1909-1914.
18. Pruitt LA, Jackson RD, Bartels RL, et al. Weight-training effects on bone mineral density in early postmenopausal women. *J Bone Miner Res* 1992;7:179-185.
19. Nichols JF, Nelson KP, Sartoris DJ. Bone mineral responses to high-intensity strength training in active older women. *Journal of Aging and Physical Activity* 1995;3:26-38.
20. Grove KA, Londeree BR. Bone density in postmenopausal women: High impact vs low impact exercise. *Med Sci Sports Exerc* 1992;24:1190-1194.
21. Welsh L, Rutherford OM. Hip bone mineral density is improved by high-impact aerobic exercise in postmenopausal women and men over 50 years. *Eur J Appl Physiol* 1996;74:511-517.
22. Nelson ME, Fisher EC, Dilmanian FA, et al. A 1-y walking program and increased dietary calcium in postmenopausal women: effects on bone. *Am J Clin Nutr* 1991;53:1304-1311.
23. Dalsky GP, Ehsani AA, Kleinheider KS, et al. Effect of exercise on lumbar bone density. *Gerontologist* 1986;26Supp:16A.
24. Bloomfield SA, Williams NI, Lamb DR, et al. Non-weightbearing exercise may increase lumbar spine bone mineral density in healthy postmenopausal women. *Am J Phys Med Rehabil* 1993;72:204-209.

25. Krølner B, Toft B, Nielsen SP, et al. Physical exercise as prophylaxis against involutional vertebral bone loss: a controlled trial. *Clin Sci* 1983;64:541-546.
26. Tsukahara N, Toda A, Goto J, et al. Cross-sectional and longitudinal studies on the effect of water exercise in controlling bone loss in Japanese postmenopausal women. *J Nutr Sci Vitaminology* 1994;40:37-47.
27. Bolotin HH, Sievänen H, Grashuis JL, et al. Inaccuracies inherent in patient-specific dual-energy X-ray absorptiometry bone mineral density measurements: Comprehensive phantom-based evaluation. *J Bone Miner Res* 2000;16:417-426.
28. Albanese CV, Civitelli R, Tibollo FG, et al. Endocrine and physical determinants of bone mass in late menopause. *Exp Clin Endocrinol Diabetes* 1996; 104:263-270.
29. Smith EL, Gilligan C, McAdam M, et al. Deterring bone loss by exercise intervention in premenopausal and postmenopausal women. *Calcif Tissue Int* 1989;44:312-321.
30. Cummings SR, Black DM, Nevitt MC, et al. Bone density at various sites for prediction of hip fractures. *Lancet* 1993;341:72-75.
31. Kerr D, Morton A, Dick I, et al. Exercise effects on bone mass in postmenopausal women are site-specific and load dependent. *J Bone Miner Res* 1996;11:218-225.
32. Heinonen A, Oja P, Sievänen H, et al. Effect of two training regimes on bone mineral density in healthy perimenopausal women: a randomised control trial. *J Bone Miner Res* 1998;13:483-490.
33. Umemura Y, Ishhiko T, Yamauchi T, et al. Five jumps per day increase bone mass and breaking force in rats. *J Bone Miner Res* 1997;12:1480-1485.
34. Rubin CT, Lanyon LE. Regulation of bone formation by applied dynamic loads. *J Bone Joint Surg* 1984;66-A:397-402.
35. Rubin CT, Lanyon LE. Regulation of bone mass by mechanical strain magnitude. *Calcif Tissue Int* 1985;37:411-417.
36. O'Connor PJ, Lanyon LE, Macfie H. The influence of strain rate on adaptive bone remodeling. *J Biomech* 1982;15:767-781.
37. Turner CH, Owan I, Takano Y. Mechanotransduction in bone: role of strain rate. *Am J Physiol* 1995;269 (Endocrinol Metab. 32):E438-E442.
38. Rubin CT. Skeletal strain and the functional significance of bone architecture. *Calcif Tissue Int* 1984;36:S11-S18.
39. Hatori M, Hasegawa A, Adachi H, et al. The effects of walking at the anaerobic threshold level on vertebral bone loss in postmenopausal women. *Calcif Tissue Int* 1993;52:411-414.
40. Turner RT. Mechanical signaling in the development of postmenopausal osteoporosis. *Lupus* 1999;8:388-92.
41. Nguyen TV, Sambrook PN, Kelly PJ, et al. Prediction of osteoporotic fractures by postural instability and bone density. *Br Med J* 1993;307:1111-1115.
42. Sinaki M, Wahner H, Offord K, et al. Efficacy of nonloading exercises in prevention of vertebral bone loss in postmenopausal women: a controlled trial. *Mayo Clin Proc* 1989;64:762-9.
43. Adami S, Gatti D, Braga V, et al. Site-specific effects of strength training on structure and geometry of ultradistal radius in postmenopausal women. *J Bone Miner Res* 1999;14:120-124.
44. Jarvinen TLN, Kannus P, Sievanen H. Have the DXA-based exercise studies seriously underestimated the effects of mechanical loading on bone? (Letter). *J Bone Miner Res* 1999;14:1634-1635.
45. Kannus P. Preventing osteoporosis, falls, and fractures among elderly people. *Br Med J* 1999;318:205-206.
46. Bassey EJ, Ramsdale SJ. Weight-bearing exercise and ground reaction force: a 12-month randomized controlled trial of effects on bone mineral density in healthy postmenopausal women. *Bone* 1995;16:469-476.
47. Lau EMC, Woo J, Leung PC, et al. The effects of calcium supplementation and exercise on bone density in elderly Chinese women. *Osteoporos Int* 1992;2:168-173.
48. Kerr D, Ackland T, Masalen B, et al. Resistance training over 2 years increases bone mass in calcium replete postmenopausal women. *J Bone Miner Res* 2001;16:175-181.
49. Cavanaugh DJ, Cann CE. Brisk walking does not stop bone loss in postmenopausal women. *Bone* 1988;9:201-204.
50. Martin D, Noteklovitz M. Effects of aerobic training on bone mineral density of postmenopausal women. *J Bone Miner Res* 1993;8:931-936.
51. Revel M, Mayoux-Benhamou MA, Rabourdin JP, et al. One-year psoas training can prevent lumbar bone loss in postmenopausal women: a randomized controlled trial. *Calcif Tissue Int* 1993; 53:307-311.
52. Tsukahara N, Toda A, Goto J, et al. Cross-sectional and longitudinal studies on the effect of water exercise in controlling bone loss in Japanese postmenopausal women. *J Nutr Sci Vitaminology* 1994; 40:37-47.
53. White MK. The effects of exercise on the bones of postmenopausal women. *Int Orthop* 1984;7:209-214.

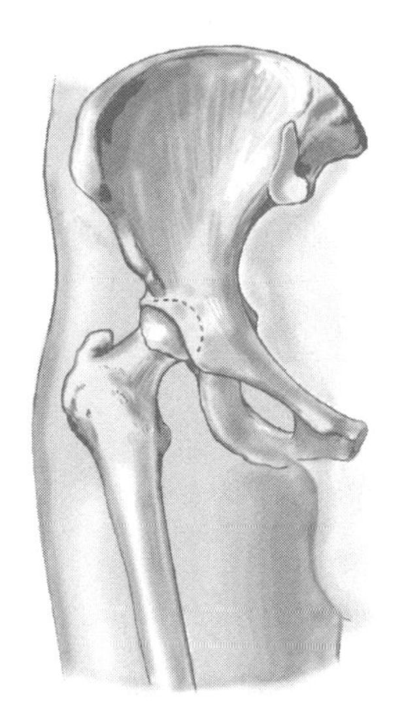

Chapter 13

Physical Activity, Targeted Bone Loading, and Bone Mineral in Men

The problem of osteoporosis in men has only recently gained the attention of more than a handful of clinicians and researchers. Historically, relatively few studies have examined the role of physical activity on bone mineral accrual or loss in men. The authors of a major textbook on osteoporosis reported that the data on physical activity and bone health in men are so limited "that it would not be prudent to draw conclusions about the effects of exercise on bone in the typical adult male"[1].

This lack of research attention is not justified, as the incidence of hip fractures rise exponentially with age in men as they do in women, albeit 5 to 10 years later [1]. Although aged-matched women are twice as likely as men to sustain osteoporosis-related hip fractures, the incidence of all fractures is higher in men than in women from adolescence until middle adulthood [2].

All of the lifestyle and biological factors that we reviewed in part II, including age, genetics, body mass, nutrition, hormones, and smoking, as well as physical activity, influence the bone mineral status of men. We remind you of the important role of the "nonactivity" determinants of bone mineral, as they can all potentially confound studies of physical activity and bone health.

In this chapter we discuss the studies of exercise and bone health in men according to study design and thus the quality of the evidence available. We begin with the few controlled exercise intervention studies and two longitudinal observational studies—one in runners, another in tennis players. Then we summarize the cross-sectional studies that have compared athletes and controls. While this has been the most popular type of study in this field, these studies may not be entirely reliable, as people with stronger bones may have self-selected into sport. We also review cross-sectional studies of populations in which physical activity is considered a determinant of bone mineral density.

These studies generally measure physical activity by questionnaire and then utilize the statistical method of regression modeling to examine the role of exercise. The last part of this chapter outlines a very practical, easy to implement exercise prescription for men who wish to gain bone mineral density.

Controlled Trials of Exercise and BMD

The only randomized, controlled trial in this field to date was part of a large "exercise and aging" study in men and women that studied the effect of exercise on various body systems [3]. In the study that focused on the skeleton, 50 men aged 66 years were randomized to either aerobic exercise, yoga and stretching, or control (table 13.1). All subjects had aerobic fitness and radial BMD measured by submaximal cycle ergometry and by SPA (single photon absorptiometry), respectively, at baseline and at four months (figure 13.1). After that time subjects from the yoga group and the control group were also prescribed the identical aerobic exercise program so that all three groups undertook aerobic exercise for four more months before further evaluation. All subjects were then given the option of six more months of aerobic exercise followed by final measurement. The aerobic exercise program consisted of a combination of stretching, cycle ergometry, arm ergometry, and walking three times per week for 60 minutes.

No subjects gained bone within the first four months, but a significant (p = 0.0009) increase was reported from four months to eight months (figure 13.2). Men who were randomized to exercise and who completed all 14 months of aerobic exercise training increased their radial

TABLE 13.1

Controlled intervention studies of exercise and measures of bone in men

Author	Study design	Outcome measures	Age (yrs)	No. of subjects	Findings
Blumenthal et al. 1991 [3]	Randomized controlled trial of aerobic exercise, yoga, and control	Aerobic fitness and BMC at baseline and 4 mo, 8 mo, and 14 mo	> 60 (mean 67)	50	BMD by SPA: significant increases in BMD associated with increases in aerobic fitness.
Fujimura et al. 1997 [25]	Controlled intervention study with 16-wk strength training intervention	Bone markers: serum OC and BAP, plasma PICP, urinary DPD, and total body and regional BMD	23–31	17	Resistance training increased bone formation while transiently suppressing bone resorption. No change in BMD.
Menkes et al. 1993 [4]	Controlled intervention study with 16-wk strength training exercise intervention	Total, spinal, and FN BMD by DXA	59 ± 2	11 subjects, 9 controls	Significant 3.8% increase in BMD of femoral neck compared with control group; 2% increase in BMD of lumbar spine (NS). No significant change in total body BMD. Osteocalcin increased by 19% after training. 26% increase in skeletal ALP after 16 wks of training.
Ryan et al. 1994 [5]	Controlled exercise intervention (16 wk); resistance training	BMD by DXA, bone markers	61 ± 1 yr	21 subjects, 16 controls	16-wk strength training intervention. 2.8 ± 1.6% increase in femoral neck BMD.
Yarasheski et al. 1997 [7]	Randomized, controlled intervention	DXA, (total body and regional), IGF by RIA osteocalcin	67 ± 1 yr	18 subjects, 18 controls	Short-term exercise intervention increased Ward's triangle BMD.

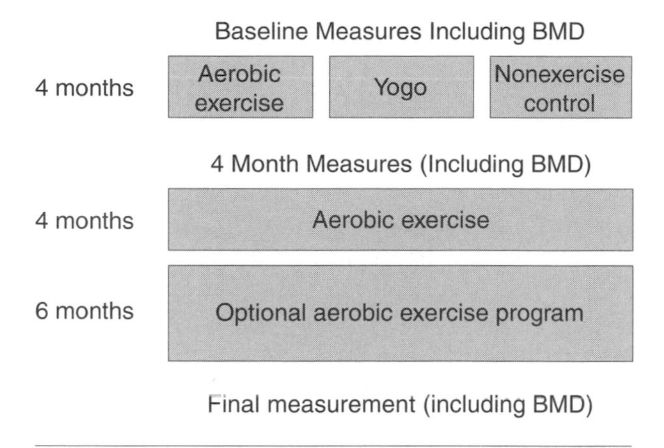

FIGURE 13.1 Study design in a randomized crossover trial of 66-year-old men performed at Duke University [3].

BMD by 19% [3]. The percent change in BMD was correlated with an increase in peak oxygen uptake (aerobic fitness) ($r = 0.35$, $p < 0.05$). On the other hand, men who were randomized to control initially, and who then completed only the four-month intervention and not the optional six-month training following, decreased BMD by 6% compared with baseline [3]. A strength of this study was that the men were normal, non-athletic, relatively sedentary older people.

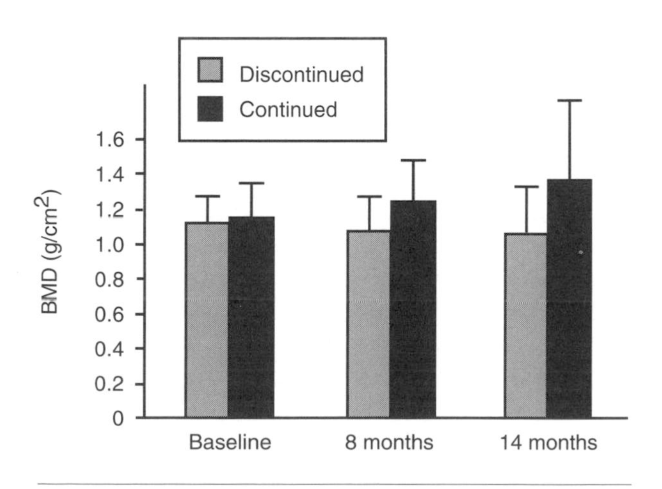

FIGURE 13.2 Changes in BMD at 8 months and 14 months in 66-year-old men who either continued (black bars) or dropped out (gray bars) of an aerobic exercise program.

The magnitude of these changes are remarkable given that much lower changes, if any, have been observed in other populations (children, women). One explanation might be the 4% greater baseline BMD of the men that continued in the exercise program as compared with those that did not. This might suggest a genetic predisposition to higher bone mass or the cumulative effect of positive factors throughout life that supported bone mineral accrual, slower rates of loss, or both.

Two controlled (but not randomized) exercise intervention studies evaluated the role of resistance training, not aerobic exercise [4, 5]. In the first study the 11 60-year-old volunteers performed chest presses, overhead presses, latissimus pull-downs (figure 13.3), upper back rows, leg presses, and leg extensions at 85% of 1-RM 3 times per week for 16 weeks.

Exercisers gained 45% in strength and 3.8% and 2.0% in BMD at the femoral neck and lumbar spine, respectively (figure 13.4). Total body BMD did not alter. Biochemical markers of bone formation—osteocalcin and skeletal alkaline phosphatase—increased by 19% and 26%, respectively (figure 13.5). Tartrate-resistant acid phosphatase—a measure of resorption—did not increase significantly. These data suggest a strength-training-induced gain in regional BMD arising from increased bone formation.

FIGURE 13.3 Latissimus pull-down.

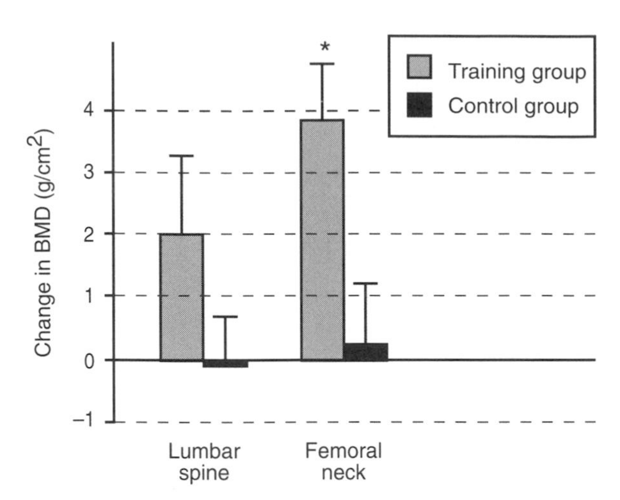

FIGURE 13.4 Strength training and BMD in men. The training group increased in BMD at both the lumbar spine and femoral neck while achieving a 45% strength gain [4].

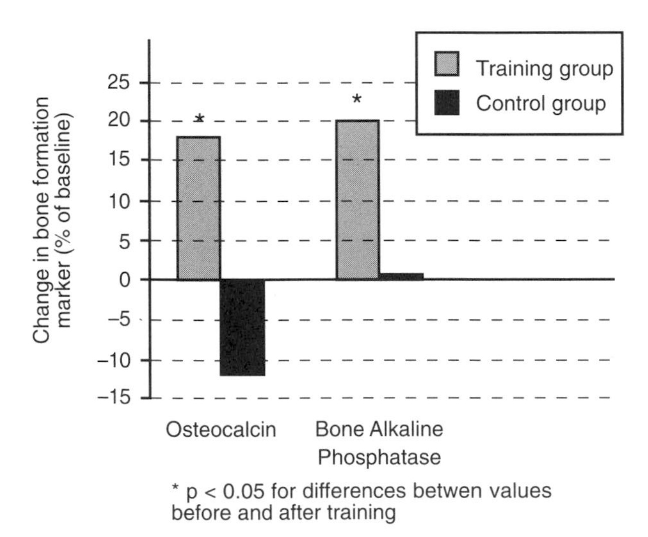

FIGURE 13.5 Markers of bone formation and strength training in 60-year-old men. Both osteocalcin and bone alkaline phosphatase increased in the training group but not in the control group [4].

Four months is a short intervention with respect to bone, since at the tissue level the lacunae formed by osteoclasts in the remodeling cycle (four months) are first refilled by nonmineralizing osteoid [6] (see chapter 2), which could result in short-term or transient increases in BMD as measured by photon absorptiometry. If this were the case, it could be argued that the regional increases in BMD reported by Menkes may represent a transient rise rather than a true structural adaptation as a result of the training. On the other hand, the lag time of the development of new osteoid [6] may have led to an underestimate of the benefits of the intervention.

The researchers from the University of Maryland then evaluated endocrine factors in a follow-up study [5] in a larger population. In 21 volunteer strength training subjects and 16 controls (mean age 60 years), a 16-week resistance training program identical to that described earlier resulted in a 2.8% increase in BMD at the femoral neck only. No changes occurred in growth hormone, IGF-I, or testosterone. In contrast with their earlier study, no changes in bone turnover markers occurred. The intervention augmented femoral neck BMD, but researchers were bemused by the inconsistency of other results. They speculated that lumbar spine BMD may not have changed because of insufficient loading of the region (subjects undertook one set of repetitions for the upper limb and spine exercises and two sets for the lower limb exercises) [5]. The authors noted that serum hormone levels may not adequately reflect the local bone tissue milieu, and thus the lack of association between serum hormone levels and BMD does not rule out the possibility that endocrine control mechanisms influence BMD. Furthermore the researchers were not able to isolate free (biologically active) IGF-I. They recommended that future research assess anabolic hormones before and after a longer-duration training program. The authors were unable to identify any specific factors to explain the discrepancy in findings of bone markers between their two studies [4, 5].

Yarasheski and colleagues [7] from Missouri, reported very similar results to Ryan and colleagues. A 16-week high-intensity resistance training program led to a significant ($p < 0.05$) increase in BMD at one of the subregions of the proximal femur without increases in serum GH or IGF-I [7].

Longitudinal Observational Studies

Four longitudinal observational studies of exercise and BMD in men have been conducted. We summarize the findings in those before we outline a recent study that prospectively examined fracture risk and sports activity using the co-twin model.

The first longitudinal observational study in men was performed in Hawaii in 1983 [8]. Single-photon absorptiometry was used to measure calcaneal BMC in 50-year-old marathon runners (n = 20) and controls at the beginning and end of a nine-month marathon training program (table 13.2). Previous level of activity can influence baseline bone mineral substantially, but none of these subjects had previous running experience. Runners who trained consistently for the marathon (> 16 km/wk in each of the nine months of training) had a significant increase in BMC over the controls; the increase was not significant for inconsistent runners (< 16 km/wk in at least one of the nine months of training). In the consistent runners, change in BMC and distance run were significantly correlated (r = 0.89).

Eleven years later, Hutchinson and colleagues [9] examined the effect of daily physical activity, rather than any specific sport, on calcaneal BMD (measured by SXA) in 26- to 51-year-old men. In a novel study design the subjects were limited to those with shoe size 9 to 10.5 to minimize anthropometric differences in bone structure and biomechanical issues of scaling and size. Calcaneal BMD measurements were undertaken at

TABLE 13.2

Longitudinal studies of exercise and measures of bone in men

Author	Study design	Outcome measures	Age (yrs)	No. of subjects	Findings
Bennell et al. 1997 [10]	Controlled trial	BMD at various sites and total body by DXA	17–26	68 athletes, 27 controls	Changes in BMD were independent of exercise status except at lumbar spine, where power athletes gained more than endurance athletes and controls.
Cohen et al. 1995 [28]	Controlled trial	Lumbar spine and hip BMD and BMC by DXA	19.5+/–2.4	17 oarsmen, 8 controls	7 months training was associated with 3% and 4% increase in BMD and BMC at the lumbar spine. No change at the hip.
Hutchinson et al. 1995 [9]	Longitudinal	Calcaneal BMD by SXA	26–51	35 subjects, 8 controls	Exercise subjects broken down into high loaders (HL) and low loaders (LL). BMD +/– 12% in HL than LL and BMD correlated with minutes of high-impact activity in high loaders. BMD not related to walking steps.
Kontulainen et al. 1999 [11]	Longitudinal (4-yr follow-up)	DXA	26 at baseline	13 subjects, 13 controls	The players' baseline mean 13-25% BMC difference was maintained despite clearly reduced training.
Michel et al. 1992 [29]	5-yr longitudinal	Lumbar spine BMD by QCT	55–77	14 runners, 14 controls	Both groups lost bone; changes in BMD correlated with changes in running (min/wk) (i.e., running tended to reduce age-related bone loss).
Williams et al. 1984 [8]	9-month longitudinal	Heel BMC	38–68	20 marathon runners, 10 controls	Runners with longer, more consistent distances gained more BMC than those with shorter, more inconsistent distances.

weeks 1 and 9 in both heels and at week 5 in the right heel only. The calcaneal BMD measure used for analysis was the mean of all calcaneal BMD measurements, which is not the traditional method of analyzing longitudinal data (see chapter 4).

The strength of the Hutchinson et al. study is in the prospective nature of the physical activity data collection. Subjects completed a detailed log of physical activity for nine weeks. Walking steps were measured with a digital stepmeter. Men were grouped according to the vertical ground reaction force of their daily physical activity. "High loaders" were subjects who performed activities that generated single-leg peak vertical ground reaction forces of twice their body weight or more. "Low loaders" reported exercise or daily activities that typically generated ground reaction forces of less than 1.5 times their body weight. Calcaneal BMD was 12% higher in high loaders than in low loaders at the end of the study, and in the former group, minutes of high-load exercise correlated with calcaneal BMD (figure 13.6). Walking steps did not correlate with BMD (figure 13.7).

This study provided evidence to support the concept that site-specific high-loading activity is a key factor in the relationship between physical activity and BMD. The differences reported were very likely due to longer-term habitual differences in activity, rather than to changes in activity during the nine-week period. In fact, the authors excluded any subject who had had a substantial change in daily exercise pattern during the study compared with six-month and one-year histories.

University of Melbourne researchers followed a group of track athletes between the ages of 17 and 27 years old prospectively for 12 months. Men in this age group have reached skeletal maturity. The study showed little change in BMD in power athletes, endurance athletes, or controls [10]. Bone turnover markers did not reveal a trend for increases in formation, or decreases in turnover, in these elite athletes who had a long history of training at a high level. These data suggest that bone mineral is relatively stable in men by the end of the second decade, a finding that is consistent with longitudinal studies in younger boys who are followed into adolescence (see chapter 10).

The most recent longitudinal observational study evaluated the effect of detraining on bone in adult (age 26 at baseline) elite male tennis players [11]. Researchers from the UKK Institute in Finland followed 13 male tennis players and 13 controls for four years. At baseline, the ten-

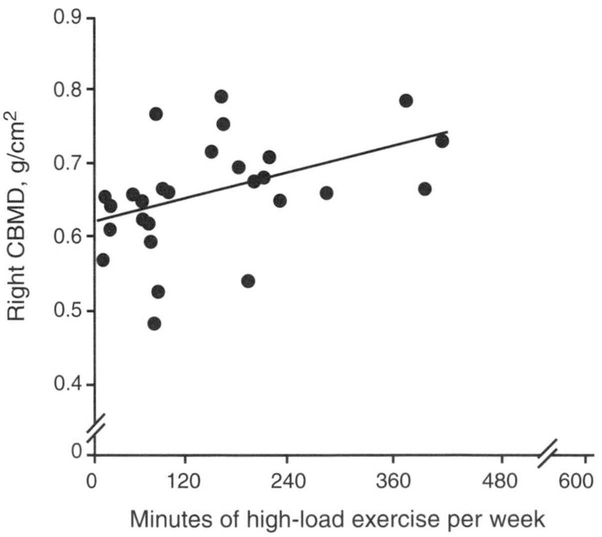

FIGURE 13.6 In high-loading subjects whose activities generated peak vertical ground reaction forces of twice their body weight or more, minutes of high-load exercise correlated with calcaneal BMD (r = 0.41, p < 0.03).

Reprinted, by permission, from TM Hutchinson, RT Whalen, TM Cleek, et al., 1995, "Factors in daily physical activity related to calcaneal mineral density in men," *Med Sci Sports Exerc* 27: 745-750. Copyright Lippincott Williams & Wilkins.

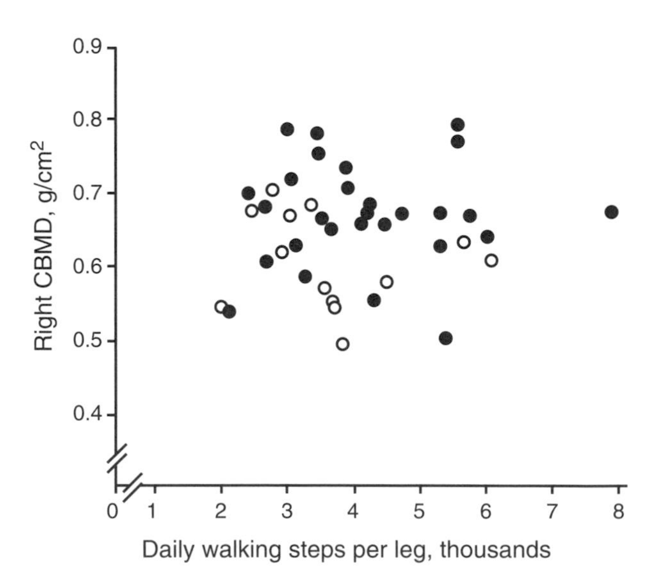

FIGURE 13.7 Calcaneal BMD and daily walking steps. There was no association between walking steps and BMD in either the high-loading or low-loading group, or all subjects combined.

Reprinted, by permission, from TM Hutchinson, RT Whalen, TM Cleek, et al., 1995, "Factors in daily physical activity related to calcaneal mineral density in men," *Med Sci Sports Exerc* 27: 745-750. Copyright Lippincott Williams & Wilkins.

nis players had significantly greater dominant–nondominant arm BMC differences than the controls. Differences in tennis players ranged from 13% to 25% depending on the upper limb site measured, whereas in controls the differences ranged from 1% to 5% (figure 13.8). A mean 2.3 years after baseline testing, all tennis players retired from high-level competition and halved their training frequency, decreasing mean training hours from 7.6 to 3.3 hours per week. Despite this decrease in loading, subjects maintained their side-to-side difference in BMC (figure 13.9) [11].

In women, detraining reduced the exercise-induced increase in bone mass to the pretraining values [12]. As Kontulainen and colleagues' study is the first in men, we can only speculate as to whether the BMC maintenance with detraining is due to a gender difference or, as the authors contend, to the fact that the men in this tennis study all started playing tennis at or before puberty [11]. Further long-term longitudinal studies are needed to support, or refute, these conclusions.

Biological research in twins provides many benefits in study design (see chapter 5). A recent Finnish twin cohort study showed prospectively that men participating in vigorous physical activity had 60% less risk of osteoporotic hip fracture than had men not participating (hazard ratio = 0.38, 95% Confidence Interval 0.16-0.91, p = 0.03) [13]. The magnitude of this protective benefit of exercise is similar to that seen in women [14, 15].

Cross-Sectional Studies of Athletes and Controls

Cross-sectional athlete–control studies have been reviewed elsewhere [16, 17] and are summarized in table 13.3. We highlight two cross-sectional studies that, despite the inherent limitations of the study design, provide useful data to base longitudinal and intervention studies on.

A study that showed enormous athlete–control differences in BMC was that of Virvidakis and colleagues from Greece [18]. In this cross-sectional study of 15- to 20-year-old male Olympic weightlifters, the mean distal and proximal forearm BMCs of subjects were 51% and 41% above the age-matched control means, respectively (see figure 10.6). This was substantially more than 2 standard deviations above the mean. In regression models, body weight and lifting record (maximal amount of weight that could be lifted) were the best predictors of BMC. These boys started training at an average age of 14 years

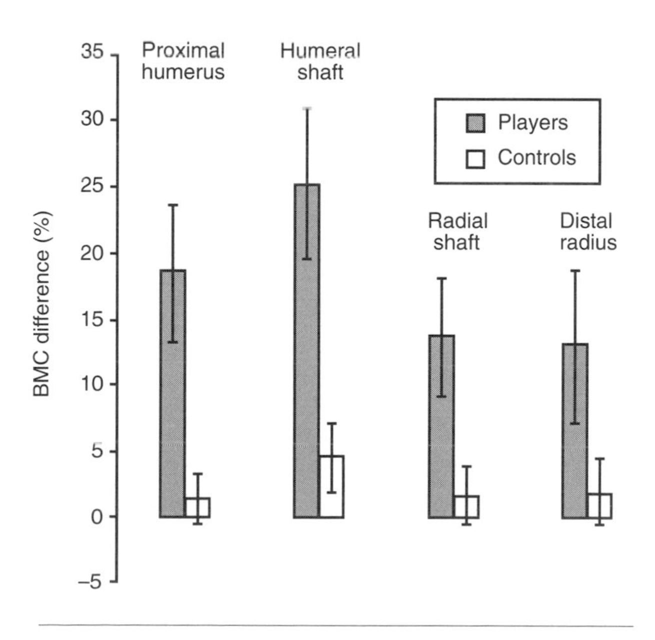

FIGURE 13.8 BMC of male tennis players and controls. Side-to-side difference in tennis players ranges from 13% to 25% depending on the site measured. In controls, the difference ranges from 1% to 5%.

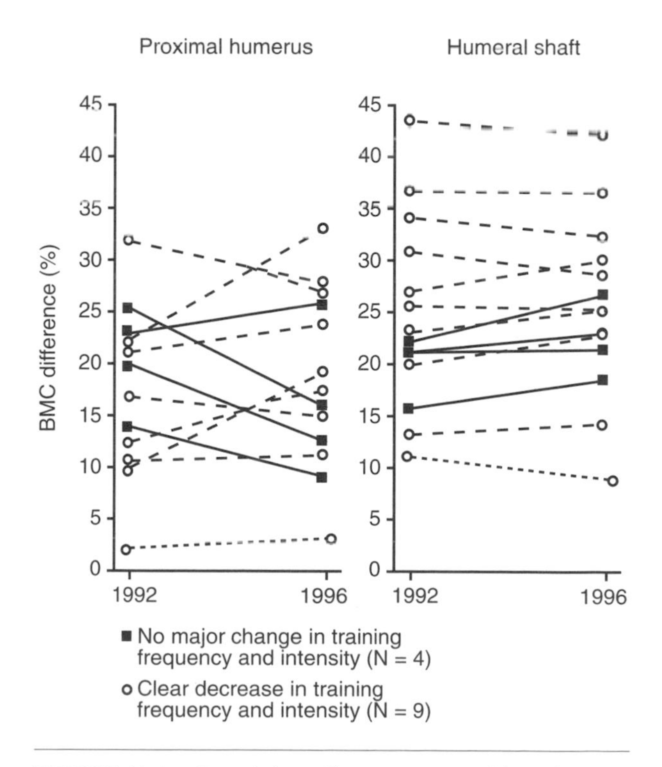

FIGURE 13.9 Detraining effect on BMC. Although players had ceased tennis on average more than two years earlier, BMC differences persisted.

TABLE 13.3

Cross-sectional studies of male athletes and controls in which BMD or BMC was measured

Authors	Group specification (no. in each group)	Age (yrs)	Years trained	Athletes/Controls (%) BMC	Athletes/Controls (%) BMD	Comments
Aloia et al. 1978 [31]	Marathon runners (30), normals (16)	30–50 (mean = 42 ± 1.4)		+4 by SPA at wrist		
Bilanin et al. 1989 [32]	Long-distance runners (13), nonrunners (11)	28 ± 4	7 ± 5	+3% at tibia and –10 at L2-4 by DPA; –1% at radius by SPA		
Block et al. 1989 [22]	Water polo players (20), weightlifters (19), controls (20)	25 ± 3			Water polo players and weightlifters had +10% T12 to L3 BMD by QCT and +9% hip by DPA.	
Colletti et al. 1989 [33]	Strength trainers (12), controls (50)	24 ± 3	>1		DPA: +11% at LS, +15% at trochanter, +16% at femoral neck, no difference at midradius.	
Conroy et al. 1993 [34]	Weightlifters (25), controls (11)	17.4 mean			13–30% higher spine and femoral neck BMD than age-matched controls.	Strength accounted for 30–65% of the variance.
Granhed et al. 1987 [35]	Weightlifters/ powerlifters (8)	28 ± 6		BMC at L3 in powerlifters 36% higher than in age-matched controls		Loads on L3 calculated to be between 19 and 36 kN.
Hamdy et al. 1994 [36]	Weightlifters (11), runners (12), cross-trainers (8), recreational exercisers (9)	19–42	> 3		From the total body scan: upper limb BMD greatest in weight trainers (+13% compared with runners). From the regional scans: no differences in proximal femur or lumbar spine BMD.	
Heinonen et al. 2001 [37]	Triple jumpers (8), controls (8)	22 ± 3	7.6 ± 3.6		Femoral neck BMD 31% greater in triple jumpers than controls. SPA: midradius +11%	

Huddleston et al. 1980 [21]	Tennis players (35)	70–84	25–72			Dominant/nondominant arm comparison.
Kannus et al. 1994 [30]	Tennis players (20), controls (20)	19–34	14	DXA: proximal humerus +14%; humeral shaft +25%; radius shaft +10%; distal radius +11%		Dominant/nondominant arm comparison; the corresponding differences in the controls were significantly lower (0–5%).
Lane et al. 1986 [38]	Long-distance runners (8), controls (8)	58			QCT: +44% at L1	
MacDougall et al. 1992 [39]	Runners (53), controls (22)	20–45			Lower leg BMD greater in moderate runners than controls but no greater in high mileage runners.	Cross-sectional area of tibia and fibula was greater in runners than controls.
Montoye & Taylor 1980 [40]	Tennis players (60)	64 ± 4	40 ± 16	SPA: radius +25%; ulna −1.4%; humerus +9.3%		
Nilsson & Westlin 1971 [41]	A variety of athletes in different sports[1] (88), nonexercising controls (15)	18–25			SPA: distal femur 27–47% in the various sporting groups	
Orwoll et al. 1989 [42]	Swimmers (58), nonexercising controls (78)	60 ± 13 (40–85 yr)	13 ± 11	SPA: proximal radius +4%	QCT: T12-L1 +14%	Positive associations were in men only, not in a parallel group of women swimmers.
Suominen 1988 [27]	Long-distance athletes (10), power athletes (8), population controls (52)	46–60	27 ± 10 for distance athletes; 34 ± 5 for power athletes		Calcaneus SPA: distance athletes +43%, power athletes +35%	
Virvidakis et al. 1990 [18]	Competitive weightlifters from 14 countries and age-matched Greek controls	15–20		SPA: distal forearm +51%; proximal forearm +41%		

[1]Weightlifters (n = 11), throwers (4), runners (25), soccer players (15), swimmers (9), exercising controls (24). Mean ± SD or range is given.

and had trained for an average of four years at the time of the study (personal communication from the author reported by Parfitt [19]).

Well before the advent of technology such as densitometry to measure bone mineral, Jones and colleagues [20] compared the X-rays of dominant and nondominant upper arms of competitive tennis players and provided circumstantial evidence (figure 13.10) that bone growth gains achieved during the key childhood years persist into adulthood [20]. Subjects were in their mid-20s at the time of this X-ray study, but the men had been playing tennis since age 9 and the women since age 10. The men had a mean between-arm difference in cortical cross-sectional area of 41%, of which 34% was present on the periosteal surface and 7% on the endosteal (figure 13.11). Results in women were in a similar vein, but of smaller magnitude. We agree with Parfitt who contends that "this difference is far greater than can be accounted for by handedness, or by periosteal reactions to repetitive trauma" [19].

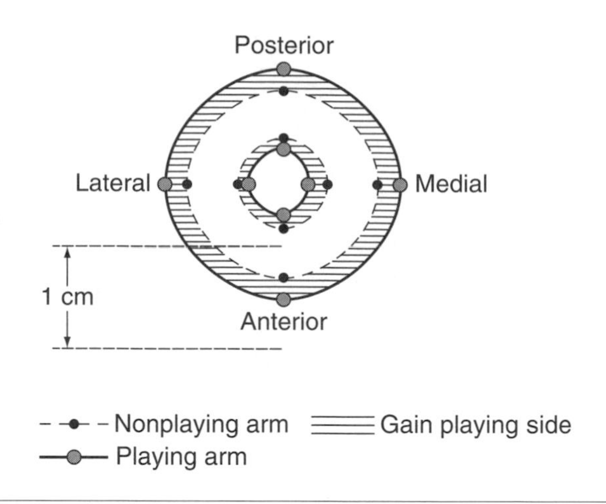

FIGURE 13.11 Pictorial representation of the gain in both periosteal bone (outer circumference) and endosteal bone (inner circumference) in the playing arm over the nonplaying arm of tennis players.

Reprinted, by permission, from HH Jones et al., 1997, "Humeral hypertrophy in response to exercise," *J Bone Joint Surg* 59(A):207.

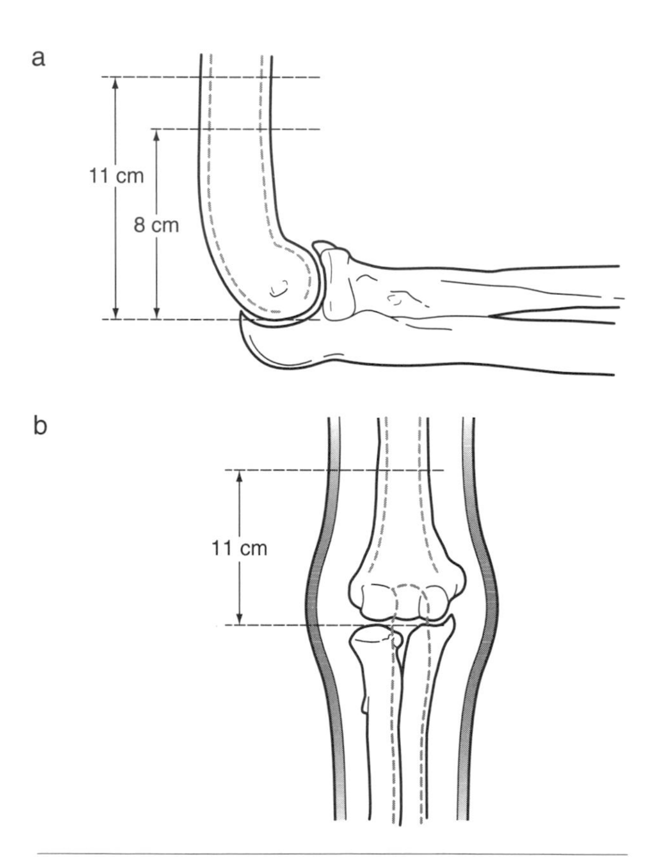

FIGURE 13.10 Line drawing representing the outline of bone as measured from X-rays in professional tennis players. *(a)* Lateral view demonstrating that the playing arm (solid line) is longer than the nonplaying arm (dotted line). *(b)* Anteroposterior view, demonstrating that the playing arm has much greater dimension than the nonplaying arm.

Reprinted, by permission, from HH Jones et al., 1977, "Humeral hypertrophy in response to exercise," *J Bone Joint Surg* 59(A):204.

One study that evaluated the association between sport and bone mineral in men aged 70-84 years [21] found that in lifetime tennis athletes the bone mass of the radius of the playing arm was 11% greater than that of the nonplaying arm and also greater than that of the dominant arms of nonathletes. Kannus and colleagues [30] found similar increases in radial bone gain in young adult male tennis players (age 25 years at measurement). These findings suggest that playing tennis during a lifetime may produce a clinically relevant site-specific increase in bone mass.

Studies Examining Physical Activity as a Determinant of BMD

In general, cross-sectional studies that evaluated physical activity as a determinant of bone mineral provide less valuable information than randomized, controlled intervention studies. However, the cross-sectional study design sometimes provides specific information that is not yet available from other sources. For example, in a study of young athletic men aged 18-30 years, Block [22] showed that paraspinous muscle area (as measured by CT) was the most robust predictor of BMD by QCT at the lumbar spine, and of BMD by DPA at the hip, explaining up to 27% of the BMD depending on the site. Exercise explained 18% of the variability of total spinal (i.e., combined cortical and trabecular bone) and trabecular BMD [22].

Cross-sectional studies can also be hypothesis generating; that is, they identify a question that requires a more sophisticated study design to be answered reliably. As an example, Michel and colleagues found that in men over 50 years, a strong positive correlation existed between exercise and lumbar spine bone density for those up to age 65 [23], but a few subjects with very high levels of exercise (> 400 min/wk) had low levels of BMD, suggesting that there may be a threshold at which exercise becomes deleterious.

Sone and colleagues [24] studied 965 Japanese men aged 27-83 years and reported a positive association between self-reported exercise in the past (none, sometimes, often) and lumbar spine and femoral neck BMD, explaining about 6-7% of the mean difference. Exercise at present explained 4% of the lumbar spine BMD only. This study suffered from several weaknesses, however. The measure of physical activity was not sophisticated, and the large age range of subjects prevented the authors from detecting age-dependent relationships, if any existed.

Mechanism of Bone Gain

Recent studies have used bone markers to evaluate the mechanism underpinning bone gain in men. A study that used weight training as the intervention demonstrated increased markers of bone formation and decreased resorption (figure 13.12) [25]. A study that compared the effect of either aerobic (40-60 minutes of running at 60-85% $\dot{V}O_2max$) or anaerobic (sprint training and leg press strength training) training found bone turnover marker changes consistent with a net increase in bone formation in the aerobic group. In contrast, anaerobic training was associated with an overall increase in bone turnover, with less pronounced increase in bone formation [26]. These preliminary data add weight to the argument that prescribing exercise is akin to prescribing medication.

Exercise Prescriptions

Far more data are available from studies of exercise prescription in women than in men. However, a study in 23- to 31-year-old Japanese men found that the following protocol led to changes in bone markers that suggested bone formation was occurring without an increase in bone resorption (table 13.4). In the four-month study, total body and regional BMD did not change [25].

In men aged 60 years who were measured to have normal hormone levels, Ryan and colleagues [5] found that a strength training program was effective in increasing BMD in the femoral neck, but not in other skeletal regions.

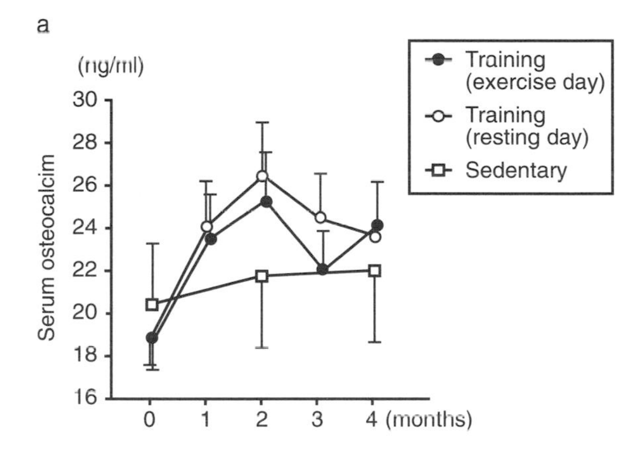

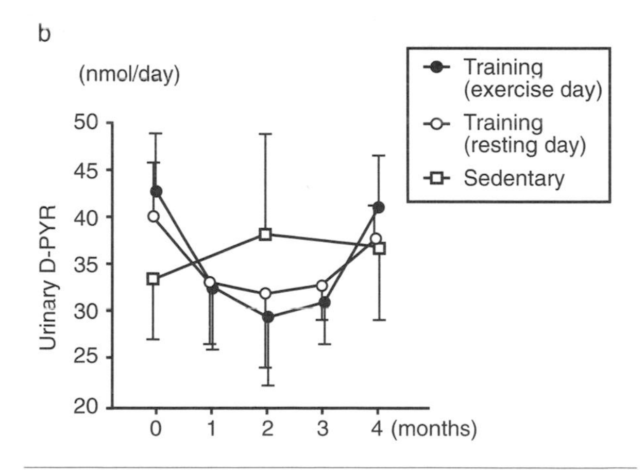

FIGURE 13.12 Biomarker changes with resistance training [25]. *(a)* Serum osteocalcin, a marker of bone formation, increased with four months of resistance training. *(b)* Urinary deoxypyridinoline, a marker of bone resorption, decreased in the exercising subjects.

TABLE 13.4

Exercise prescription (3 times/ wk) in young men that led to increase in biomarkers of bone formation without evidence of increased turnover

Exercise x/wk	Example 1-RM weight (lb/kg)	Weeks	Intensity (% 1-RM)	Reps	Sets
Leg extension (2)	60 lb/30 kg	1–4 5–16	60–80% 60–80%	10 10	2 3
Leg curl (2)	30 lb/15 kg	1–4 5–16	60–80% 60–80%	10 10	2 3
Bench press (3)	150 lb/70 kg	1–4 5–16	60–80% 60–80%	10 10	2 3
Sit-up (3)	50 lb/22 kg	1–4 5–16	60–80% 60–80%	10 10	2 3
Back extension (3)	30 lb/15 kg	1–4 5–16	60–80% 60–80%	10 10	2 3
Wrist curl (3)	N/A	1–4 5–16	60–80% 60–80%	10 10	2 3
Leg lunge (2)	N/A	1–4 5–16	60–80% 60–80%	10 10	2 3
Latissimus pull-down (1)	100 lb/45 kg	1–4 5–16	60–80% 60–80%	10 10	2 3
Arm curl (1)	50 lb/20 kg	1–4 5–16	60–80% 60–80%	10 10	2 3
Half squat (1)	150 lb/70 kg	1–4 5–16	60–80% 60–80%	10 10	2 3
Back press (1)		1–4 5–16	60–80% 60–80%	10 10	2 3

Free weights were used except for leg extension, leg curl, bench press, and latissimus pull-down, which were all done on a machine [25].

SUMMARY

- Far fewer studies of exercise intervention have been conducted in men than in women, but it appears that the skeleton responds similarly in both sexes.
- In men, high-impact training and strength training produce site-specific increases in BMD.
- Measures of bone turnover suggest that bone is accrued by increased new bone formation, rather than by reduced bone turnover (remodeling).
- Detraining may not be associated with rapid bone loss, at least not in elite athletes.
- Fracture rates are the ultimate test of bone health. A recent prospective study showed a 60% reduction in hip fracture rates in men who performed vigorous sporting activity compared with those who did not [13].

References

1. Orwoll ES, Klein RF. *Osteoporosis in Men. Epidemiology, pathophysiology and clinical characterization.* In: Marcus R, Feldman D, Kelsey J, ed. Osteoporosis. 3rd ed. San Diego: Academic Press, 1996: 745-784.
2. Melton LJ, Cummings SR. Heterogeneity of age-related fractures: implications for epidemiology. *Bone Miner* 1987;2:321-31.
3. Blumenthal JA, Emery CF, Madden DJ, et al. Effects of exercise training on bone density in older men and women. *J Am Gerentol Soc* 1991;39:1065-1070.
4. Menkes A, Mazel S, Redmond RA, et al. Strength training increases regional bone mineral density and bone remodeling in middle-aged and older men. *J App Physiol* 1993;74:2478-2484.
5. Ryan AS, Treuth MS, Rubin MA, et al. Effects of strength training on bone mineral density: Hormonal and bone turnover relationships. *J Appl Physiol* 1994;77:1678-1684.
6. Frost HM. Some effects of the basic multicellular unit-based remodeling on photon absorptiometry of trabecular bone. *Bone Miner* 1989;7:47-65.
7. Yarasheski KE, Campbell JA, Kohrt WM. Effect of resistance exercise and growth hormone on bone density in older men. *Clin Endocrinol* 1997;47:223-239.
8. Williams JA, Wagner J, Wasnich R, et al. The effect of long-distance running upon appendicular bone mineral content. *Med Sci Sports Exerc* 1984;16:223-227.
9. Hutchinson TM, Whalen RT, Cleek TM, et al. Factors in daily physical activity related to calcaneal mineral density in men. *Med Sci Sports Exerc* 1995;27:745-750.
10. Bennell KL, Malcolm SA, Khan KM, et al. Bone mass and bone turnover in power athletes, endurance athletes and controls: A 12 month longitudinal study. *Bone* 1997;20:477-484.
11. Kontulainen S, Kannus P, Haapasalo H, et al. Changes in bone mineral content with decreased training in competitive young adult tennis players and controls: a prospective 4-year follow-up. *Med Sci Sports Exerc* 1999;31:646-652.
12. Dalsky GP, Stocke KS, Ehsani AA, et al. Weight-bearing exercise training and lumbar bone mineral content in postmenopausal women. *Ann Int Med* 1988;108:824-828.
13. Kujala UM, Kaprio J, Kannus P, et al. Physical activity and osteoporotic hip fracture risk in men. *Arch Intern Med* 2000;160:705-8.
14. Joakimsen RM, Magnus JH, Fonnebo V. Physical activity and predisposition for hip fractures: A review. *Osteoporos Int* 1997;7:503-513.
15. Gregg EW, Cauley JA, Seeley DG, et al. Physical activity and osteoporotic fracture risk in older women. Study of Osteoporotic Fractures Research Group. *Ann Intern Med* 1998;129:81-8.
16. Suominen H. Bone mineral density and long term exercise. An overview of cross-sectional athlete studies. *Sports Med* 1993;16:316-30.
17. Chilibeck PD, Sale DG, Webber CE. Exercise and bone mineral density. *Sports Med* 1995;19:103-122.
18. Virvidakis K, Georgiou E, Korkotsidis A, et al. Bone mineral content of junior competitive weightlifters. *Int J Sports Med* 1990;11:244-246.
19. Parfitt AM. The two faces of growth: benefits and risks to bone integrity. *Osteoporos Int* 1994;4:382-398.
20. Jones HH, Priest JD, Hayes WC, et al. Humeral hypertrophy in response to exercise. *J Bone Joint Surg* 1977;59(A):204-208.
21. Huddleston AL, Rockwell D, Kulund DN, et al. Bone mass in lifetime tennis athletes. *JAMA* 1980;244:1107-1109.
22. Block JE, Friedlander A, Brooks GA, et al. Determinants of bone density among athletes engaged in weight-bearing and non-weight-bearing activity. *J App Physiol* 1989;67:1100-1105.
23. Michel BA, Bloch DA, Fries JF. Weight-bearing exercise, overexercise, and lumbar bone density over age 50 years. *Arch Intern Med* 1989;149:2325-2329.
24. Sone T, Miyake M, Takeda N, et al. Influence of exercise and degenerative vertebral changes on BMD: a cross-sectional study in Japanese men. *Gerontology* 1996;42 Suppl 1:57-66.
25. Fujimura R, Ashizawa N, Watanabe M, et al. Effect of resistance exercise training on bone formation and resorption in young male subjects assessed by biomarkers of bone metabolism. *J Bone Miner Res* 1997;12:656-662.
26. Woitge HW, Friedmann B, Suttner S, et al. Changes in bone turnover induced by aerobic and anaerobic exercise in young males. *J Bone Miner Res* 1998;13:1797-804.
27. Suominen H. Physical activity and bone. *Ann Chir Gynaecol* 1988;77:184-188.
28. Cohen B, Millett PJ, Mist B, et al. Effect of exercise training programme on bone mineral density in novice college rowers. *Br J Sports Med* 1995; 29:85-88.
29. Michel BA, Lane NE, Bjorkengren A, et al. Impact of running on lumbar bone density: a 5-year longitudinal study. *J Rheumatol* 1992;19:1759-1763.
30. Kannus P, Haapasalo H, Sievänen H, et al. The site-specific effects of long-term unilateral activity on bone mineral density and content. *Bone* 1994;15:279-284.
31. Aloia JF, Cohn SH, Abesamis C, et al. Skeletal mass and body composition in runners. *Metabolism* 1978;27:1793-1796.

32. Bilanin JE, Blanchard MS, Russek-Cohen E. Lower vertebral bone density in male long distance runners. *Med Sci Sports Exerc* 1989;21:66-70.

33. Colletti AL, Edwards J, Gordon L, etal. The effects of muscle building exercise on bone mineral density of the radius, spine and hip in young men. *Calcif Tissue Int* 1989;45:12-14.

34. Conroy BP, Kraemer WJ, Maresh CM, et al. Bone mineral density in elite junior Olympic weightlifters. *Med Sci Sports Exerc* 1993;25:1103-1109.

35. Granhed H, Jonson R, Hansson T. The loads on the lumbar spine during extreme weight lifting. *Spine* 1987;12:146-149.

36. Hamdy RC, Anderson JS, Ke KEW, et al. Regional differences in bone density of young men involved in different exercises. *Med Sci Sports Exerc* 1994;26:884-888.

37. Heinonen A, Sievänen H, Kyrolainen H, et al. Mineral mass, bone size and estimated mechanical strength of lower limb in triple jumpers. *Osteoporos Int* 2001;(in press).

38. Lane NE, Bloch DA, Jones HH, et al. Long distance running, bone density, and osteoarthritis. *JAMA* 1986;255:1147-1151.

39. MacDougall JD, Webber CE, Martin J, et al. Relationship among running mileage, bone density, and serum testosterone in male runners. *J Appl Physiol* 1992;73:1165-1170.

40. Montoye HJ, Taylor HL. Measurement of physical activity in population studies: a review. *Human Biol* 1984;56:192-216.

41. Nilsson BE, Westlin NE. Bone density in athletes. *Clinics in Orthopaedic and Related Research* 1971;77:179-182.

42. Orwoll ES, Ferar J, Oviatt SK, et al. The relationship of swimming exercise to bone mass in men and women. *Arch Intern Med* 1989;149:2197-2200.

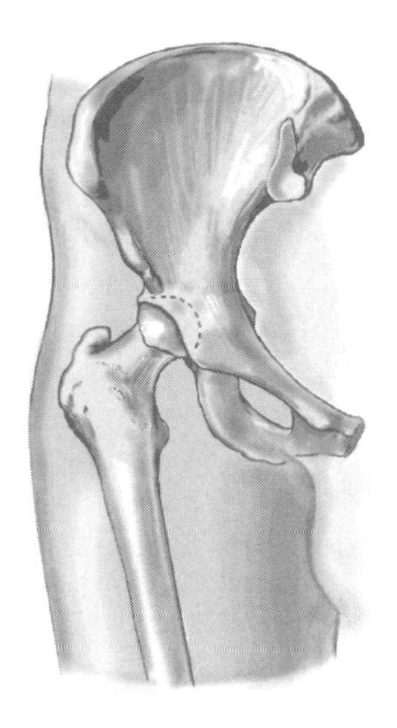

Chapter 14

Exercise and Fall Prevention

with Nick Carter

A book on bone health and physical activity would be incomplete without a chapter focusing specifically on exercise as a measure to prevent falls. Falls are an increasingly important cause of serious fracture and its corollaries, impairment, disability, and in a substantial number of cases, death [37]. Fall-related fractures can also occur in people with normal bone, so the problem is not restricted to those with osteoporosis (see chapter 15).

As exercise may provide one important part of the solution to this major public health problem, this chapter provides both the exercise scientist and the exercise therapist with a summary of the physiology of falling and the evidence for exercise prescription in fall prevention. The chapter concludes with examples of exercise prescriptions that have been proven effective in reducing falls.

Fall-related injuries and the resulting deaths in older adults are a major, and increasing, health problem worldwide [1-3, 37]. Approximately 30% of individuals over 65 years of age fall at least once per year and about half of these may do so recurrently [1, 3, 4]. Nonfatal falls may result in individuals, considerably reducing their activity for fear of future falls. Almost half of fallers are unable to get up without help [5].

A fall may result in a fracture, particularly in an older person. About 90% of hip fractures result from falls [6] and in individuals who sustain hip fracture, the outcome is fatal in 12-20% of cases [7]. In nonfatal cases, long-standing pain, disability, and functional impairment often ensue with tremendous socioeconomic consequences. Annual costs associated with fall-related fractures in the US are estimated at $10 billion [8]. Furthermore, the incidence of hip fractures continues to rise steadily, even using age-adjusted figures [1, 2]. Regular exercise has been proposed as one method of preventing falls and fall-related fractures in older adults [9].

Falling, Fracture, and Age-Related Physiological Changes Among Older Adults

The incidence of falls increases with age [9] because of the deterioration of the three sensory systems that control posture: vestibular, visual, and somatosensory (figure 14.1). The vestibular system provides input as to the head position in relation to gravity, and it also senses how fast, and in which direction, the head is accelerating. The visual system provides information about the body's location relative to its environment. The somatosensory system provides information about the body's position in space (proprioception) and detects movement (kinesthesia). With aging, the sensory cells in the otolith of the ear, joint position sense, and peripheral vision (important in sway stabilization) all deteriorate.

Age-related changes in muscle and bone also predispose people to both falls and fractures. Muscle strength and mass decline 30-50% between the ages of 30 and 80 [10]. Normal muscle mass and function prevent instability and correct imbalance and may protect the proximal femur by attenuating impact forces on the hip that occur in sideways falls in older adults [11].

There are in excess of 130 different risk factors for falling [12]. The simplest way to categorize risk factors for falling is to divide them into intrinsic, host factors (increased personal liability to fall) and extrinsic, environmental factors (increased opportunity to fall) [13]. This distinction is important because it addresses the extrinsic factors that have received little attention in medical strategies to prevent falls [14].

The act of falling comprises three stages: initiation, descent, and impact. As different factors can act at each stage of the fall process, this categorization provides several areas of focus for medical intervention. This approach has, for example, led to some researchers focusing on hip

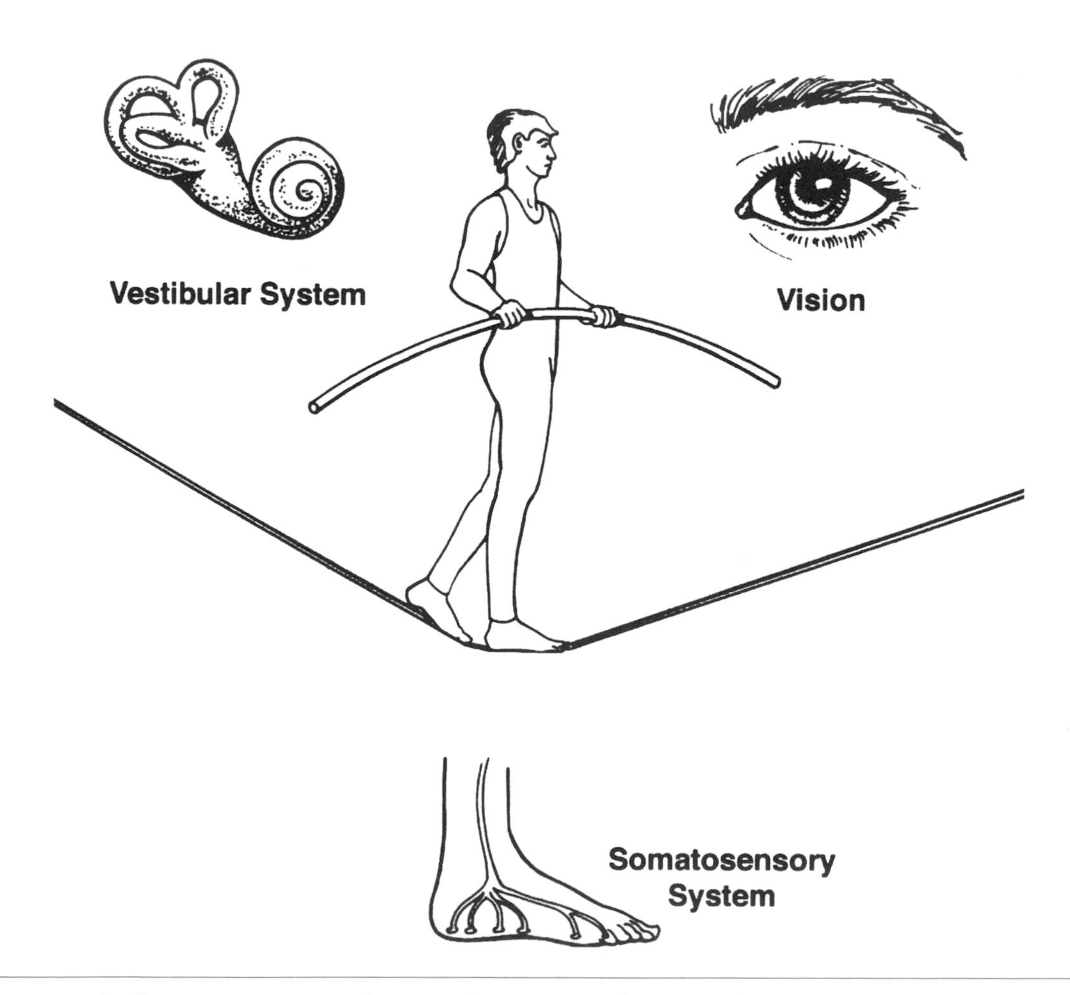

FIGURE 14.1 The three sensory systems that control posture: vestibular, visual, and somatosensary.

Finer Point: Impairment Versus Disability

In view of the complexity of the interaction among risk factors for falling, we advocate using the rehabilitation model of impairment and disability. The World Health Organization's definition of impairment is any loss or abnormality of psychological, physiological, or anatomical structure or function. Disability is defined as any restriction or lack of ability (resulting from impairment) to perform an activity in the manner or within the range considered normal for a human being [16]. Stroke, for example, increases the risk of falling, but it is muscle weakness or sensory loss (impairment) and poor balance (disability) resulting from the stroke that produces the increased risk, not the stroke itself. The terms *impairment* and *disability* are widely used in rehabilitation medicine, and they are particularly useful for understanding the mechanisms of falling.

protectors to reduce fall impact and thus fracture risk [9, 15].

A strength of the rehabilitation model outlined in the Finer Point sidebar is that it reveals multiple sites for interventions that may reverse fall risk factors. Clearly reversible risk factors (e.g., multiple drug therapy) can be attended to directly. The rehabilitation model indicates that irreversible risk factors (e.g., stroke, osteoarthritis) can theoretically be tackled at the impairment level (therapy to improve strength and proprioception) and at the disability level (therapy to improve gait). Thus, risk of falling may be modifiable, even though the underlying medical condition may not be.

Using this model, table 14.1 summarizes the results of studies that have documented the specific impairments and disabilities that predispose people to falls, and the relative risk for falling that each risk factor imparts. The broad range of risk may be due to the consideration of continuous variables (e.g., strength) as dichotomous (normal or abnormal) without defining what constitutes abnormal. Figure 14.2 schematically represents an integrated model for the interaction between intrinsic impairments and disabilities associated with aging and disease, and extrinsic and environmental hazards in the initiation of falls and fractures.

TABLE 14.1

Impairments and disabilities as risk factors for falls

Risk factor for falling	Relative risk for falls reported in various studies (values > 1 represent increased risk)
Impairment	
Lower limb strength	0.5–10.3
Upper limb strength	1.5–4.3
Lower limb range of motion	1.9
Sensation	0.6–5.0
Vestibular function	4.0
Vision	1.3–1.6
Cognition	1.2–5.0
Disability	
Static balance	1.5–4.1
Dynamic balance/gait	1.6–3.3

Can Exercise Decrease the Incidence of Falls?

This question has two parts, as exercise may reduce fall risk or reduce the incidence of actual falls. Let us look at the data for fall risk first. Figure 14.2 illustrates the numerous areas in which intervention strategies may modify risk factors for falls. Exercise intervention can alter intrinsic risk factors for falling (table 14.2). Strength, flexibility, balance, and reaction time—the factors most amenable to modification—should be addressed in exercise intervention trials to lower fall risk in the elderly.

Multifactorial hazard reduction interventions have been effective in reducing falls [14,

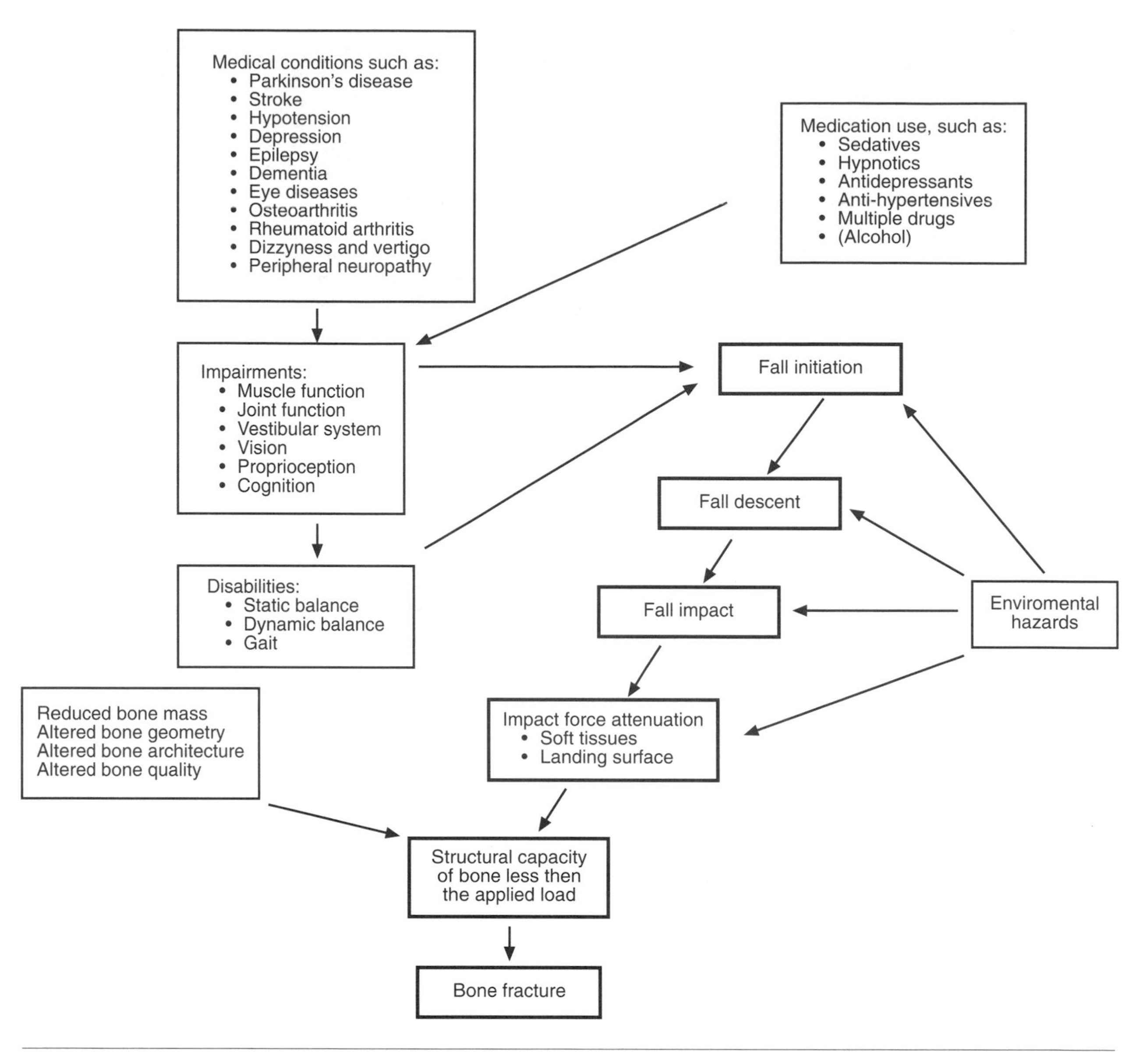

FIGURE 14.2 An integrated model for the interaction between various factors that influence fall risk. Intervention at the levels of Impairment and Disabilities can be effective.

TABLE 14.2

Intervention studies of exercise to modify risk factors for falls

Risk factor for falling	Average % improvement (range between studies)
Muscle strength	6 to 174
Range of motion	0.5 to 18
Balance	–7 to 53
Gait	12 to 48
Reaction time	0 to 4

18, 20], as have interventions aimed at reducing the numbers of medications elderly people use [5]. The results of a systematic review [5] of 13 randomized, controlled studies of subjects 60 years or older that had falls as an outcome and exercise as an intervention are summarized in table 14.3.

Table 14.3 reveals that the three oldest studies did not find exercise to reduce the risk of falling in older adults, while the eight more recent studies (since Wolf et al. 1996) [17] confirmed the value of exercise in fall prevention. Four of them demonstrated a significant reduction in falls [17-20] while in the remaining

TABLE 14.3

Summary of the randomized controlled trials that included exercise as an independently analyzed part of the trial in reducing or delaying falls in the older adults

Author	Subjects, age	Intervention	Falls outcome
Reinsch et al. 1992 [31]	n = 230 (C) age = 74	(In)—3 groups: exercise (n = 57), exercise/cognition (n = 72), cognition/behavioral (n = 51) *Exercise*—60 min, 3×/wk, 12 mo. Stand-up, step-up, stretching, and movement to music. *Cognition/behavioral*—health and safety curriculum to prevent falls, relaxation, video games. *Exercise/cognition*—2×/wk exercise, 1×/wk cognition (CT)—(n = 50).	(In)—exercise = 24.7%, cognition = 19.1%, ex/cog = 37.1%, (Ct) = 19.1%, (NS).
MacRae et al. 1994 [32]	n = 80 (C) age > 69	(In)—(n = 42) stand-up/step-up routine progressing to 4 sets of 10 reps 60 min 3×/wk. (Ct)—(n = 38) hourly meeting each wk focusing on health promotion and safety education.	Fallers in 12-mo period: (In) = 36%, (Ct) = 45% (NS).
Mulrow et al. 1994 [33]	n = 194 (I) age > 60	(In)—(n = 97) individually tailored one-one physiotherapy sessions 3×/wk for 4 mo, including range of motion, strength, balance, transfer, and mobility. Each session 30-40 min. (Ct)—(n = 97) same frequency friendly visits.	Total number of falls: (In) = 79, (Ct) = 60 (NS), Subjects with falls (%): (In) = 43, (Ct) = 37 (NS).
Lord et al. 1995 [34]	n = 197 (C) age = 72	(In)—(n = 100) 60-min exercise sessions, twice weekly in 4 terms of 10-12 wk. Four sections per session: warm-up, conditioning (aerobic, strength, balance, and flexibility), stretching and relaxation. (Ct)—(n = 97).	One or more falls: (In) = 34.7%, (Ct) = 35.1% (NS), Two or more falls: (In) = 10.7%, (Ct) = 12.8% (NS).
Wolf et al. 1996 [17]	n = 200 (C) age = 80	(In)—2 groups: Tai Chi (TC) (n = 72) 15 min twice daily at home for 4 mo. Computerized balance training (BT)—(n = 64). (Ct)—(n = 64) education sessions once per week.	Risk ratio of time to one or more falls as compared with controls: (TC) – 0.525 (47.5% reduction in fall incidence) ($p < 0.01$). (BT) = 1.136 (NS).
Campbell et al. 1997 [18]	n = 232 (C) age = 80	(In)—(n = 116) individually tailored program of exercise. Physiotherapist visited 4 × in first 2 mo. Exercises 3×/wk, 30 min each, lower limb strength and balance plus encouraged walking outside 3×/wk. (Ct)—(n = 116) equal care and frequent social visits.	Total falls: (In) = 88, (Ct) = 152. Rate of falls per year: (In) = 0.87 [SD 1.29], (Ct) = 1.34 [SD 1.93], difference 0.47 ($p < 0.05$).
Buchner et al. 1997 [19]	n = 100 (C) age = 78	All subjects had at least mild deficits in strength or balance. (In)—3 groups: (S) strength and flexibility (n = 25), (E) endurance and flexibility (n = 25), (SE) strength and endurance (n = 25). Strength training: upper and lower limb. 3 sessions/wk for 60 min. Endurance training: stationary cycle 75% max. heart rate. (Ct)—(n = 25).	3 intervention groups analyzed as one group—falls in the first year: (In) = 42%, (Ct) = 60%. Relative risk in the intervention group 0.53 ($p < 0.05$).
McMurdo et al. 1997 [35]	n = 118 (C) age = 65	(In)—45-min weightbearing exercise to music, 3×/wk for 3 × 10-wk terms for 2 yr. (Ct)—calcium supplementation (1000 mg/day).	Falls: (In) = 15, (Ct) = 31 (NS at 2 years). Difference between groups from 12 to 18 months ($p = 0.011$).
Campbell et al. 1999 [21]	n = 93 (C) age > 65	(In)—3 groups: gradual psychotropic withdrawl over 14 wk plus home-based program of exercises (see Campbell 1997) (n = 24), drug withdrawal	Subjects with falls: (In)—drug = 30% ($p = 0.05$), exercise = 39% (NS),

(continued)

TABLE 14.3 *(continued)*

Author	Subjects, age	Intervention	Falls outcome
		only (n = 24), exercise only (n = 21). Groups analyzed as exercise or drug withdrawal. (Ct)—(n = 22).	(Ct) = 51%.
Campbell et al. 1999 [20]	n = 152 (C) age = 84	2-yr follow-up of the above noted 12-mo study [18]. (In)—(n = 71) individually tailored program of exercise. Physiotherapist visited 4 × in first 2 mo of the original study. Exercises 3×/wk, 30 min each, lower limb strength and balance plus encouraged walking outside 3×/wk. (Ct)—(n = 81) equal care and frequent social visits.	Total falls over 2 years: (In) = 138, (Ct) = 220. Relative hazard for falls for the exercise group at 2 years = 0.69 (95% confidence interval [CI] for intervention group compared with controls 0.49, 0.97).
Steinberg et al. 2000 [23]	n = 252 (C) 75% aged 50–74 25% aged > 75	12 mo follow-up. (In)—3 groups: Exercise to improve balance and strength, frequency and duration of exercises not defined (n = 69). Home safety advice to modify environmental hazards (n = 61). Medical assessment to optimize health (n = 59). (Ct)—(n = 63) education and awareness of fall risk factors.	Fall events per 100 person months. (In) (exercise) = 6.37, (Ct) = 7.05. Time to first fall, adjusted hazard ratio 0.67 (95% CI 0.42, 1.07).
Rubenstein et al. 2000 [24]	n = 59 (C) age = 75	12 wk follow-up. (In)—(n = 31) strength, endurance, mobility, and balance training for 90 min, 3×/wk for 12 wk, (Ct)—(n = 28) usual activities for the follow-up period.	(In) 38.7% reported falling, (Ct) 32.1% reported falling (NS). Falls adjusted for activity (In) 6/1000 hours activity, (Ct) 16.2 (p < 0.05).
Lehtola et al. 2001 [36]	n = 131 (C) age = 79–75	Additional 4 mo follow-up after 6 mo intervention. (In)—(n = 92) an exercise class including Tai Chi once weekly plus walking with sticks, and home exercises each at least 3×/wk for 6 mo. Ct)—(n = 39) usual activities for the follow-up period.	Relative hazard for falls for the exercise group in 10 mo = 0.60 (95% CI for [In] compared with [Ct] 0.43, 0.84).

n = number of subjects; I = institution-dwelling; C = community-dwelling; In = intervention group; Ct = control group; p = significance level; NS = not significant.

four [21-24] a numerical reduction was evident but statistically not significant.

In the Wolf et al. study [17], a program of tai chi resulted in a 48% reduction of falls in participants (mean age 76 years) compared with controls. Such a reduction was not seen in the subjects who followed a computerized balance training program. It is of interest that while the computerized balance training group developed greater stability on balance platform measures, the tai chi group showed little change in this parameter [25].

In the study of Campbell and colleagues [18] a physiotherapist-led, but individualized, program of predominantly lower limb strength and balance exercises for 30 minutes, three times per week plus additional walking resulted in a significantly reduced annual rate of falls among women aged 80 years and older, compared with controls. After one year, the exercise group had 32% less likelihood of having 4 falls compared with controls. The benefit of exercise for the reduction of falls continued in the two-year follow-up [20].

Buchner and colleagues [19] in turn, reported that in 75 community-dwelling elderly subjects who underwent strength, endurance, and flexibility training, fewer persons fell in the first year (42%) compared with controls (60%) ($p < 0.05$). These data were originally presented comparing three exercise groups, each with 25 subjects (strength and flexibility, endurance and flexibility, and strength and endurance) with controls as part of the Frailties and Injuries: Cooperative Studies of Intervention Techniques (FICSIT) meta-analysis (in which seven independent, randomized, controlled trials assessed intervention efficacy in reducing falls) [26].

Analysis by these individual groups did not demonstrate a significant reduction in the incidence of falls.

The meta-analysis of the seven FICSIT trials showed a reduction in the falls for treatment arms including exercise (10% fewer falls) and balance (17% fewer falls)[26]. Repeat meta-analysis, however, excluding interventions with a nonexercise component, revealed that although the effects of balance training remained (24% fewer falls), the pooled estimate for overall exercise became nonsignificant at 13% fewer falls. There was no significant effect of the other exercise domains (resistance, endurance, and flexibility) on the incidence ratios for falls [26].

In five studies in addition to the 13 studies described, the exercise intervention arm combined exercise with the correction of intrinsic risk factors (smoking/alcohol/nutrition [27]), drug treatment [14] or extrinsic risk factors (environmental hazards [27-30]). As the effect of exercise cannot be separated from the other components of the multifactorial intervention, these studies cannot be analyzed further in terms of exercise and fall prevention.

Guidelines for Exercise Prescription

Table 14.4 summarizes the exercise dimensions in the positive and negative outcome studies of exercise in the prevention of falls in older adults. Clearly, the paucity of data and the limited power of studies undertaken to date preclude any conclusions from being drawn about the most appropriate dimensions of exercise prescription for fall prevention. Exercise interventions in the meta-analysis of the seven FICSIT trials, however, pooled estimates of the effect of individual training types across the studies. Pooling indicated lower fall incidence ratios as a result of balance, resistance, and flexibility training than as a result of endurance training. It must be noted, however, that confidence intervals overlapped [26].

Unfortunately, these published data do not permit us to make a precise evidence-based exercise prescription. Further well-designed and adequately powered studies are urgently required [37, 38]. It is of interest, however, that all randomized exercise interventions published in 1996 and later support the concept that exercise can prevent falls among older adults [3, 5].

TABLE 14.4

Summary of the disparity of outcomes among different studies

Exercise dimension	Studies in which falls were reduced	Studies in which falls were not reduced
Type of activity		
Endurance	[19, 24]	[34, 35]
Strength	[18–20, 24]	[21, 23, 31–35]
Balance	[17, 18, 20, 24]	[21, 23, 31–34]
Flexibility	[18–20]	[31–35]
Duration per session	15 min [17] 30 min [18] 60 min [19] 90 min [24]	30 min [21] 40 min [33] 45 min [35] 60 min [31, 32, 34]
Frequency of exercise	3×/wk [18, 19, 24] 14×/wk [17]	2×/wk [34] 3×/wk [21, 31–33, 35]

The dimensions of exercise needed to reduce falls remain unclear [5].

SUMMARY

- Falls and related fractures are a major health problem for older individuals and for society [37].
- Involutional changes in sensory and musculoskeletal structure and function among older adults render them at increased risk of falls and injuries.
- Many intrinsic and extrinsic risk factors for falls have been identified.
- Exercise can modify the intrinsic fall risk factors and thus prevent falls in older adults; however, the optimal exercise prescription to prevent falls has not yet been defined.
- A physically active lifestyle is associated with a reduced risk of osteoporotic fracture.
- Future exercise trials need to focus on (1) the oldest individuals with osteoporosis as a target population, (2) better control for confounding variables, (3) identifying an optimal exercise program for each specific group of older adults, and (4) using falls and fractures as primary outcomes. All of these are likely to require multicenter collaboration.

References

1. Kannus P, Niemi S, Parkkari J, et al. Hip fractures in Finland between 1970 and 1997 and predictions for the future. *Lancet* 1999;353:802-805.
2. Kannus P, Palvanen M, Niemi S, et al. Epidemiology of osteoporotic pelvic fractures in elderly people in Finland: sharp increase in 1970-1997 and alarming projections for the new millennium. *Osteoporos Int* 2000;11:443-8.
3. Gardner MM, Robertson MC, Campbell AJ. Exercise in preventing falls and fall related injuries in older people: a review of randomised controlled trials. *Br J Sports Med* 2000;34:7-17.
4. Tinetti ME, Lui WL, Claus E. Predictors and prognosis of inability to get up after falls among elderly persons. *JAMA* 1993;269:65-70.
5. Carter N, Kannus P, Khan KM. Exercise and the prevention of falls in older people: a systematic literature review examining the rationale and the evidence. *Sports Med* 2001;31:427-438.
6. Parkkari J, Kannus P, Palvanen M, et al. Majority of hip fractures occur as a result of a fall and impact on the greater trochanter of the femur: A prospective controlled hip fracture study with 206 consecutive patients. *Calcif Tissue Int* 1999;65:
7. Peck WA, Riggs BL, Bell NH, et al. Research directions in osteoporosis. *N Eng J Med* 1988;85:275-282.
8. Cummings SR, Rubin SM, Black D. The future of hip fractures in the United States. *Clin Orthop* 1990;252:163-166.
9. Kannus P. Preventing osteoporosis, falls, and fractures among elderly people. Promotion of lifelong activity is essential. *Br Med J* 1999;318:205-206.
10. Frischknecht R. Effect of training on muscle strength and motor function in the elderly. *Reprod Nutr Dev* 1998;38:167-174.
11. Kannus P. Preventing osteoporosis, falls, and fractures among elderly people. *Br Med J* 1999;318:205-206.
12. Myers AH, Young Y, Langlois JA. Prevention of falls in the elderly. *Bone* 1996;18:
13. Davis JW, Ross PD, Nevitt MC, et al. Risk factors for falls and serious injuries among older Japanese women in Hawaii. *J Am Gerentol Soc* 1999;47:792-798.
14. Tinetti ME, Baker DI, McAvay G, et al. A multifactorial intervention to reduce the risk of falling among elderly people living in the community. *N Engl J Med* 1994;331:821-827.
15. Kannus P, Parkkari J, Niemi S, et al. Prevention of hip fracture in elderly people with use of a hip protector. *N Engl J Med* 2000;343:1506-1513.
16. Schuntermann MF. The international classification of impairments, disabilities and handicaps (ICIDH) - results and problems. *Int J Rehab Res* 1996;19: 1-11.
17. Wolf SL, Barnhart MX, Kutner NG, et al. Reducing frailty and falls in older persons: An investigation of Tai-Chi and computerized training. *J Am Geriatr Soc* 1996;44:489-497.
18. Campbell AJ, Robertson MC, Gardner MM, et al. Randomised controlled trial of a general practice programme of home based exercise to prevent falls in elderly women. *Br Med J* 1997;315:1065-1069.
19. Buchner DM, Cress ME, deLauteur BJ, et al. The effect of strength and endurance training on gait, balance, fall risk and health services use

in community-living older adults. *J Gerontol* 1997;52A:M218-M224.

20. Campbell A, Robertson M, Gardner M, et al. Falls prevention over 2 years: a randomized controlled trial in women 80 years and older. *Age Ageing* 1999;28:513-518.
21. Campbell AJ, Roberton MC, Gardner MM, et al. Psychotropic medicine withdrawl and a home-based exercise program to prevent falls: a randomized controlled trial. *JAGS* 1999;47:850-853.
22. McMurdo ME, Rennie L. A controlled trial of exercise by residents of old people's homes. *Age Ageing* 1993;22:11-15.
23. Steinberg M, Cartwright C, Peel N, et al. A sustainable programme to prevent falls and near falls in community dwelling older people: results of a randomised trial. *J Epidemiol Com Health* 2000; 54:227-232.
24. Rubenstein LZ, Josephson KR, Trueblood PR, et al. Effects of a group exercise program on strength, mobility, and falls among fall-prone elderly men. *J Gerontol A Biol Sci Med Sci* 2000;55:M317-21.
25. Wolf SL, Barnhart MX, Ellison GL, et al. The effect of Tai Chi Quan and computerized balance training on postural stability in older subjects. *Phys Ther* 1997;77:371-381.
26. Province MA, Hadley EC, Hornbrook MC, et al. The effects of exercise on falls in elderly patients. *JAMA* 1995;273:1341-1347.
27. Vetter NJ, Lewis PA, Ford D. Can health visitors prevent fractures in elderly people? *Br Med J* 1992;304:888-890.
28. Wagner EH, La Croix AZ, Buchner DM, et al. Effects of physical activity on health status in older adults. Observational studies. *Ann Rev Public Health* 1992;13:451-468.
29. Rubenstein LZ, Robbins AS, Schulman BL, et al. Falls and instability in the elderly. *J Am Gerentol Soc* 1988;36:266-278.
30. Hornbrook MC, Stevens VJ, Wingfield DJ, et al. Preventing falls among community-dwelling older persons: results from a randomized trial. *Gerontologist* 1994;34:16-23.
31. Reinsch S, Macrae P, Lachenbruch PA, et al. Attempts to prevent falls and injury: A prospective community study. *Gerontologist* 1992;32:450-456.
32. MacRae PG, Feltner ME, Reinsch S. A 1-year exercise program for older women: effects on falls, injuries and physical performance. *J Ageing Phys Activity* 1994;2:127-142.
33. Mulrow CD, Gerety MB, Kanten D, et al. A randomized trial of physical rehabilitation for very frail nursing home residents. *JAMA* 1994;271:519-524.
34. Lord SR, Ward JA, Williams P, et al. The effect of a 12-month exercise trial on balance, strength, and falls in older women: a randomized controlled study. *J Am Geriatr Soc* 1995;43:1198-1206.
35. McMurdo E, Mole P, Paterson C. Controlled trial of weight bearing exercise in older women in relation to bone density and falls. *Br Med J* 1997;314:569.
36. Lehtola S, Hanninen L, Paatalo M. The incidence of falls during a six-month exercise trial and four month follow up among home dwelling persons aged 70-75 years [Finnish]. *Liikunta Tiede* 2000;6:41-47.
37. Lord SR, Sherrington C, Menz HB. *Falls in older people: Risk factors and strategies for prevention.* Cambridge: Cambridge University Press, 2001:1-260.
38. Khan KM, Lu-Ambrose T, Donaldson MG, et al. Physical activity to prevent falls in older people: time to intervene in high risk groups using falls as an outcome. *Br J Sports Med* 2001;35:159-160.

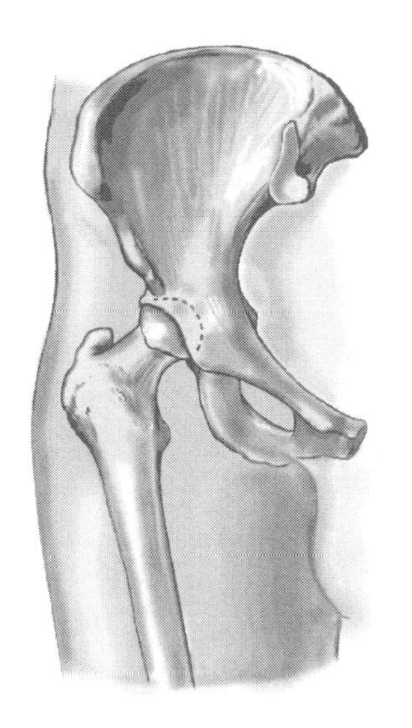

Chapter 15

Exercise Prescription for People With Osteoporosis

In the 1990s the general public became aware of the condition of osteoporosis and its manifestations: fracture and fall-related fracture. Although now of epidemic proportions, this condition was not even familiar to many doctors 20 years earlier. The awareness of osteoporosis has resulted in an explosion of clinical research and steady increases in the availability of bone density testing. In addition, pharmaceutical companies now provide large-scale education and marketing programs emphasizing the prevalence and significance of this condition. In this chapter we define osteoporosis and discuss briefly its relevance to and impact on medicine and society today. The chapter then provides practical guidelines for prescribing exercise safely for people with osteoporosis.

Osteoporosis may be defined conceptually as a condition of generalized skeletal fragility such that fractures occur with minimal trauma, often no more than is applied by routine daily activity. The operational definitions of osteoporosis relate to BMD scores measured by DXA.

severe osteoporosis—Bone mineral density at least 2.5 standard deviations below adult peak mean together with evidence of low-energy fractures (figure 15.1).

osteoporosis—BMD at least 2.5 standard deviations below adult peak mean without evidence of low-energy fractures.

osteopenia—BMD between 1 and 2.5 standard deviations below adult peak mean without evidence of low-energy fractures.

Normal BMD is defined as that which is better than 1 standard deviation below the adult peak mean. These criteria are used to define osteoporosis because they reflect fracture risk (figure 15.2). Although these definitions are not ideal, they provide a useful framework for current clinical practice.

Before tackling the subject of osteoporosis, we should mention that there have been very few exercise intervention studies performed in women with osteopenia [1-4] or osteoporosis

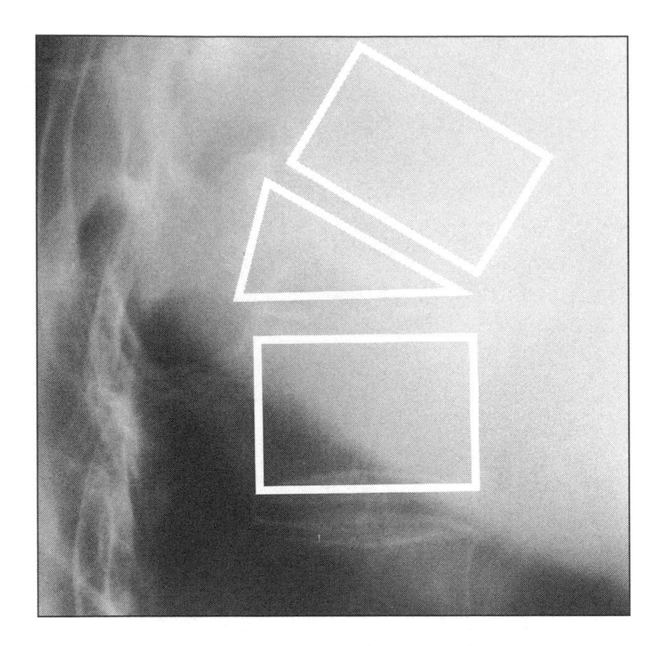

FIGURE 15.1 Lateral X-ray of the spine showing a profound vertebral crush fracture in a woman who leaned over to pick up a shopping bag.

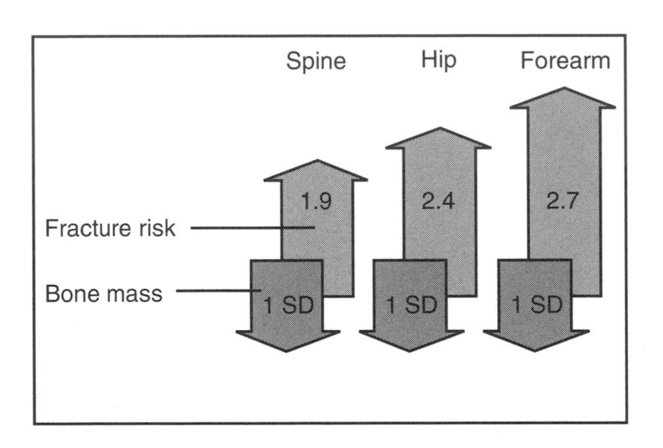

FIGURE 15.2 Graphic depicting the increased risk of fracture at the spine, hip, and forearm as a result of a 1-standard-deviation decrease in BMD.

[17, 24, 25, 28]. One such 12-month intervention examined BMD as well as functional activity level and balance as outcome measures in women aged 50-70 years [2]. The intervention consisted of weight-bearing exercises, including walking, stepping up and down from benches, aerobic dancing, and flexibility exercises, performed three times a week for one hour. Spinal BMD stabilized in these women; femoral neck BMD did not change relative to controls. These results are consistent with those reported in postmenopausal women who had normal bone mineral density (see chapter 12).

The Problem of Osteoporosis

Osteoporosis and osteoporotic fractures are unarguably major public health problems, as shown by the direct expenditures for the treatment of osteoporotic fractures in table 15.1. In addition to their economic cost, they have a debilitating effect on independence and quality of life [5, 6]. In sum, osteoporosis and related fractures are devastating psychologically, socially [5], and economically. Moreover, the problem will worsen, as it has been estimated that 30 to 40 years from now hip fractures will be three to four times more prevalent than they are currently [7, 8]. The reason for the secular trend of increased hip fractures remains unexplained. One candidate is the increasingly sedentary nature of everyday life that has resulted from technological advances and labor-saving devices [9].

A logical strategy for tackling osteoporotic fracture is to identify, and modify, proven risk factors for fracture. Epidemiological studies of osteoporotic fractures have identified some significant risk factors [10, 16]. Some risk factors, such as age, female gender, and Caucasian race, are unalterable [11]. Others, such as low bone mineral density (BMD), may only be modifiable to a small extent, particularly in the elderly [12]. Researchers have identified some fracture risk factors that respond well to intervention, for example, recurrent falling (see chapter 14) [19, 20].

Calcium, vitamin D, and *physical activity* are discussed fully elsewhere, but they remain important parts of treatment of osteoporosis. Physicians and patients must not overlook the importance of these factors when seeking to reduce fracture risk in osteoporosis.

Exercise is the only single therapy that can simultaneously ameliorate low BMD, augment muscle mass, promote strength gain, and improve dynamic balance—all of which are independent risk factors for fracture [12, 13]. For this reason we are convinced that exercise is a unique and vastly underused therapy for people with osteoporosis and increased risk of falling. The rest of this chapter outlines how to prescribe exercise for these people. We emphasize that exercise for osteoporosis should only be prescribed by those appropriately trained and certified.

Exercise Prescriptions

Few well-designed trials have been performed to guide exercise prescription in women

TABLE 15.1

Cost estimates of osteoporotic fractures around the globe

Country and author	Type of fractures	Cost in local currency	Cost converted to U.S. dollars (for comparison)	Cost (U.S. dollars) per head of population studied
Australia Randell 1995 [21]	Osteoporotic fractures	800 million	530 million	31
United States Cummings et al. 1995 [16]	Hip fractures (1984)	7 billion	7 billion	28
United States Cummings et al. 1995 [16]	Hip fractures (2040—projected)	16 billion	16 billion	64
Finland Kannus et al. 1999 [8]	Hip fractures	350 million	70 million	14
Canada Goeree et al. 1996 [7]	Treatment of osteoporotic fractures	400 million	264 million	10

Finer Point: Commonly prescribed treatments for osteoporosis

Most of the medications used to treat osteoporosis act by slowing down or stopping further bone loss. Thus, they stop bone resorption (chapter 2) and are called antiresorptive agents. Estrogen, calcitonin, and bisphosphonate drugs are all examples of antiresorptive agents.

Estrogen is commonly used in hormone replacement therapy, where it is often used with the hormone progesterone, and it has been shown to prevent fractures in postmenopausal women. It acts in a variety of ways to improve calcium absorption and maintain bone mass. Estrogens can cause side effects in some women and there is concern among the medical and scientific community that estrogens may increase the risk of breast or uterine cancer in some women.

Progesterone is one of the two important female sex hormones. When menstrual cycles are disturbed, even in subtle ways, progesterone levels can drop and this may affect bone mass. There remains a great deal of debate about the role of progesterone in osteoporosis treatment and prevention. However, progesterone treatment can improve bone density in premenopausal women with menstrual cycle disturbance.

SERMs (selective estrogen receptor modulators) are synthetic drugs that have been designed to provide the benefit of estrogen without the risk. Raloxifene (Evista) has an antiresorptive effect on bone: like estrogen, it blocks the effect of estrogen on breast and uterine tissue. This means it does not increase the risk of breast or uterine cancer. Thus, it is often prescribed for osteoporosis in women with past gynecological or breast cancers or women with a strong family history of these types of cancers. The most common side effects are leg cramps and worsening hot flashes. Like estrogen, raloxifene may increase the risk of blood clots.

Bisphosphonates are potent inhibitors of bone turnover. This category of medications includes Alendronate (Fosamax), Etidronate (Didronel), and Risedronate (Actonel). Bisphosphonates bind permanently to the surfaces of bones and limit osteoclast activity. Some patients have gastrointestinal discomfort when taking bisphosphonates, even when special precautions are taken to prevent this from happening.

Calcitonin is a hormone that also regulates bone metabolism and it can be prescribed for treatment of postmenopausal osteoporosis. It is used either by injection or by nasal spray—at present it is not available in tablet form. Calcitonin also has pain-relieving properties so it is often used to treat painful vertebral osteoporotic fractures. Side effects include local nasal irritation and headaches.

Phytoestrogens, or plant-based estrogens, are the subject of much discussion. They are present in soybeans and wild yams. Although not as powerful antiresorptive agents as pharmaceutical estrogen products they have been reported to reduce hot flashes. Further research is needed to clarify their effect on bone and other tissues.

with osteoporosis (table 15.2). Nevertheless, after an appropriate assessment, principles of exercise prescription can be combined with clinical experience to provide a useful exercise regimen.

Assessment Before Exercise Prescription

1. **Determine participant's lifestyle goals.** Find out from the participant what goals she has for the exercise program, for example, to increase bone mineral, improve balance, improve quality of life, and/or improve independence. The aim will influence both your choice of exercise prescription and your measures of progress.

2. **Conduct a health appraisal.** Because the client may have a number of frailties, assess her health (if you are trained) or have her obtain a health evaluation from her physician or physical therapist. This assessment should include the biomechanical competence of the spine, the musculoskeletal system, the neuromuscular system, and cardiovascular and respiratory function. In this age group it is also important to identify any other conditions and know of any medications the person may be taking that might limit her participation in an activity program.

3. **Discuss activity preferences and interests.** An exercise program must be tailored to an individual's interests if it is to be successful. Obtain some background on the person's sporting and activity preferences and profile. A lifestyle questionnaire can assist you in developing a program that is both appropriate and interesting to the individual (see appendix B.1).

4. **Develop an action plan.** It is important at this stage to match the participant's goals with her needs and preferences. An assessment form for this purpose is provided in appendix B.2. If there is a poor match between goals, preferences, and abilities, refocus the participant on each element and inform her of the benefits and limitations of her choices. Become aware of the facilities and programs available to the person both in the home and in the local community (e.g., a heated pool may be easily accessible). The Action Plan Worksheet can assist you at the program planning stage (see appendix B.3). It includes reasonable lifestyle changes and achievable short-term goals. These are both key to the success of any program. If you are untrained in exercise prescription, seek the counsel of a kinesiologist or someone who is trained in this area of specialization.

5. **Have contingency plans and follow up.** Be alert to personal barriers to achieving the goals set up in the action plan and have a contingency plan. For example, if the participant chooses a combined land and water exercise program, ask her if she has a swimsuit and whether a pool is accessible by bus. Have an alternative plan should the program be offered on days or at times that are inconvenient to the participant. In addition, have a means by which to assess the individual's progress and success. This can be as simple as sending out a "How is it going?" postcard or allowing time to discuss the program during the next scheduled office visit.

6. **Other tips.** Encourage the participant to involve family or friends in the program goals. Even a casual question from friends such as, How is your activity program coming along? may provide the impetus needed to maintain compliance in this age group. Studies suggest that adherence to programs is improved with group participation. Suggest to your patient/client that she use a buddy system, if possible. Also, direct her to a program in which strong social support can be established.

TABLE 15.2

Studies of exercise interventions in women with evidence of osteoporosis or osteopenia

Reference	Study design	Rand	Blinded	N	Age/ Sex	Definition of OP	Groups and treatment	Compliance	Duration of program	Outcome measures	Results
Carter 2001 [17]	RCT	Y	Y-examiner N-subjects	97	65-75 all F	WHO definition of OP at either lumbar spine or hip or both (i.e., T-score < 2.5)	• Osteofit program for flexibility, strength, posture and balance, 20 wks	> 80% attendance at exercise classes; 11% drop out during 20 wk; no adverse effects	20 weeks	• Balance by computerized dynamic posturography • Figure 8 run • Quadriceps strength • Quality of life • Low back pain	Improved dynamic balance
Malmros 1998 [24]	RCT	Y	Y-examiner N-subjects	53 (27 Ex 25 C)	62-71 all F	• OP but not defined • 1 spinal crush • Back pain in last 3 yrs	Example: Land based group (n = 6) exercise program including balance (standing on 1 leg), strengthening (isometric trunk flexion/extension), stretching, and relaxation 1 hr, 2 ×/wk Home exercise after 5 wks C = no treatment	100% attendance at exercise start 9% drop out during 10 wk exercise No adverse effects	10 wks Follow up at 12 weeks after exercise	• Pain—VAS • Use of analgesia • Modified Oswetry questionnaire—daily level of function • QOL questionnaire • Balance—Chattex • Isometric back extension and abdominal strength • Isometric quadriceps strength with twitch interpolation	Exercise group had a significant reduction in use of analgesia and pain and an increase in QOL
Kronhed 1998 [28]	Controlled study	N	N	34 (19 Ex 15 C)	Mean = 55 Ex, 59 C; 18 men, 12 women (4 not specified)	Low bone mass by single photon absorptiometry at the wrist (Z score < −1)	• Physiotherapist-administered balance and strength exercise training in Vadstena, Sweden • Control group continued with "usual physical activity"	4 drop outs of 19 in exercise group; 25-95% compliance in exercise group Mean about 50%	12 months	• BMD at the hip and lumbar spine • Muscle strength • Flexibility • Balance • Aerobic activity	Improved BMD, balance, and aerobic capacity in the exercise group

(continued)

TABLE 15.2 *(continued)*

Reference	Study design	Rand	Blinded	N	Age/ Sex	Definition of OP	Groups and treatment	Compliance	Duration of program	Outcome measures	Results
Mitchell 1998 [25]	RCT	Y	N-examiner or subjects	30 (16 Ex 15 C)	Mean = 59 Rx, 63 C All F	• Diagnosed with OP but not further defined • No mention of having to have pain or disability	Example: Group based low impact weight bearing aerobic exercise at 60-75% of max HR 2 ×/wk Home exercise—20 min brisk walk 1 ×/wk C = no treatment	Average = 87% (79-100%) attend 13% drop out	12 wks	• Weight • % body fat • Steady state heart rate • Functional reach • Sit and reach • Isokinetic quadriceps strength	Significant improvements in most measures
Bravo et al. 1997 [26]	Pre-post one group	No	N/A	86	50-70 all F	• Osteopenic BMD < 1g/cm^2 L2-4 or < 0.9 g/cm^2 femoral neck • No mention of of having to have pain or disability	Example: Group water-based exercise program—jumping in waist-high water interspersed with muscular exercise (groups of 12-30) 60 mins, 3 ×/wk	75% (22-99%) attendance 25% drop out	12 months	• BMD • Fitness measures (e.g., flexibility, coordination, agility, strength) • Psychological well being	• Stabilized femoral neck BMD but not LSp • Significant improvements in flexibility, agility, strength/endurance and CV endurance
Bravo 1996 [2]	RCT	Y	Not stated	142 (61 Ex 63 C)	50-70 All F; post-meno-pausal	• Osteopenic BMD < 1g/cm^2 L2-4 or < 0.9 g/cm^2 at femoral neck; higher than WHO definition • No mention of having to have pain or disability	• 25 min rapid walking • 15 min step-up/down 60-70% max HR • 15 min isotonic exercise for UL, scapular, abdominals, back • 5 min cool down including balance and coordination • Education seminars 3×/wk C = bimonthly educational seminar on OP	73% attendance 13% lost to follow-up (equal numbers in both groups)	12 months	• BMD • Fitness measures (e.g., flexibility, coordination, agility, strength • Psychological well being	• Stabilized spinal but not femoral BMD • Improved fitness • Improved psychological well being • Decreased back pain

Preisinger 1996 [23]	RCT	Y	N-examiner or subjects	92 (71 Ex 31 C)	45-75 All F; all post-menopausal	• Only 36% had OP • Moderate idiopathic mid or low back pain during walking, standing, sitting > 1 hr, lifting weights, bending down, carrying, shopping, or other vigorous ADL	Example: Individual stretches, exercises for posture, motor control, coordination 3 ×/wk Supervised for first 6 wks then home exercises C = no exercise, but could have other pain relieving techniques	44% compliance for exercisers 3 ×/wk	4 yrs	• BMD—SPA of FA • Q for back complaints, #	• Significantly less bone loss and less back complaints in "compliant" exercisers
Ayalon 1987 [1] Simkin 1987 [4]	Quasi-experimental	N	N-examiner or subjects	40 (14 Ex 26 C)	53-74 All F; all post-menopausal	Morphological changes in the lumbar vertebrae	• Limb loading exercises of the distal forearm—hanging from a ladder, pulling and twisting a hand against a partner, pushing against a wall, arm wrestling • Rationale was to provide tension, torsion and tension, compression and bending forces respectively 3 ×/wk 50 min sessions individual or group	73% attendance	5 months	• Distal forearm—BMD-Compton scattering and BMC single photon absorptiometry	• Significant increase of 3.8% in BMD while controls lost 1.9%. No change in BMC

(continued)

TABLE 15.2 *(continued)*

Reference	Study design	Rand	Blinded	N	Age/ Sex	Definition of OP	Groups and treatment	Compliance	Duration of program	Outcome measures	Results
Sinaki and Mikkelsen 1984 [27]	Quasi-experimental	N	N-subjects Y-examiner	59 (53 Ex 6 C)	49-60 all post-menopausal	• Radiographic diagnosis of spinal OP—decreased BMD, biconcavity of vertebral bodies with overextension of discs, collapse of vertebral bodies • Kyphosis and decreased height	• Back extension exercises performed prone and sitting (n = 25) • Back flexion exercises—sit-ups and stretching of erector spinae (n = 9) • Combined back flexion and extension exercises (n = 19) • Controls—instruction in posture and some patients given isometric abdominal exercises (n = 6)	Not stated	1-6 yrs average 1.4 to 2 yrs for individual groups	X-ray appearance of spinal vertebrae	84% of patients in the extension exercise group showed no further evidence of wedging or compression of vertebrae compared with 11% of the flexion exercise group, 47% in the combined group, and 33% in the no exercise group

ADL = activities of daily living; C = control; CV = cardiovascular; Ex = exercise; F = female; HR = heart rate; N = no; N/A = not applicable; OP = osteoporosis; QOL = quality of life; Rand = randomized; RCT = randomized, controlled trial; VAS = visual analogue scale; Y = yes.

Patients Limited by Pain

For patients with debilitating osteoporosis and severe pain, exercise program options will be limited. It may be necessary to begin exercise prescription with a warm pool-based program (e.g., hydrotherapy), which, although non-weight-bearing, can improve flexibility and muscle strength. If the person cannot tolerate active exercises at all, functional electrical stimulation may improve spinal muscle strength in preparation for active strengthening as pain diminishes. Even simple analgesics taken shortly before planned activities can be helpful. Moreover, the hormone calcitonin, which can be injected or administered via nasal spray, can reduce bone-induced pain, for example, in acute osteoporotic fractures. A physician could prescribe this to help permit a patient to exercise again. Because this medication is a hormone normally found in the body, side effects are minimal, consisting primarily of nasal irritation resulting from the spray.

Program for Relatively Pain-Free Patients

Most types of activity programs are preferable to a sedentary lifestyle and should be encouraged. If a person chooses to exercise alone, the following program is recommended. Exercise sessions should begin with an 8- to 15-minute warm-up of gentle stretching and joint motion. Cardiovascular exercises should follow for 5 to 10 minutes. Targeted heart rate should be 60% of the calculated maximum heart rate (220 minus age) for a beginner or a deconditioned woman and 70-75% of that value for those in more intermediate health. Range of motion exercises for the shoulder and hip joints should follow (figures 15.3 and 15.4). Gradual weight training with light free weights and rubber tubing can then be incorporated. Effective upper arm exercises include pushing against a wall (figure 15.5) or pulling and twisting against a partner [4].

FIGURE 15.3 Wall arch. Face a wall with feet six inches from the wall and six inches apart and stretch arms to touch the wall while taking a deep breath. The stomach should be flat. Reaching up with one arm while stretching down with the other improves shoulder range of motion.

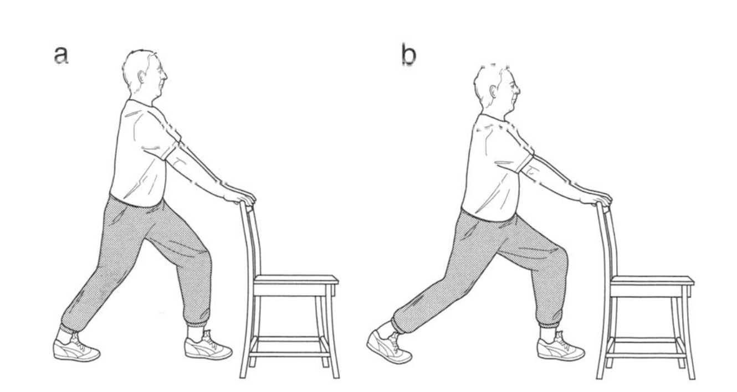

FIGURE 15.4 Leg and hip stretch. *(a)* To stretch calf and thigh muscles and hip joint, perform a calf stretch as shown. Lift heel up and bend the knee of the posterior leg. *(b)* Lean forward on bent knee for stretch. Repeat with other leg.

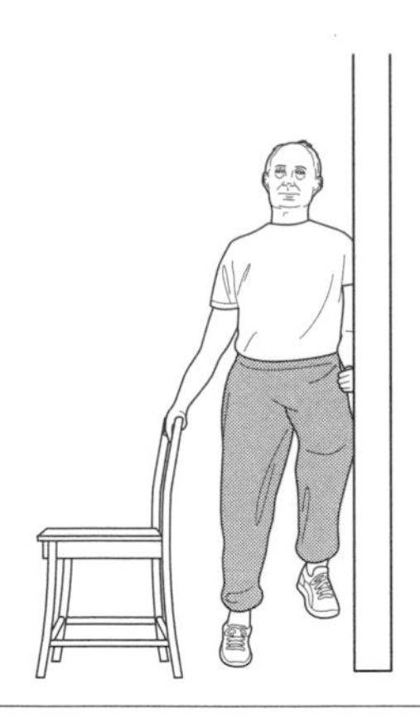

FIGURE 15.5 Wall push. Stand sideways with left side next to a wall, holding the back of a chair with the right hand for support. Bend the left knee and left arm. Push against the wall and count for 3-10 seconds. Repeat using the right side.

Quadriceps strength can be improved with a wall slide exercise (figure 15.6) or by practicing standing from a seated position (figure 15.7). Back posture correction can be done standing (figure 15.8), in a chair (figure 15.9), or prone (figure 15.10). Abdominal strengthening is illustrated in figures 15.11 and 15.12. Leg strength exercises include side-lying knee lifts (figure 15.13), prone leg lifts (figure 15.14) that also strengthen the lower back, and leg raises on all fours (figure 15.15).

FIGURE 15.6 Wall slide/wall sit. Stand as illustrated and flatten the stomach. Slide up and down against the wall, bending the knees while keeping the back and stomach flat.

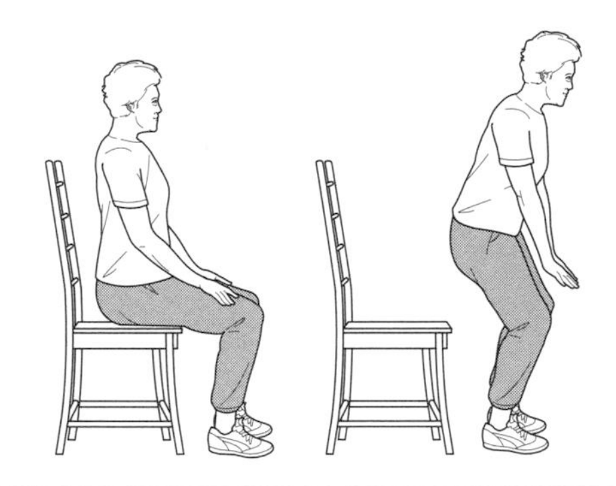

FIGURE 15.7 Chair rise. Sit on the edge of a chair and repeatedly rise to a standing position. The knees and feet should remain hip-width apart at all times. The weight should remain forward on the toes in both the standing and seated position. The legs supply the power. This exercise is best done in two or three sets during the day. A set is equal to 5 to 20 chair rises, as tolerated.

FIGURE 15.8 Standing back bend. To restore spinal posture make two fists and place them on the lower back below the waist. Arch backward slowly while taking a deep breath. Repeat.

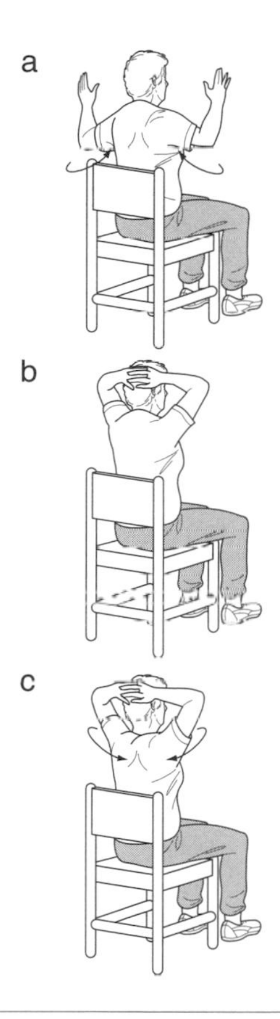

FIGURE 15.9 Seated posture correction. Sit or stand as tall as possible with the chin in, not up, stomach tight, chest forward, and good leg posture. *(a)* W exercise: Place arms in a W position with shoulders relaxed, not hunched. Bring elbows back, pinching shoulder blades together. Hold for a slow count of three. Relax for a slow count of three. *(b)* This can also be done with hands behind the head. *(c)* Next, push against the back of the scalp while pinching the shoulder blades together.

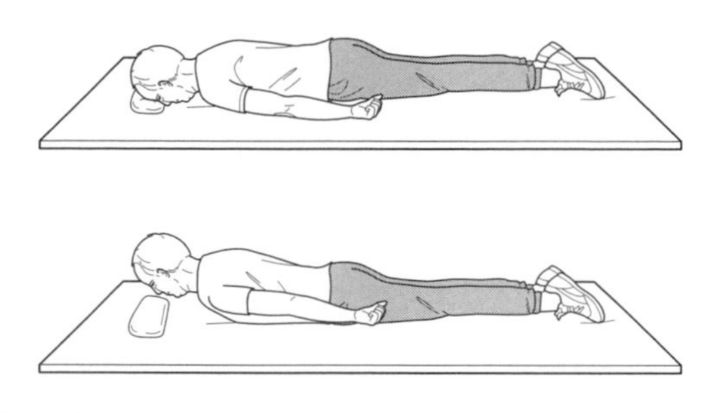

FIGURE 15.10 Prone trunk lift. Additional extension strength can be gained lying prone and keeping the stomach tight, feet down, and head in a neutral position, with hands at sides, and shoulder blades pinched together. Hold for one to three seconds. Relax, repeat.

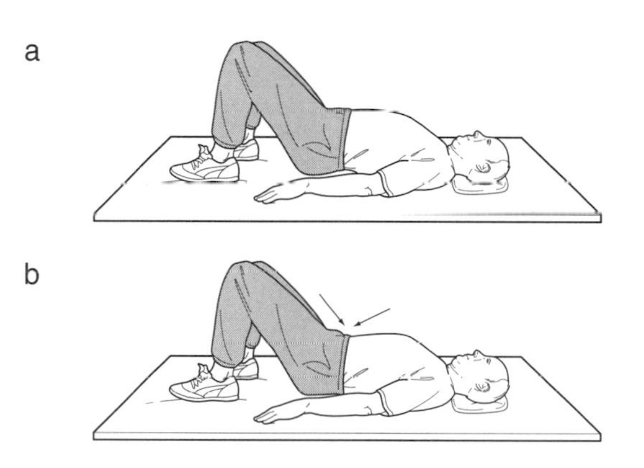

FIGURE 15.11 Abdominal strengthening. *(a)* Lie back with knees bent and feet flat on the floor with a small pillow under the head. *(b)* Perform pelvic tilt by tightening the abdominal muscles and "pulling" the pelvis toward the ribs. This pulls the ribs down and tilts the pelvis forward, flattening the lower back toward the floor.

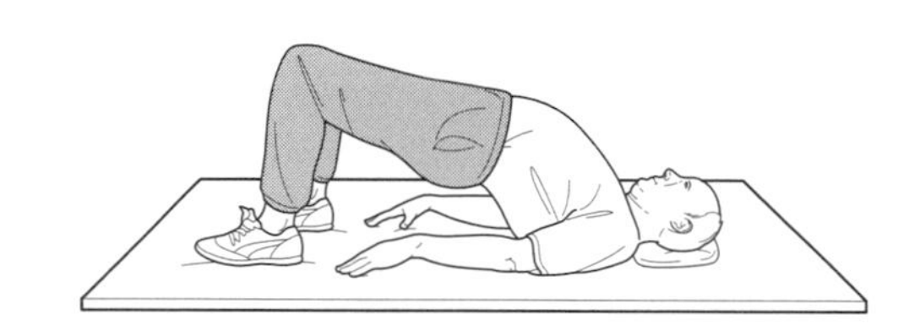

FIGURE 15.12 The bridge. To begin, lie supine with feet flat and knees bent, keeping the arms at the sides. First press the head and shoulders down. To add difficulty, lift then lower the hips and trunk. Relax, repeat. This should only be performed if cleared by an appropriate authority.

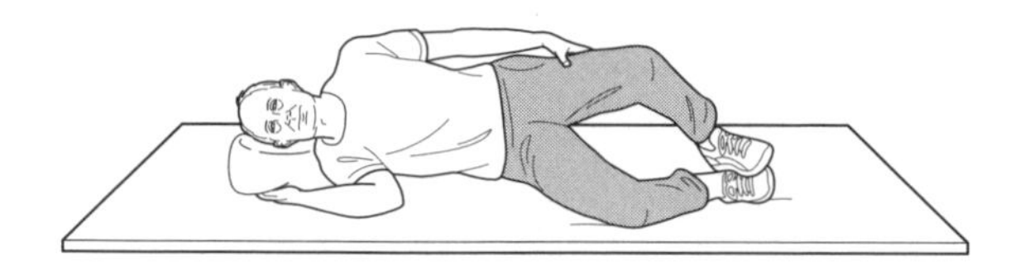

FIGURE 15.13 Side-lying knee lift. This is done lying on the side with knees bent, pelvis forward, and stomach flat. Lift the top knee, keeping feet together.

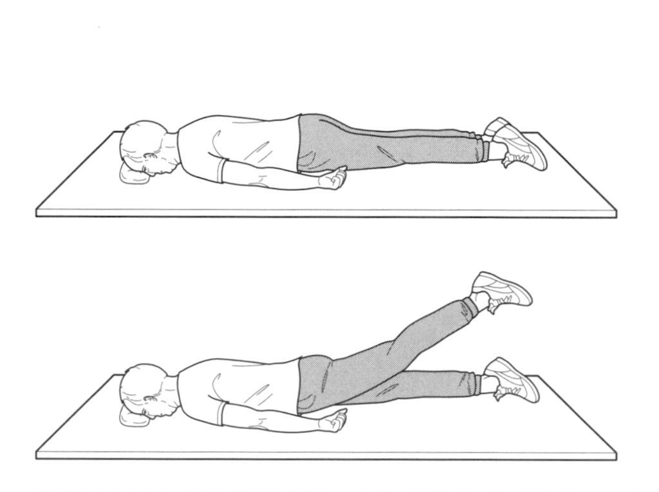

FIGURE 15.14 Prone leg lift. To strengthen lower back and buttocks, lie on the stomach, hands at the sides. Fold a towel beneath the head for comfort. Bend the left leg slightly and keep the thigh off the floor. Keep the foot relaxed. Repeat on the right side.

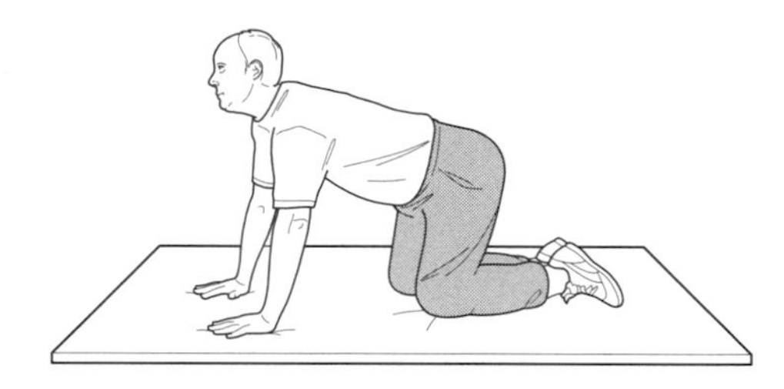

FIGURE 15.15 All-fours leg raise. Starts in the all-fours position, on hands and knees, keeping stomach and back flat. Lift one leg without changing trunk position, hold three to five seconds, then repeat with the other leg.

The exercise program should conclude with a seven- to eight-minute cool-down period. Stretches can be done sitting (figure 15.16) or on all fours (figure 15.17).

Remember that any exercise prescription is only as good as the willingness of the person to participate. When assessing a client, try to discover the types of activities the person enjoys. Many popular activities require some degree of strength, flexibility, endurance, balance, and coordination. For example, a good line dancing class emphasizes posture and the attributes listed, is fun, and does not require a partner. This is certainly likely to provide the patient with greater benefit than a dutiful 45-minute walk in the rain [14].

While these guidelines have not been proven in a scientific manner, they represent current "best practice" guidelines that are safe and appear appropriate from clinical experience. Further studies should improve our ability to prescribe effective exercise in the future.

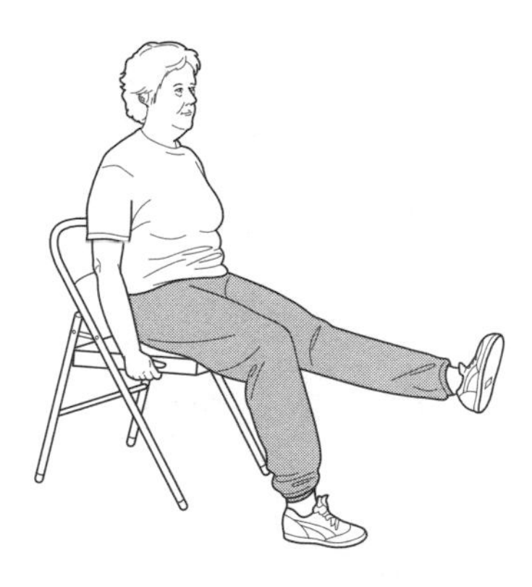

FIGURE 15.16 Sitting stretch. Hamstring stretch can be done sitting with a straight back, tight stomach, and arms resting on the thighs. Slowly straighten the knee, being sure not to let the pelvis rock backward. After the knee is as straight as possible, dorsiflex the ankle fully before lowering the leg again and relaxing. Repeat with the other leg.

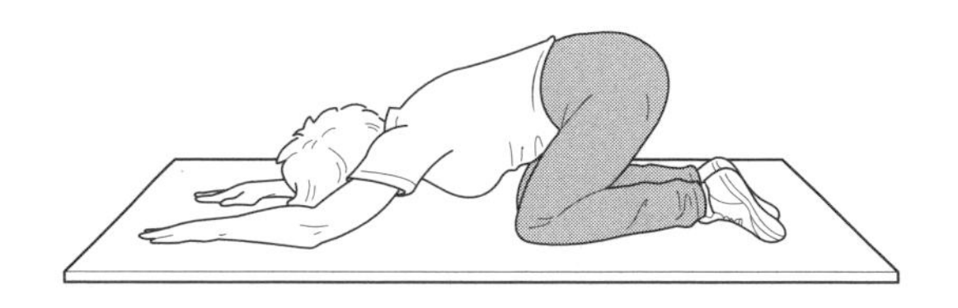

FIGURE 15.17 Cat stretch. This excellent stretch begins on all fours. Sit back slightly with hands and arms out in front. Stretch the arms forward and allow the back to flatten as you relax.

An Effective Exercise Program: Osteofit

Osteofit is an exercise-based program devised by the staff of the BC Women's Hospital and Health Centre Osteoporosis Program in Vancouver, Canada. This community-based program for women and men with osteoporosis aims to reduce participants' risk of falling and improve their functional ability, thereby enhancing quality of life. It differs from typical senior exercise classes by specifically targeting posture, balance, gait, coordination, and hip and trunk stabilization rather than general aerobic fitness.

A typical class consists of a warm-up, the workout, and a relaxation component, all of which are outlined here. During the last five minutes of Osteofit classes instructors offer Osteofit Tips—education topics related to bone health.

Warm-up. The general 10- to 15-minute warm-up is done to music and commences with gentle range of motion exercises for the major joints, which are performed either seated or standing. Static stretching exercises are usually not included. The warm-up ends with walking and simple dance routines with a tempo of between 110 and 126 beats per minute so that participants can remain in control.

Workout. The workout consists of strengthening and stretching exercises intended to improve posture by combating medially rotated shoulders, chin protrusion (excessive cervical extension), thoracic kyphosis, and loss of lumbar lordosis. Exercises to improve balance and coordination may progress from heel raises and toe pulls, to the mildly challenging two-legged heel-toe rock and the more challenging tandem walks (figure 15.18) and obstacle courses. Hip stabilization is trained using leg exercises (e.g., hip abduction and extension) or balance exercises. Trunk stabilization is addressed when the participant is cued and positioned to do all standing exercises with resistance for the arms (e.g., biceps curls) and shoulders (e.g., lateral arm raises). The abdominal muscles are strengthened in their function as stabilizers rather than as prime movers. Exercises to improve functional ability include chair squats (figure 15.19) and getting up from and down to the floor.

Exercises are arranged so that upper and lower body activities are alternated to reduce the risk of tendon pain. If the class includes more than one set of an exercise, the sets are separated by a short rest period. Repetitions are kept to between 8 and 16, and weights are relatively light so that participants do not work to fatigue with each set. The exercises are arranged so that the less strenuous exercises, such as hamstring stretching, are at the end of the workout.

Relaxation. The last few minutes of the class are devoted to relaxation techniques such as deep breathing, progressive muscle tensing and relaxing, and visualizations to a background of soft music and/or nature sounds.

In a study of the efficacy of Osteofit in women aged 65-75 with osteoporosis, the women who completed the Osteofit Program had (1) increased ability to undertake activities of daily living, (2) decreased back pain, (3) increased general health, and (4) decreased fear of falling [17]. Importantly, no patient suffered an exercise-related injury. These data suggest that Osteofit is safe and effective for a population that is at high risk of osteoporotic fracture.

FIGURE 15.18 A participant in the Osteofit Exercise Program performs the tandem walk to improve balance

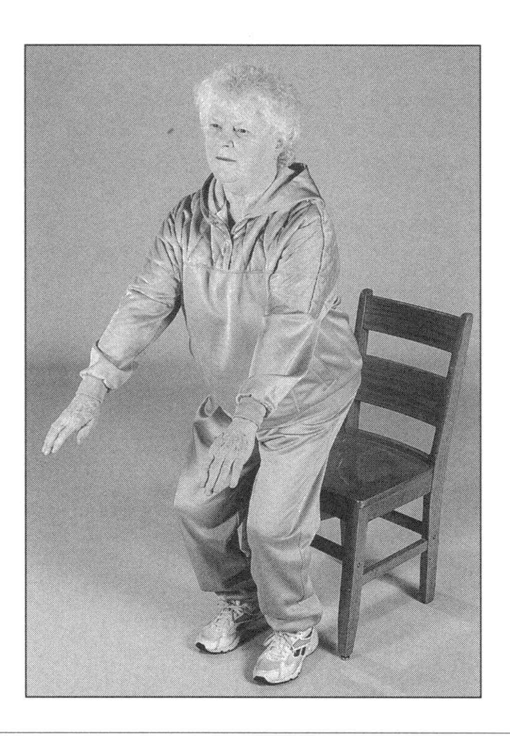

FIGURE 15.19 Chair squats are an exercise that help improve daily function.

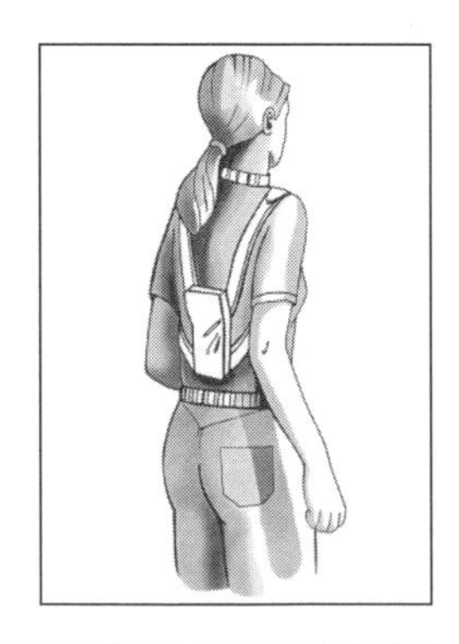

FIGURE 15.20 A spinal orthosis, worn like a backpack but containing only 750 g (1.5 pounds) of weight, has provided symptomatic relief to some patients.

Nonpharmacological Management of Osteoporosis

Some encouraging data suggest that physical activity intervention in osteoporosis may be facilitated with the use of certain back orthoses (figure 15.20)—devices to support the thoracic spine or improve postural awareness. However, others believe that orthoses are ineffective because patients do not use their own muscles to strengthen their backs and improve their posture. We remain in the latter camp but encourage appropriate trials and await their outcome with interest.

Exercises Contraindicated in Osteoporosis

Several exercises are not suitable for people with osteoporosis as they can generate large forces on relatively weak bone. Dynamic abdominal exercises (e.g., situps) and excessive trunk flexion can cause vertebral crush fractures (figure 15.21). Twisting movements such as a golf swing can also cause fractures [15]. Exercises that involve abrupt or explosive loading, or high-impact loading, are also contraindicated. Daily activities such as sitting and bending to pick up objects can cause vertebral fracture if not performed correctly (figure 15.22).

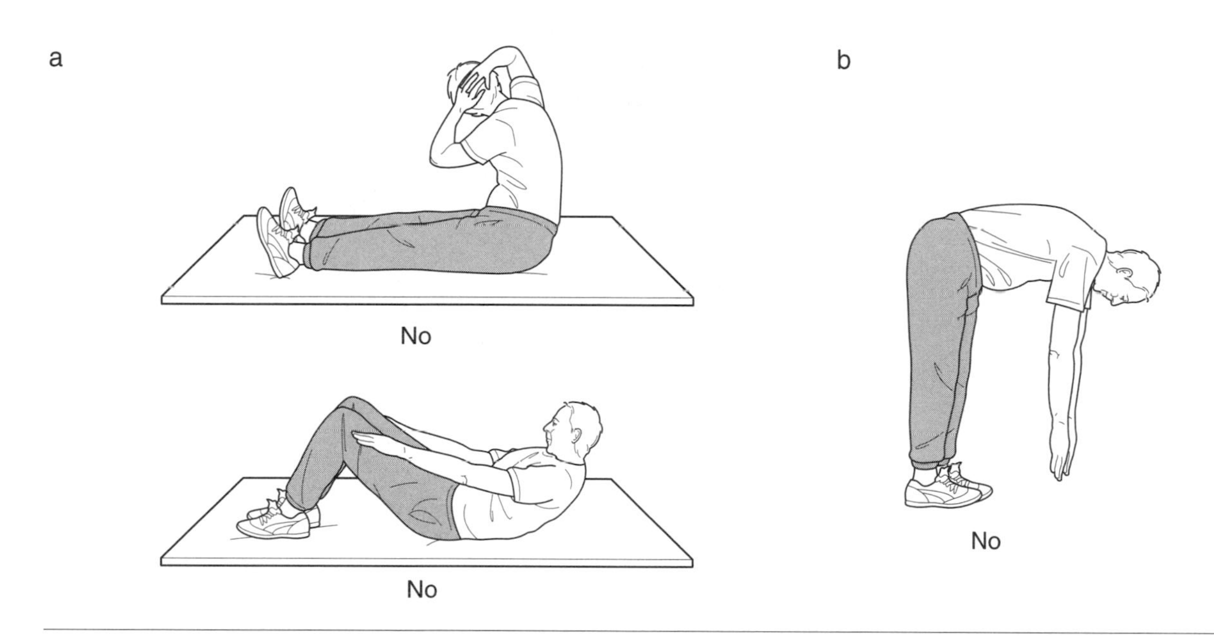

FIGURE 15.21 People with osteoporosis should avoid *(a)* sit-ups and *(b)* excessive forward flexion.

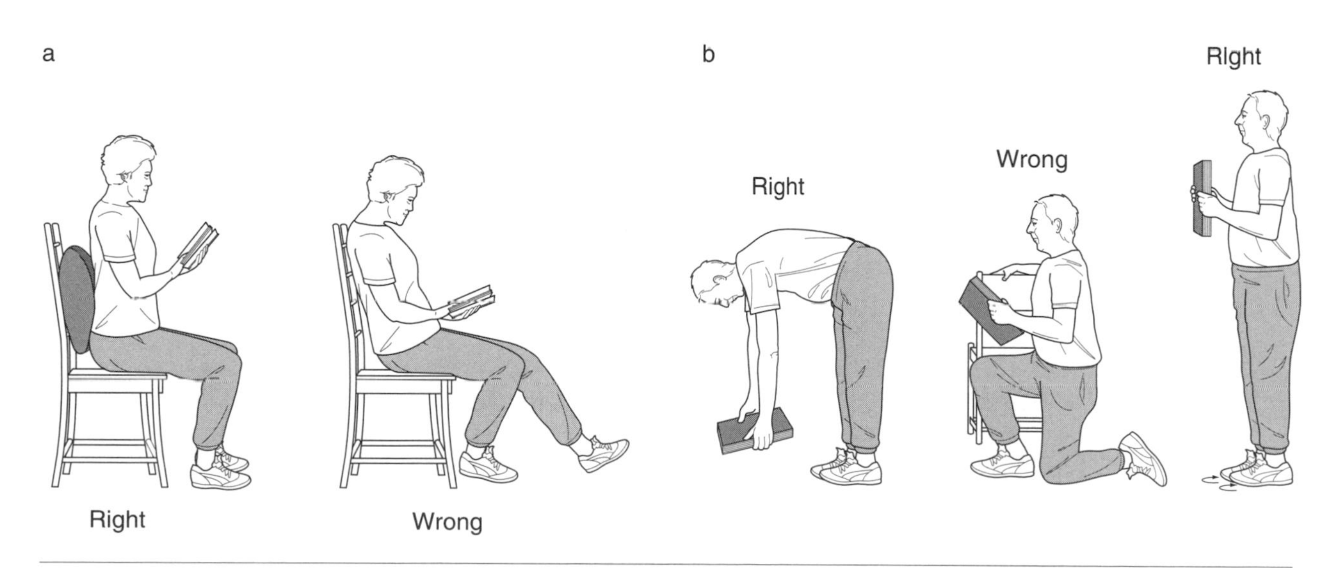

FIGURE 15.22 Correct posture for activities of daily living. *(a)* Sitting with back supported by a pillow. *(b)* Correct bending and lifting technique.

Tips for Daily Living

Clinicians who see a large number of patients with osteoporosis who fracture doing everyday activities have the following practical suggestions:

- Be sure to attend to fall risks (see chapter 14), as falling remains the key risk factor, even in a person with osteoporosis. In practice, be aware of environmental and home hazards (minimize or, preferably, avoid scatter rugs and don't have electric cords running across hallways).
- Address intrinsic fall risks by attending to vision, balance, and strength.
- Let things drop; your back is worth more.
- When getting up from a bed, roll onto the side and then tilt upward; don't curl straight up, which puts large forces through the lumbar spine on the way.
- You may need to install grab bars for hand support near the toilet and place the toilet paper where you can reach without twisting.
- Never get dressed while standing on one leg. Sit to put on pants and then stand to pull them up.
- Keep frequently used heavier refrigerator items on an easy to reach shelf at waist level.

CASE REPORT

Putting It All Into Practice

Barbara Beatty, a Vancouver physical therapist who has an international reputation in treating patients with severe osteoporotic fractures, provides a wonderful case report of how a practical exercise prescription that fully assesses the patient's needs can virtually provide a new life for a patient.

The patient, Mrs. O., and her husband lived in their dream retirement home, perched on cliffs overlooking the Pacific Ocean. Their home was 75 stairs below the garage level. There was no level ground anywhere near the home. Six months after a vertebral fracture, Mrs. O. was virtually housebound. She would climb stairs by resting on each landing. Her husband carried a chair so she could sit and rest and wait for back spasms to subside. She walked with a stationary walker in her home. She and her husband were thinking they would have to sell their home and move back to the city so she could attend medical appointments, start an exercise program, and perhaps go shopping on level ground. She felt guilty about how dependent she was on her husband.

Assessment suggested that she needed to learn to negotiate stairs effectively. Barbara Beatty arranged for a wheeled walker, and Mrs. O. began a gradually progressed walking program back and forth across the 40-foot deck of the home. Once she was able to get out of the house, Mrs. O. began taking tai chi classes. Over the next 12 months she mastered the stairs and walked 30 minutes daily without a walker back and forth on her deck overlooking the ocean. Tai chi taught her to stand indefinitely on one foot. She gradually substituted 40 minutes of tai chi for her deck walks. After resuming driving, she drove regularly to a flat area to walk and bird watch with her friend. She summed up her improvement by stating, "I have my life back!"

SUMMARY

- Osteoporosis is defined according to bone mineral density. A BMD t score more than 2.5 standard deviations below the young normal mean ($t < -2.5$) corresponds with osteoporosis.
- Osteoporosis and osteoporotic fractures are a major and increasing problem in the Western world. In the United States, the medical cost of hip fractures alone is $4 billion.
- Exercise is not the most powerful therapy to increase bone mineral, but it is the only therapy that can simultaneously modify several key risk factors for osteoporotic fracture, including muscle strength, BMD, and dynamic balance. It may also improve quality of life.
- As yet, no direct evidence exists that exercise reduces the risk of osteoporotic fracture. However, until exercise studies are funded to the same extent as some of the large pharmaceutical trials, clinicians ought to embrace the theoretical basis behind exercise prescription in osteoporosis and prescribe safe, appropriate exercises.
- Effective exercise programs for people with osteoporosis are summarized and an example prescription is provided. People with osteoporosis should avoid certain exercises, such as sit-ups and rapid twisting movements.

References

1. Ayalon J, Simkin A, Leichter I, et al. Dynamic bone loading exercises for postmenopausal women: effect on the density of the distal radius. *Arch Phys Med Rehabil* 1987;68:280-283.
2. Bravo G, Gauthier P, Roy P-M, et al. Impact of a 12-month exercise program on the physical and psychological health of soteopenic women. *J Am Geriatr Soc* 1996;44:756-762.
3. Prince RL, Smith M, Dick IM, et al. Prevention of postmenopausal osteoporosis: a comparative study of exercise, calcium supplementation and hormone-replacement therapy. *N Engl J Med* 1991;325:1189-1195.
4. Simkin A, Ayalon J, Leichter I. Increased trabecular bone density due to bone loading exercises in postmenopausal osteoporotic women. *Calcif Tissue Int* 1987;40:59-63.
5. Gold DT. The clinical impact of vertebral fractures: Quality of life in women with osteoporosis. *Bone* 1996;18:185S-189S.
6. Coelho R, Silva C, Maia A, et al. Bone mineral density and depression: a community study in women. *J Psychosom Res* 1999;46:29-35.
7. Goeree R, O'Brien B, Pettit D, et al. An assessment of the burden of illness due to osteoporosis in Canada. *J Soc Obstet Gynaecol Can* 1996;18 (suppl, July):
8. Kannus P, Niemi S, Parkkari J, et al. Hip fractures in Finland between 1970 and 1997 and predictions for the future. *Lancet* 1999;353:802-805.
9. Kannus P, Palvanen M, Niemi S, et al. Epidemiology of osteoporotic pelvic fractures in elderly people in Finland: sharp increase in 1970-1997 and alarming projections for the new millennium. *Osteoporos Int* 2000;11:443-8.
10. Wasnich RD. *Epidemiology of osteoporosis*. In: Favus MJ, ed. Primer on the metabolic bone diseases. 3rd ed. Philadelphia, PA: Lippincott-Raven, 1996: 249-251.
11. Ross PD. Risk factors for osteoporotic fracture. *Endocrinol and Metab Clin North Am* 1998;27:289-301.
12. Nelson ME, Fiatarone MA, Morganti CM, et al. Effects of high-intensity strength training on multiple risk factors for osteoporotic fractures. A randomized control trial. *JAMA* 1994;272:1909-1914.
13. Kannus P. Preventing osteoporosis, falls, and fractures among elderly people. *Br Med J* 1999;318:205-206.
14. Beatty B. Osteoporosis: Optimal exercise behaviour. *Canadian Physiotherapy Association Gerontology Newsletter* 1997;Spring:
15. Ekin JA, Sinaki M. Vertebral compression fractures sustained during golfing: report of three cases. *Mayo Clin Proc* 1993;68:566-70.
16. Cummings SR, Nevitt MC, Browner WS, et al. Risk factors for hip fracture in white women. Study of Osteoporotic Fractures Research Group. *N Engl J Med* 1995;332:767-73.
17. Carter ND, Khan KM, McKay HA, et al. Osteofit—a community-based strength and balance training

program improves dynamic balance: A single blind, randomized, controlled trial in 65-75 yr old women with osteoporosis. *J Am Geriatr Soc* (submitted).

18. Preisinger E, Alacamlioglu Y, Pils K, et al. Therapeutic exercise in the prevention of bone loss: a controlled trial with women after the menopause. *Am J Phys Rehabil* 1995;74:120-123.
19. Tinetti ME, Baker DI, McAvay G, et al. A multifactorial intervention to reduce the risk of falling among elderly people living in the community. *N Engl J Med* 1994;331:821-827.
20. Lord SR, Ward JA, Williams P, et al. The effect of a 12-month exercise trial on balance, strength, and falls in older women: a randomized controlled study. *J Am Geriatr Soc* 1995;43:1198-1206.
21. Randell A, Sambrook PN, Nguyen TV, et al. Direct clinical and welfare costs of osteoporotic fractures in elderly men and women. *Osteoporos Int* 1995;5:427-432.
22. Knott, M, Voss DE. 1968. *Proprioceptive neuromuscular facilitation, patterns and techniques.* (2nd ed.) New York: Harper and Row.
23. Preisinger E, Alacamlioglu Y, Pils K, et al. Therapeutic exercise in the prevention of bone loss: a controlled trial with women after the menopause. *Am J Phys Rehabil* 1995;74:120-123.
24. Malmros B, Mortenson L, Jensen MB, et al. Postive effects of physiotherapy on chronic pain and performance in osteoporosis. *Osteoporosis Int* 1998;8:215-221.
25. Mitchell SL, Grant S, Aitchison T. Physiological effects of exercise on post-menopausal osteoporotic women. *Physiotherapy* 1988;84:157-163.
26. Bravo G, Gauthier P, Roy PM, et al. A weight-bearing, water-based exercise program for osteopenic women: its impact on bone, functional fitness, and well-being. *Arch Phys Med Rehabil* 1997;78:1375-80.
27. Sinaki M, Mikkelsen BA. Postmenopausal spinal osteoporosis: flexion versus extension exercises. *Arch Phys Med Rehabil* 1984;65:593-6.
28. Kronhed AC, Moller M. Effects of physical exercise on bone mass, balance skill and aerobic capacity in women and men with low bone mineral density, after one year of training-a prospective study. *Scand J Med Sci Sports* 1998;8:290-298.

part IV

Intense Physical Activity and Bone Health

The first three parts of this book focused on the positive effects of physical activity on bone health. This part addresses potential deleterious effects—after all, every therapy has potential side effects. Before we tackle the specific topics in this part, we feel this is an appropriate place to raise an issue people are often reluctant to talk about: the lack of reporting of adverse events.

With some notable exceptions [1], exercise intervention studies usually do not report adverse outcomes. Despite our optimistic hope that this is because exercise *never* causes problems, we fear that injuries are not being reported. If subjects drop out because of injury, this should be noted and reported, even if it is a benign injury such as a muscle strain. Until we are confident that we truly know the rates of injuries in exercise interventions, all of us who prescribe exercise must be especially careful and safety conscious. This is particularly relevant when patients and subjects already have compromised bone health (e.g., osteopenia, osteoporosis, osteoarthritis, or other medical conditions).

Part IV is concerned with the long-term side effects of high levels of activity, rather than with the acute injuries that may occur as a result of incorrect exercise prescription. There is increasing awareness that athletes who overtrain or undereat may develop skeletal problems that do not beset those who train in moderation. Chapter 16 addresses the association between athletic activity and bone mineral when menstrual disturbance is present. Chapter 17 is concerned with fractures that result from overuse, rather than from single traumatic events. Recent research has added a great deal to our understanding of the factors that influence the development of stress fractures in athletes.

Reference

1. Heinonen A, Kannus P, Sievänen H, et al. Randomised control trial of effect of high-impact exercise on selected risk factors of osteoporotic fractures. *Lancet* 1996;348:1343-1347.

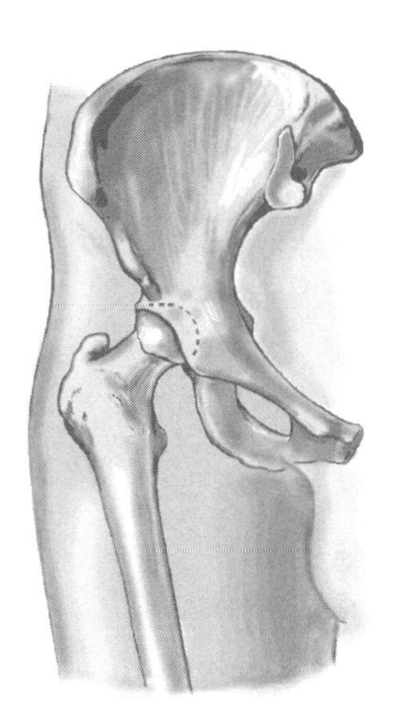

Chapter 16

Skeletal Effects of Menstrual Disturbance

While this chapter reports the potential ill effects of athletic overactivity, we believe the significance of this problem can be compared to passengers' concern for the martini glasses falling over when the *Titanic* hit the iceberg. A larger problem looms. Inactivity is an enormous, and increasing, problem in our society, and it threatens to sink the health of current and future generations [1].

High levels of physical activity have been blamed for various skeletal problems. Researchers have claimed that intense athleticism can delay menarche and limit peak bone mineral accrual in girls and adolescent females [2]. Also, some experts fear that athletic training can retard linear growth in females [3]. We begin this chapter by discussing these pediatric and adolescent issues.

Experts in many circles are concerned that menstrual disturbance may be associated with suboptimal increases in bone mass and perhaps even bone loss. This raises the following questions:

What are the effects of menstrual disturbance on bone mass?

Will all athletes with menstrual disturbance develop osteopenia (low bone mass)?

Is an athlete with menstrual disturbance at greater risk for osteoporosis in later life?

In this chapter, we address each of these issues regarding menstrual disturbance and bone mineral in both girls and women. The chapter concludes with a discussion of the recently discovered physiological mechanisms associated with bone loss in active women and a brief discussion of the treatment of amenorrhea.

Stress fracture, an injury that appears to occur more frequently in women with menstrual disturbance than in those without, is not discussed in this chapter. As it is a condition that also affects men and eumenorrheic women, it is the subject of chapter 17.

Delayed Menarche and Bone Mass

The timing of menarche, the first menstrual period, is controlled by genetic and hormonal variables in addition to being influenced by social and lifestyle variables in a way that is not yet fully understood [4]. These latter variables include nutrition, socioeconomic status, and family size [5].

Menarche occurs later in athletes than in nonathletes, particularly in certain sports such as ballet, gymnastics (figure 16.1), and running [6]. Delayed menarche or primary amenorrhea could theoretically be associated with a lower rate of bone mineral accretion during adolescence and therefore decreased peak bone mass [7, 8]. The relationship between age of menarche and bone density in female athletes is unclear, however, with some investigators finding statistically significant, but moderate to weak, negative correlations at a number of bone sites [2, 9, 10] and others finding no significant correlations (see also chapter 7, which reviews the influence of normal endocrine function on bone mineral) [11-13].

These data are not definitive since most of these studies were cross-sectional, the sample sizes were small, and the influence of confounding variables was not taken into account. In some cohorts of healthy adolescents and pre- and postmenopausal women, a significant negative correlation between menarcheal age and bone mass occurred [7, 14, 15]. In contrast, the University of Saskatchewan's longitudinal bone mineral accrual study showed no relationship between bone mineral accrued in the two years around the adolescent height spurt and age at menarche [16]. Further, no association was noted between the absolute amount of total body bone mineral content at final measurement and the age at menarche in this healthy population [16].

The contribution of factors such as genetics on age of menarche and bone mass remains a fertile area for research. Dr. Robert Malina, a fellow of the American College of Sports Medicine and respected professor and researcher in growth and development, recently presented the conclusions of two comprehensive discussions of exercise and reproductive health of women:

> *"Although menarche occurs later in athletes than non-athletes, it has yet to be shown that exercise delays menarche in anyone." [17]*

> *"The general consensus is that while menarche occurs later in athletes than in non-athletes, the relationship is not causal and is confounded by other factors." [18]*

We feel these conclusions accurately summarize the current evidence.

Intense Physical Training and Linear Growth

Research has suggested that menstrual disturbance may be associated with submaximal linear growth [3]. If intense physical activity were to compromise linear growth, it would diminish eventual adult height. An early longitudinal

FIGURE 16.1 While gymnasts have a later age of menarche than nonathletes, no convincing evidence exists that exercise itself is a cause of delayed menarche.

study compared the growth proportions of ballet dancers with same-age controls and found that pre-, peri- and postpubertal female ballet dancers had a decreased upper to lower body ratio ("eunuchoid proportions"). This characteristic is consistent with the hypogonadotrophic hypogonadism associated with undernutrition [19]. As athletes may have been selected based on a specific body type, Warren compared this ratio and the upper arm span of dancers with their siblings [19]. Although significant differences existed between the groups for these parameters (dancers had smaller ratios and wider arm spans), no difference existed in relative height between dancers, their siblings, and their mothers.

Longitudinal studies lend strong support to the idea that children self-select into sports suited to their body shape and size. For example, the standardized heights of children competing in gymnastics are closely associated with standardized mid-parent height values (figure 16.2) [20]. This suggests that genetically shorter individuals are

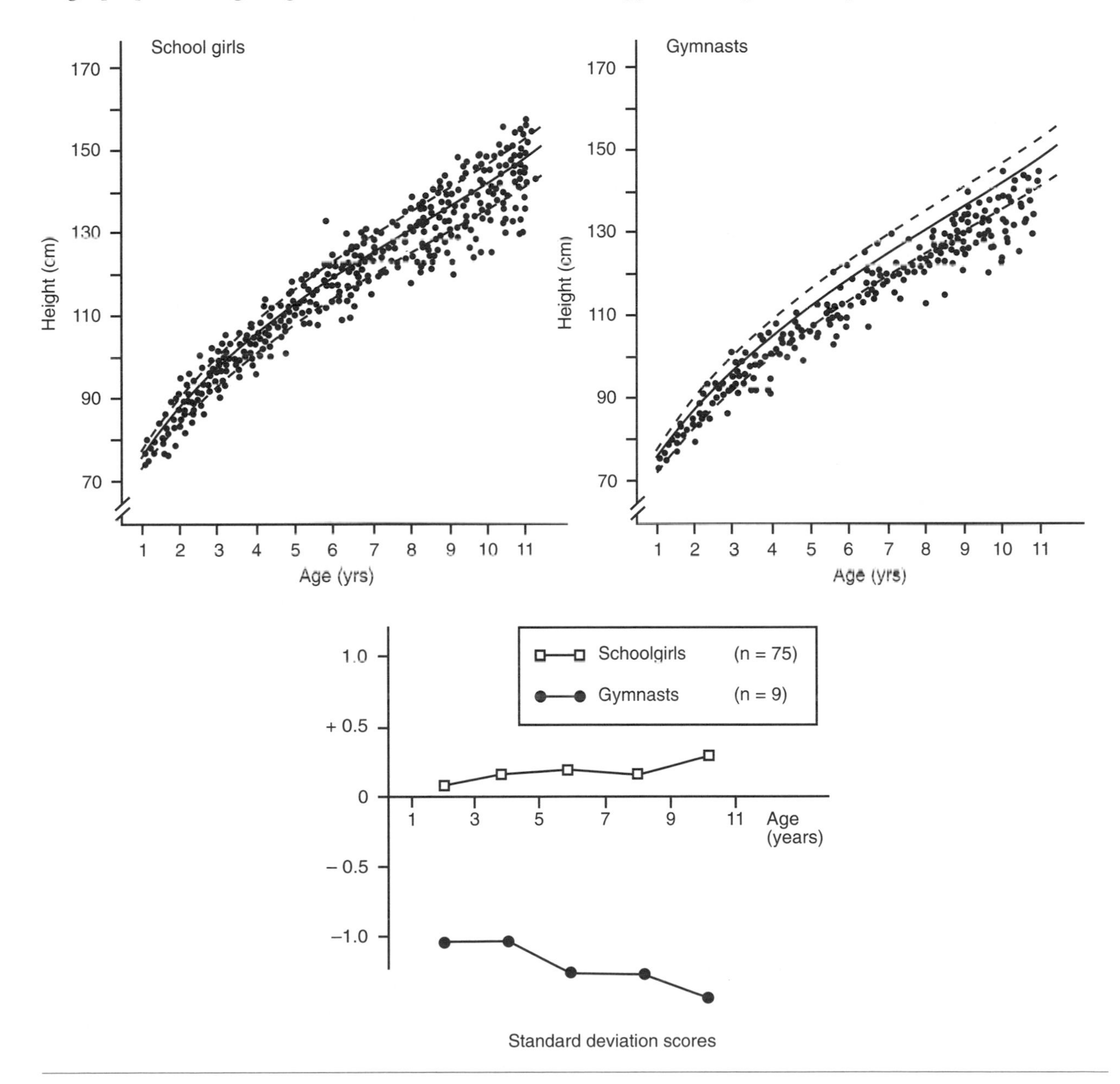

FIGURE 16.2 A longitudinal study of gymnasts found them to be smaller than their nonathletic counterparts from a young age.

Reprinted, by permission, from Peltenburg et al., 1984, "A retrospective growth study of female gymnast and girl swimmers," *Int J Sports Med* (5):262-267.

selected into aesthetic sports, where performance is closely linked to a small body habitus. This is not unlike the phenomenon of the taller-than-average children of national team basketball players being chosen by coaches for higher-level basketball leagues based on their large stature. Taller or early maturing children tend to drop out of gymnastics at a young age [20], just as shorter children tend to drop out of basketball.

Further support for self-selection comes from two recent Australian studies. These studies monitored the growth rate and hormone levels of boys and girls in elite gymnastics and found that athletes and size-matched, normally active control children did not differ in either of these factors [21, 22].

Despite these recent data, the issue remains controversial. A number of important questions related to the relationship of intense training to normal bone size and mass remain to be answered. Until these questions are addressed with well-designed long-term longitudinal studies in which genetics are controlled for, it seems inappropriate to suggest that delayed or late menarche results in either low bone mass or short stature. It also seems prudent for coaches, clinicians, and parents to advise young athletes to consume an adequate and healthy diet, and for pubertal female athletes to modify training intensity so as to avoid altered menstrual function.

Dr. Robert Malina concluded:

> *It has been suggested that intensive training may delay the timing of the growth spurt and stunt the spurt in female gymnasts. Unfortunately, the data are not sufficiently longitudinal to warrant such a conclusion, and many confounding factors are not considered, especially the rigorous selection criteria for gymnastics (and dance), marginal diets, short parents and so on. Female gymnasts show the growth and maturation characteristics of short normal, slow maturing children with short parents. Male gymnasts also present with consistently short stature and late maturation, but these trends are not attributed to intensive training. [5, 23, 24]*

Evidence seems to be insufficient at present to blame lack of height on intensive training.

Menstrual Disturbance and Bone

We now focus on menstrual disturbances and their associations with BMD. We define terms, outline the prevalence of this phenomenon in athletes, briefly summarize the physiological aberrations that cause the absence of periods, and review the literature that examined the association between menstrual flow and BMD. The term *menstrual disturbance* is used to include amenorrhea, oligomenorrhea, and menstrual cycles with shortened luteal length (a shortened period of time between ovulation and the onset of menstruation, less than 10 days, whereas normal is 14 [25]) unless otherwise specified. *Eumenorrhea* is defined as 10-13 cycles per year, *oligomenorrhea* as 3-6 cycles per year, and *amenorrhea* as fewer than 3 cycles per year or no cycles for the past six months [26]. Amenorrhea is commonly categorized as either primary (the woman has never menstruated) or secondary (menstruation has been present for a variable period of time in the past and has ceased). Note that particularly in the early postmenarcheal years, anovulatory cycles are not uncommon.

Physiology of Menstrual Dysfunction

There are two main hypotheses about the mechanism of altered menstrual cycles [17]. The first is that the adrenal axis inhibits activation of the hypothalamic gonadotropin releasing hormone (GnRH) pulse generator (figure 16.3), which is consistent with the findings of mildly elevated cortisol in women with athletic amenorrhea [27]. In this type of amenorrhea, the athlete does not generate luteinizing hormone (LH) pulses and there is loss of frequency and amplitude of LH release. Consequently, no follicular or luteal development occurs [27] and levels of estrogen and progesterone are low. Because LH and FSH remain low, the endocrine pattern is termed hypogonadotrophic hypogonadism.

Stimulating the pituitary in amenorrheic athletes with exogenous GnRH causes a rapid and substantial LH/FSH response [27], suggesting that the failure to release gonadotropin is due to decreased activity of the hypothalamic GnRH pulse generator. As strenuous and prolonged exercise inevitably activates the adrenal axis, this model suggests that certain "stressful" forms of exercise training are also likely to be associated with a negative effect on BMD via hypercortisolism.

The second hypothesis has been known as the energy drain hypothesis [28]. According to this hypothesis caloric intake does not meet the demands of exercise training. Amenorrheic athletes report caloric intakes that are low for their activity levels [29, 30]. Thyroid hormone levels are

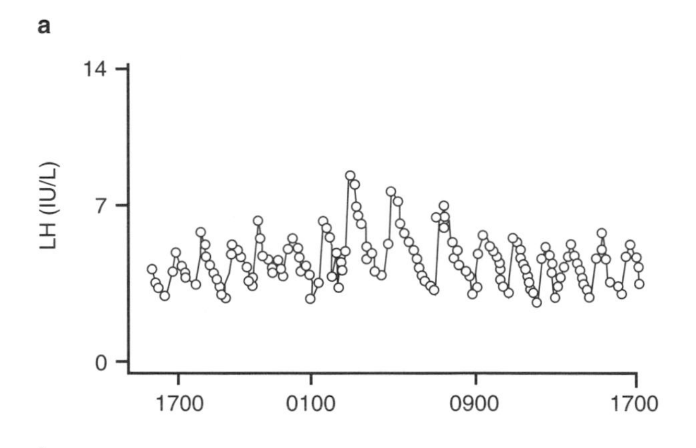

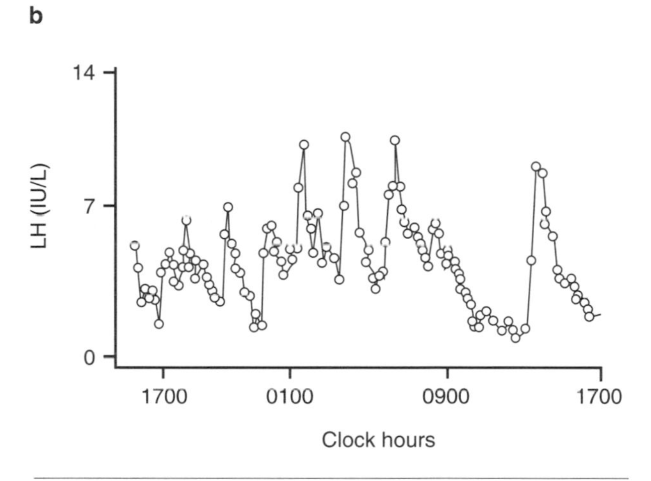

FIGURE 16.3 Luteinizing hormone (LH) during the menstrual cycle: *(a)* normal LH pulsatility and *(b)* lack of pulsation frequency and low amplitude.

low (as in critically ill, malnourished patients) [31], but normal thyroid stimulating hormone levels exclude hypothyroidism [30]. The data that support this model suggest that the caloric deficiency, not the exercise per se, results in the failure of bone mineral to accrue. The clinical implication is that athletes may be able to improve their nutrition and regain bone health without having to reduce training loads. However, further prospective studies are needed to validate this hypothesis.

Menstrual Disturbance and Bone Mass

Research evaluating the effect of current menstrual status on bone density mostly consists of cross-sectional studies in which athletes currently experiencing amenorrhea or oligomenorrhea are compared with their eumenorrheic counterparts and/or sedentary controls. These studies are summarized in appendix A, table A.4 and have been reviewed in detail elsewhere [15]. A major problem in this area of research is that factors known to influence bone density, such as body weight and BMI, training habits, diet, and soft tissue composition, often differ among menstrual groups or are generally not documented. Other important methodological issues include small subject numbers; different definitions of menstrual status; inadequate documentation of estrogen levels or ovulation; and differences in subject age, type of sport, activity level, technique used to assess bone density, and measurement site.

Results of Longitudinal and Cross-Sectional Studies. Several longitudinal studies have reported vertebral bone loss of about 3-4% per year in athletes with amenorrhea [32-34]. Taafe and colleagues [35] reported longitudinal (8- and 12-month) changes in BMD in 20-year-old gymnasts and runners, some of whom had eumenorrhea or oligomenorrhea at baseline. "Bone density changes did not differ by menstrual status," although the power to detect such differences was not reported and would have been small [35].

Many studies have reported a significant association between current menstrual status and BMD in sedentary and exercising subjects [15]. In summary, those studies found that currently amenorrheic running athletes had lower BMD than oligomenorrheic athletes who in turn had lower BMD than eumenorrheic athletes [36-38]. Prior menstrual history has also proven to be a strong, and perhaps the best [39], predictor of vertebral BMD [11, 40-42]. More specifically, both Keay [43] and Rencken [44] found that three factors—duration of amenorrhea, delayed menarche, and body weight—explained about a third of the variation in vertebral BMD among amenorrheic dancers and runners, respectively. Biochemical studies confirmed that subjects were hormone deficient [44]. Recent evidence suggests that decreased bone formation (as opposed to increased bone turnover) is the cause of low bone mass (see "Mechanism of Bone Loss" on page 209).

Affected Sites. Although the low bone mass and bone loss associated with menstrual disturbance is most commonly seen at the lumbar spine (trabecular bone), it is also evident at other sites. Amenorrhea in runners is associated with low femoral neck and femoral shaft BMD compared with controls [11]. This has been corroborated subsequently in several studies of amenorrheic runners [42, 44-47]. There is yet no evidence that femoral BMD is diminished in amenorrheic women in sports that generate high ground reaction forces such as gymnastics [10] and ballet [48].

At one time it was thought that a history of menstrual disturbance was not associated with decreased forearm BMD [49-54]. However, more recent studies have reported lower radial cortical BMD in amenorrheic athletes [34] and dancers [2].

Although amenorrhea is the most obvious sign of reproductive hormone disturbance in young women, menstrual frequency is not a precise index of ovarian hormone status [55]. Exercise may cause subtle changes in reproductive hormone levels and ovulatory dysfunction, while menstrual periods appear fairly regular. This phenomenon has been described as a luteal phase defect or a short luteal phase.

Luteal Phase Defects and Anovulation. Women with normal menstrual cycle duration and normal flow characteristics but with a short luteal phase have decreased progesterone production [56-58]. This menstrual disturbance may be significant, as progesterone is considered by some to promote bone formation [58-61].

Much of the evidence to support this comes from Canadian studies. In a prospective study involving eumenorrheic women, two thirds of whom were runners, recurrent short luteal phase cycles and anovulation were associated with spinal bone loss of approximately 2-4% per year [57]. Serum progesterone levels and the proportion of the total menstrual cycle spent in the luteal phase were significant predictors of both lumbar spine BMD [62] and rate of change of bone mass at this site [57]. A subsequent cross-sectional study, however, failed to find a significant difference in spinal bone density between groups with short and long luteal phases [63].

A more recent study revisited the Prior et al. [57] data to see if exercise-related effects on spinal trabecular BMD in regularly cycling premenopausal women were related to luteal phase length. The researchers found independent positive effects of both luteal length ($p = 0.001$) and activity ($p = 0.041$) on BMD. The 11 runners with luteal length of 11 or more days had a nonsignificant 0.5% increase in lumbar BMD, while the 15 who averaged luteal length of less than 10 days experienced a significant 3.6% loss [61].

Although further research is needed to clarify the effects of subclinical menstrual disturbances on bone density at different sites, it may be worth monitoring ovulatory status and luteal phase length in athletes training intensely. This can be achieved noninvasively via measurement of basal body temperature (BBT) or urinary analysis of hormonal levels. Body temperature monitoring is cheap and reliable if performed correctly, but compliance is often low [58].

Reversibility of Bone Loss. Short-term (14 months) studies [32, 34, 64] showed apparent reversal of low bone mass as formerly amenorrheic athletes augmented their vertebral bone density an average of 6% during the study period. Unpublished data showed that these gains slowed to 3% for the next year and then stopped for the next two years of observation [41]. A recent paper [26] reported an average eight-year follow-up in 29 subjects from an earlier study of amenorrheic athletes [41]. After making reasonable and necessary adjustments to the data to correct for different types of instruments used to measure subjects on the two occasions, the authors reported that after chronic amenorrhea and oligomenorrhea, low vertebral BMD did not appear to respond to the resumption of apparently normal menstrual cycles [26].

These data are supported by the results of the cross-sectional study by Micklesfield and colleagues [42] referred to earlier. A history of menstrual disturbance was associated with lower spinal BMD despite menstrual periods being regular at the time of measurement. These authors followed up 19 marathon runners with a history of menstrual irregularity and found that restoration of lumbar spine BMD was slow and did not reach the same level as age-related controls [46].

Prevalence of Menstrual Disturbance in Athletes

Reports of the prevalence of menstrual disturbances in athletes vary considerably depending on the definition of menstrual disturbance and on the population surveyed in terms of sport, age, activity level, parity, and nutritional status. Nevertheless, the prevalence in athletes (1-44%) is greater than that in the general population (2-5%) [15]. Younger nulliparous women of excessive leanness who train intensely appear to be particularly at risk. In a questionnaire survey of 226 elite athletes, the prevalence of menstrual disturbance was higher in ballet dancers (52%), gymnasts (100%), lightweight rowers (67%), and distance runners (65%) than in swimmers (31%) or those playing team sports (17%) [15]. Menstrual disturbances clearly represent a significant problem in some sports in the female athletic community.

Other Factors Contributing to Osteopenia

Although it is apparent that hypogonadism can be associated with bone loss, particularly at trabecular sites, not all athletes with menstrual disturbances develop osteopenia. Other factors such as type of sport, genetic background, body composition, and calcium intake determine the impact of menstrual disturbance on bone mass.

At sites subjected to high mechanical loads during exercise, detrimental effects may in fact be partially or completely offset. Indirect support for this theory comes from findings of greater bone density in highly active women with anorexia nervosa compared with less active women with the same condition [65], and in amenorrheic athletes compared with amenorrheic nonathletes [34, 49, 54, 66]. In a cross-sectional study, 44 elite teenage ballet dancers with oligomenorrhea or delayed menarche had significantly higher bone density at weight-bearing sites compared with 18 sedentary amenorrheic girls with anorexia nervosa [49]. This implies that sports such as ballet may provide relative protection against the deleterious effects of menstrual disturbance at the significantly loaded sites.

The much higher bone mass commonly observed in gymnasts compared with controls was attenuated in amenorrheic athletes [10]. At the lumbar spine, BMD of amenorrheic gymnasts was significantly lower than that of their eumenorrheic teammates, but both lumbar spine and total body BMD were the same as that of the control subjects (figure 16.4). Similarly, low bone mass commonly associated with amenorrhea was not observed in amenorrheic rowers [67] and figure skaters [68]. Also, spinal trabecular bone density was higher in amenorrheic rowers compared with amenorrheic runners and dancers [50, 67]. This suggests that rowing, an activity that uses back musculature, may provide a more effective osteogenic stimulus at the lumbar spine than other sports, which do not load the spine to the same extent.

The fact that different forms of exercise may partially or fully compensate for the effects of menstrual disturbance at some skeletal sites may be due to the production of local or systemic factors or to higher mechanical loads exerted on the skeleton. The latter explanation fits with Frost's mechanostat model (see chapter 3). Based on this model, low estrogen levels increase the set-points for mechanical strain such that higher mechanical loads are required to maintain or increase bone mass in the presence of estrogen deficiency [69, 70]. While some forms of exercise (such as gymnastics or rowing) may provide sufficient strain to reach these increased set-points, others (such as running) may not. This is supported by studies showing that during running, vertical ground reaction forces vary from two to three times body weight, while jumping and landing activities generate considerably higher forces, up to 12 times body weight [71]. Thus, the degree to which menstrual disturbances affect bone appears to depend on the type of sport and the pattern of loading.

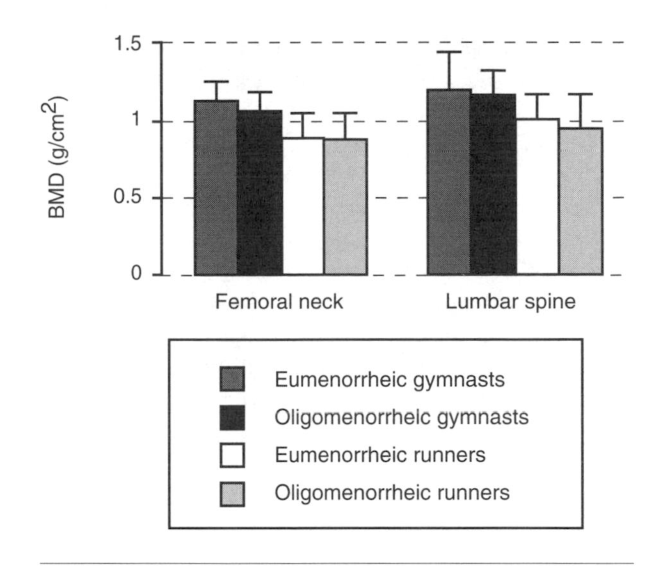

FIGURE 16.4 Despite oligomenorrhea and amenorrhea, gymnasts had higher BMD than even eumenorrheic runners [10].

Peak bone mass is determined by an interaction between genotype and environment. While genetic endowment likely determines the potential to accrue bone, other factors affect what level an individual reaches [72]. Therefore, the clinical consequences of menstrual disturbances in athletes will be partly determined by genetic influences. For example, if a sportswoman has a genetic potential for peak bone mass that is above average, but fails to reach this level or loses bone prematurely, her BMD may still fall within normal levels, resulting in no adverse skeletal consequences. Thus, menstrual disturbance may be of less concern in an athlete blessed with "good genes." Conversely, a less skeletally advantageous genetic makeup may render the presence of menstrual disturbance of greater concern to the woman. This is represented diagrammatically in figure 16.5.

Body weight has a positive effect on bone density (see chapter 6). As a result of the significant interaction among menstrual history, body weight, and bone mass [41, 73], as the severity of menstrual disturbance increases, body weight becomes a more important predictor of vertebral bone density. Similarly, amenorrheic runners with lower weight and less body fat had significantly lower bone mass than amenorrheic runners with higher relative body fat, or weight. The influence of menstrual disturbance on bone density independent of differences in body weight can be evaluated to an extent by adjusting the data for weight.

Calcium is an important threshold nutrient for bone health (see chapter 8). Can calcium intake modulate the effects of low hormone levels associated with amenorrhea in female athletes? In animals, low calcium intake or reduced calcium bioavailability may limit bone's response to exercise training [74]. Since calcium deficiency is a stimulus for bone resorption, the effect of calcium deficiency and hypogonadism may be additive [70]. Wolman and colleagues [75] reported a linear relationship between calcium intake and trabecular bone density at the lumbar spine in both amenorrheic and eumenorrheic athletes. However, BMD was significantly lower in amenorrheic athletes at all levels of calcium intake. Therefore, even though additional calcium attenuated the bone loss in amenorrheic athletes, it did not compensate for the effects of amenorrhea on the skeleton.

In summary, menstrual disturbance can have detrimental skeletal effects including a predisposition to osteoporosis, particularly at the lumbar spine. Equally important are the data that prove that type of sport, genetics, and body weight also determine the net result of menstrual disturbance. Therefore, we emphasize that menstrual disturbances are not necessarily harmful for all sportswomen, as there are considerable individual differences in predisposition and response to menstrual cycle irregularities.

Athlete Menstrual Disturbance and Osteoporosis

Although it is has been suggested that menstrual disturbance in sportswomen will result in an in-

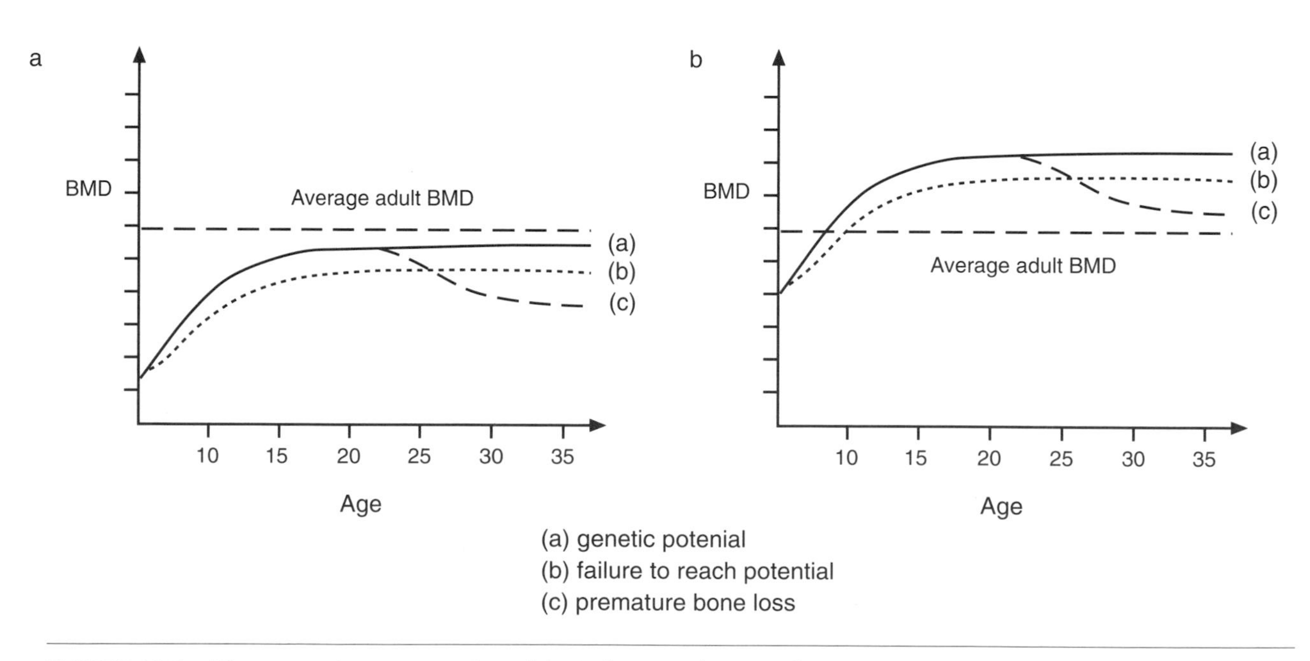

FIGURE 16.5 Diagrammatic representation of the influence of genetic factors on the clinical significance of menstrual disturbances. *(a)* If the athlete has average or below average BMD, she is more susceptible to fracture when BMD is affected by menstrual disturbance. *(b)* If, however, her genetic potential is above that of her colleagues, she is afforded a degree of protection against bone loss associated with menstrual disturbance.

creased risk for osteoporosis in later life, this cannot be confirmed or refuted given that no longitudinal studies have followed female athletes over several decades to assess patterns of change in bone mass. If improvements in bone mass can be made following resumption of menses, or as a result of pharmacologic or nonpharmacologic interventions, then the future risk of osteoporosis may not be influenced by the presence of menstrual disturbance. However, if significant deficits in bone mass remain, and bone loss with aging follows the usual pattern of loss for the female population, menstrual disturbance will have an impact on the future risk for osteoporosis.

As discussed in chapters 14 and 17, low bone mass is not the only determinant of osteoporotic fracture risk. Bone geometry, bone microstructure (see chapter 3), and the likelihood of falling [76] all contribute to fracture risk. Further longitudinal studies are required to determine whether the exercise that is also associated with menstrual disturbance improves these other risk factors for fracture. It should be noted, however, that from a clinical perspective hip fracture and not spine or forearm fracture causes the most disability. Menstrual disturbances have their greatest adverse effects at the lumbar spine and not at the proximal femur.

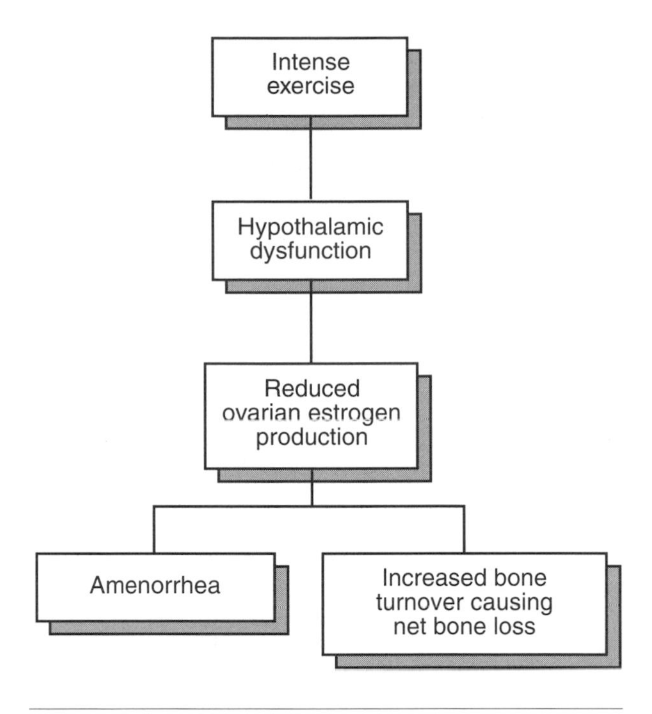

FIGURE 16.6 Flow chart showing traditional explanation of estrogen-associated bone loss with amenorrhea. This notion has been undermined by recent data [30].

Mechanism of Bone Loss

The mechanism underlying bone loss in athletic women with menstrual disturbance is an issue of great current interest, and it appears that we are witnessing a major paradigm shift. Traditional dogma explained premature bone loss in young women as essentially due to estrogen deficiency (figure 16.6). A well-designed study by Cathy Zanker, however, vigorously disputes the primary role of estrogen in exercise-associated bone loss [77-79]. Zanker and colleagues studied bone turnover in women with exercise-associated amenorrhea using biochemical markers (see chapter 4). Biomarkers proved that bone was in a state of reduced turnover, with markedly decreased bone formation, in direct contrast to bone in estrogen deficiency, which remodels excessively (high levels of bone formation and resorption with net bone loss). The state of excessive bone resorption can be treated with estrogen, but amenorrheic athletes did not display that pattern of bone turnover.

Zanker's data explain why treatment of amenorrheic women with estrogen has not led to the same gains in bone mineral as it has in postmenopausal women [80, 81]. Zanker's work forces exercise physiologists and endocrinologists to seek the mechanisms that underpin decreased bone formation. Undernutrition is an obvious candidate, as women with anorexia nervosa have a comparable clinical pattern of amenorrhea, bone loss, and biomarkers showing reduced bone formation. With refeeding they can recover bone mass and improve biochemical markers of bone formation even without resumption of menses. Also, malnutrition plays a role in other forms of osteoporosis [82]. This model suggests that the primary problem in exercise-associated bone loss is energy deficit rather than low estrogen (figure 16.7).

Interesting physiological correlations exist between undernutrition and impaired bone formation. Acute or chronic energy deficits can cause low T_3 syndrome and IGF-I deficiency. In an experimental fast, subjects had reduced bone formation and IGF-I that was rectified by infusion of IGF-I [83]. IGF-I administered to women with anorexia nervosa is associated with increased levels of serum bone formation markers [84]. The implication of this exciting new work is that

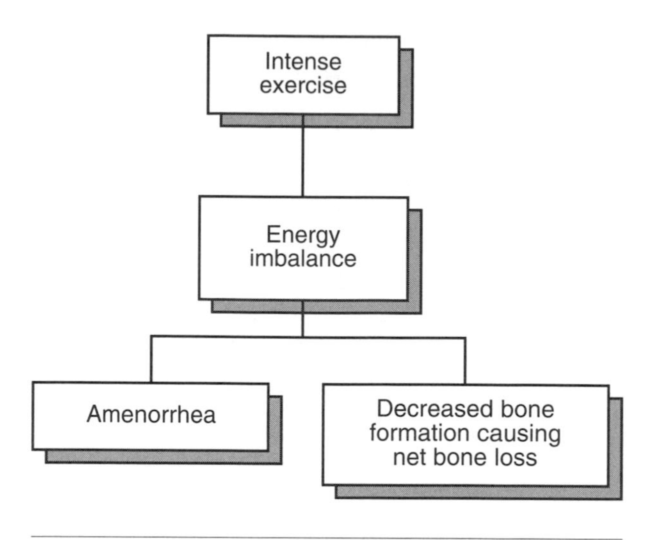

FIGURE 16.7 Zanker's model showing that energy imbalance is the key factor influencing bone's response to intense exercise.

therapy for exercise-associated bone loss should target the return of bone formation to normal, rather than antiresorptive therapy [30].

Treatment of Amenorrheic Women

For the reasons outlined earlier (in the section "Mechanism of Bone Loss"), traditional therapy for young women with amenorrhea has been estrogen and the oral contraceptive pill [85]. Individuals who have not finished growing are usually denied hormonal therapy as it can cause epiphyseal closure. Not surprisingly, according to Zanker's findings [77], treatment of amenorrhea with oral contraceptives has not proven universally successful [85, 86], and theoretical reasons now exist that may explain why [30]. We recommend that practitioners pay close attention to any potential energy imbalance in female athletes with amenorrhea.

SUMMARY

- There are no convincing data that intense exercise stunts linear growth or that exercise causes delayed menarche.
- There are no convincing data that delayed menarche is a cause of failure to achieve peak bone mass.
- Prolonged menstrual disturbance and hypogonadism appears to promote loss of bone mineral, and this is not necessarily reversible with return of normal menstrual cycles.
- Bone loss associated with hypogonadism occurs primarily at the trabecular bone of the vertebral bodies and less at cortical sites such as the femur.
- Although hypogonadism can be associated with bone loss, not all athletes with menstrual disturbances develop osteopenia. Thus, type of sport, genetics, and body weight also determine the net result of menstrual disturbance.
- We emphasize that menstrual disturbances per se are not necessarily harmful for all sportswomen as there are considerable individual differences in predisposition and response to menstrual cycle irregularities.
- It appears that energy imbalance, rather than low circulating estrogen, may be the key factor influencing bone's response to intense exercise. The clinical implication of this finding is that in a woman with exercise-associated menstrual disturbance, any treatment of bone loss should concentrate on improving energy balance, rather than medicating with estrogen.

References

1. Booth FW, Gordon SE, Carlson CJ, et al. Waging war on modern chronic diseases: primary prevention through exercise biology. *J Appl Physiol* 2000;88:774-87.
2. Warren MP, Brooks-Gunn J, Fox RP, et al. Lack of bone accretion and amenorrhea: Evidence for a relative osteopenia in weightbearing bones. *J Clin Endocrinol Metab* 1991;72:847-853.
3. Theintz G, Buchs B, Rizzoli R, et al. Longitudinal monitoring of bone mass accumulation in healthy adolescents: Evidence for a marked reduction after 16 years of age at the levels of lumbar spine and femoral neck in female subjects. *J Clin Endocrinol Metab* 1992;75:1060-1065.
4. Malina RM, Bouchard C. *Growth, maturation, and physical activity*. Champaign, IL: Human Kinetics, 1991:501 pages.
5. Malina RM, Katzmarzyk PT, Bonci CM, et al. Family size and age at menarche in athletes. *Med Sci Sports Exerc* 1997;29:99-106.
6. Malina RM, Spirduso WW, Tate C, et al. Age at menarche and selected menstrual characteristics in athletes at different competitive levels and in different sports. *Med Sci Sports Exerc* 1978;10:218-222.
7. Lu PW, Briody JN, Morely K, et al. Bone mineral density of total body, spine and femoral neck in children and young adults: A cross sectional and longitudinal study. *J Bone Miner Res* 1994;9:1451-1458.
8. Young D, Hopper JL, Nowson CA, et al. Determinants of bone mass in 10- to 26-year-old females: A twin study. *J Bone Miner Res* 1995;10:558-567.
9. Dhuper S, Warren MP, Brooks-Gunn J, et al. Effect of hormonal status on bone density in adolescent girls. *J Clin Endocrinol Metab* 1990;71:1083-1088.
10. Robinson TL, Snow-Harter C, Taafe DR, et al. Gymnasts exhibit higher bone mass than runners despite similar prevalence of amenorrhea and oligomenorrhea. *J Bone Miner Res* 1995;10:26-35.
11. Myburgh KH, Bachrach LK, Lewis B, et al. Low bone mineral density in axial and appendicular sites in amenorrheic athletes. *Med Sci Sports Exerc* 1993;25:1197-1202.
12. Fehily AM, Coles RJ, Evans WD, et al. Factors affecting bone density in young adults. *Am J Clin Nutr* 1992;56:579-586.
13. Rutherford OM. Spine and total body bone mineral density in amenorrheic endurance athletes. *J App Physiol* 1993;74:2904-2908.
14. Katzman DK, Bachrach LK, Carter DR, et al. Clinical and anthropometric correlates of bone mineral acquisition in healthy adolescent girls. *J Clin Endocrinol Metab* 1991;73:1332-1339.
15. Bennell KL, Malcolm SA, Wark JD, et al. Skeletal effects of menstrual disturbances in athletes. *Scand J Med Sci Sports* 1997;7:261-273.
16. McKay HA, Bailey DA, Mirwald RL, et al. Peak bone mineral accrual and age of menarche in adolescent girls: A 6-yr longitudinal study. *J Pediatr* 1998;133:682-687.
17. Loucks AB, Vaitukaitis J, Cameron JL, et al. The reproductive system and exercise in women. *Med Sci Sports Exerc* 1992;24:S288-S293.
18. Clapp JF, Little KD. The interaction between regular exercise and selected aspects of women's health. *Am J Obstet Gynecol* 1995;173:2-9.
19. Warren MP. The effects of exercise on pubertal progression and reproductive function in girls. *J Clin Endocrinol Metab* 1980;51:1150-1157.
20. Peltenburg AL, Erich WB, Zonderland ML, et al. A retrospective growth study of female gymnasts and girl swimmers. *Int J Sports Med* 1984;5:262-267.
21. Daly RM, Rich PA, Klein R, et al. Effects of high impact exercise on ultrasonic and biochemical indices of skeletal status: A prospective study in young male gymnasts. *J Bone Miner Res* 1999;14:1222-1230.
22. Bass S, Bradney M, Pearce G, et al. Short stature and delayed puberty in gymnasts: influence of selection bias on leg length and the duration of training on trunk length. *J Pediatr* 2000;136:149-155.
23. Malina RM, Ryan RC, Bonci CM. Age at menarche in athletes and their mothers and sisters. *Ann Human Biol* 1994;21:417-422.
24. Malina RM, Woynarowska B, Bielicki T, et al. Prospective and retrospective longitudinal studies of the growth, maturation, and fitness of Polish youth active in sport. *Int J Sports Med* 1997;18 (Supplement 3):S179-S185.
25. Bennell K. *The female athlete*. In: Brukner P, Khan K, ed. Clinical sports medicine. 2nd edition. Sydney: McGraw-Hill, 2001: 906 pages.
26. Keen AD, Drinkwater BL. Irreversible bone loss in former amenorrheic athletes. *Osteoporos Int* 1997;7:311-315.
27. Loucks AB, Mortola JF, Girton L, et al. Alterations in the hypothalamic-pituitary-ovarian and the hypothalamic-pituitary-adrenal axes in athletic women. *J Clin Endocrinol Metab* 1989;68:402-411.
28. Warren MP. Effects of undernutrition on reproductive function in the human. *Endocr Rev* 1983;4:363-377.
29. Myerson M, Gutin B, Warren MP, et al. Resting metabolic rate and energy balance in amenorrheic and eumenorrheic runners. *Med Sci Sports Exerc* 1991;23:15-22.
30. Zanker C. Bone metabolism in exercise associated amenorrhoea: the importance of nutrition. *Br J Sports Med* 1999;33:228-229.
31. Loucks AB, Laughlin GA, Mortola JF, et al. Hypothalamic-pituitary-thyroidal function in

eumenorrheic and amenorrheic athletes. *J Clin Endocrinol Metab* 1992;75:514-518.

32. Drinkwater BL, Nilson K, Chesnut CHI, et al. Bone mineral density after resumption of menses in amenorrheic athletes. *JAMA* 1986;256:380-382.
33. Cann CE, Cavanaugh DJ, Schnurpfiel K, et al. Menstrual history is the primary determinant of trabecular bone density in women runners. *Med Sci Sports Exerc* 1988;20:S59.
34. Jonnavithula S, Warren MP, Fox RP, et al. Bone density is compromised in amenorrheic women despite return of menses: A 2-year study. *Obstet Gynecol* 1993;81:669-674.
35. Taaffe DR, Snow-Harter C, Connolly DA, et al. Differential effects of swimming versus weight-bearing activity on bone mineral status of eumenorrheic athletes. *J Bone Miner Res* 1995;10:586-593.
36. Constantini NW. Clinical consequences of athletic amenorrhoea. *Sports Med* 1994;17:213-223.
37. Lloyd T, Myers C, Buchanan JR, et al. Collegiate women athletes with irregular menses during adolescence have decreased bone density. *Obstet Gynecol* 1988;72:639-642.
38. Highet R. Athletic amennorhoea: an update on etiology, complications and management. *Sports Med* 1989;7:82-108.
39. Otis CL, Drinkwater B, Johnson M, et al. ACSM position stand on the female athlete triad. *Med Sci Sports Exerc* 1997;29:i-ix.
40. Grimston SK, Sanborn CF, Miller PD, et al. The application of historical data for evaluation of osteopenia in female runners: the menstrual index. *Clin Sports Med* 1990;2:108-118.
41. Drinkwater BL, Bruemner B, Chesnut CH. Menstrual history as a determinant of current bone density in young athletes. *JAMA* 1990;263:545-548.
42. Micklesfield LK, Lambert EV, Fataar AB, et al. Bone mineral density in mature premenopausal ultramarathon runners. *Med Sci Sports Exerc* 1995;27:688-696.
43. Keay N, Fogelman I, Blake G. Bone mineral density in professional female dancers. *Br J Sports Med* 1997;31:143-147.
44. Rencken ML, Chesnut III CH, Drinkwater BL. Brief report. Bone density at multiple skeletal sites in amenorrheic athletes. *JAMA* 1996;276:238-240.
45. Pettersson U, Stalnacke B, Ahlenius G, et al. Low bone mass density at multiple skeletal sites, including the appendicular skeleton in amenorrheic runners. *Calcif Tissue Int* 1999;64:117-25.
46. Micklesfield LK, Reyneke L, Fataar A, et al. Long-term restoration of deficits in bone mineral density is inadequate in premenopausal women with prior menstrual irregularity. *Clin J Sport Med* 1998;8:155-63.
47. Tomten SE, Falch JA, Birkeland KI, et al. Bone mineral density and menstrual irregularities. A comparative study on cortical and trabecular bone structures in runners with alleged normal eating behavior. *Int J Sports Med* 1998;19:92-7.
48. Khan KM, Green R, Saul A, et al. Retired elite female ballet dancers have similar bone mineral density at weightbearing sites to nonathletic controls. *J Bone Miner Res* 1996;11:1566-1574.
49. Young N, Formica C, Szmukler G, et al. Bone density at weight-bearing and non weight-bearing sites in ballet dancers: The effects of exercise, hypogonadism and body weight. *J Clin Endocrinol Metab* 1994;78:449-454.
50. Drinkwater BL, Nilson K, Chesnut CH, et al. Bone mineral content of amenorrheic and eumenorrheic athletes. *N Engl J Med* 1984;311:277-281.
51. Cann CE, Martin MC, Genant HK, et al. Decreased spinal mineral content in amenorrheic women. *JAMA* 1984;251:626-9.
52. Lindberg JS, Fears WB, Hunt MM, et al. Exercise-induced amenorrhea and bone density. *Ann Int Med* 1984;101:647-648.
53. Nelson ME, Fisher EC, Catsos PD, et al. Diet and bone status in amenorrheic runners. *Am J Clin Nutr* 1986; 43:910-916.
54. Marcus R, Cann R, Madvig P. Menstrual function and bone mass in elite women distance runners: endocrine and metabolic features. *Ann Int Med* 1985;102:158-163.
55. Winters KM, Adams WC, Meredith CN, et al. Bone density and cyclic ovarian function in trained runners and active controls. *Med Sci Sports Exerc* 1996;28:776-785.
56. Beitens IZ, McArthur JW, Turnbull BA, et al. Exercise induces two types of human luteal dysfunction: confirmation by urinary free progsterone. *J Clin Endocrinol Metab* 1991;72:1350-1358.
57. Prior JC, Vigna YM, Schechter MT, et al. Spinal bone loss and ovulatory disturbances. *N Engl J Med* 1990;323:1221-1227.
58. Prior JC, Vigna YM. Ovulation disturbances and exercise training. *Clin Obstet Gynecol* 1991;34:180-90.
59. Karambolova KK, Snow GR, Anderson C. Surface activity on the periosteal and corticoendosteal envelopes following continuous progestogen supplementation in spayed beagles. *Calcif Tissue Int* 1986;38:239-43.
60. Petit MA, Hitchcock CL, Prior JC, et al. Ovulation and spinal bone mineral density. *J Clin Endocrinol Metab* 1998;83:3757-3758.
61. Petit MA, Prior JC, Barr SI. Running and ovulation positively change cancellous bone in premenopausal women. *Med Sci Sports Exerc* 1999;31:780-787.
62. Snead DB, Weltman A, Weltman JY, et al. Reproductive hormones and bone mineral density in women runners. *J App Physiol* 1992;72:2149-2156.
63. Barr SI, Prior JC, Vigna YM. Restrained eating and ovulatory disturbances: possible implications for bone health. *Am J Clin Nutr* 1994;59:92-7.

64. Lindberg JS, Powell MR, Hunt MM, et al. Increased vertebral bone mineral in response to reduced exercise in amenorrheic runners. *West J Med* 1987;146:39-42.
65. Rigotti NA, Nussbaum SR, Herzog DB, et al. Osteoporosis in women with anorexia nervosa. *N Engl J Med* 1984;311:1601-1606.
66. Jones KP, Ravnikar D, Tulchinsky D, et al. Comparison of bone density in amenorrheic women due to athletics, weight loss, and premature menopause. *Obstet Gynecol* 1985;66:5-8.
67. Wolman R, Clark P, McNally E, et al. Menstrual status and exercise as determinants of spinal trabecular bone density in female athletes. *Br Med J* 1990;301:516-518.
68. Slemenda CW, Johnston CC. High intensity activities in young women: Site specific bone mass effects among female figure skaters. *Bone Miner* 1993;20:125-132.
69. Cheng MZ, Zaman G, Lanyon LE. Estrogen enhances the stimulation of bone collagen synthesis by loading and exogenous prostacyclin, but not prostaglandin E2, in organ cultures of rat ulnae. *J Bone Miner Res* 1994;9:805-16.
70. Dalsky GP. Effect of exercise on bone: permissive influence of estrogen and calcium. *Med Sci Sports Exerc* 1990;22:281-285.
71. McNitt-Gray JL. Kinetics of the lower extremities during drop landings from three heights. *J Biomech* 1993;26:1037-46.
72. Kelly PJ, Eisman J, Sambrook PN. Interaction of genetic and environmental influences on peak bone density. *Osteoporos Int* 1990;1:56-60.
73. Linnell SL, Stager JM, Blue PW, et al. Bone mineral content and menstrual irregularity in female runners. *Med Sci Sports Exerc* 1984;16:343-348.
74. Lanyon LE, Rubin CT, Baust G. Modulation of bone loss during calcium insufficiency by controlled dynamic loading. *Calcif Tissue Int* 1986;38:209-216.
75. Wolman RL, Clark P, McNally E, et al. Dietary calcium as a statistical determinant of spinal trabecular bone density in amenorrheic and estrogen-replete athletes. *Bone Miner* 1992;17:415-423.
76. Nguyen TV, Sambrook PN, Kelly PJ, et al. Prediction of osteoporotic fractures by postural instability and bone density. *Br Med J* 1993;307:1111-1115.
77. Zanker CL, Swaine IL. Relation between bone turnover, oestradiol, and energy balance in women distance runners. *Br J Sports Med* 1998;32:167-171.
78. Zanker CL, Swaine IL. The relationship between serum oestradiol concentration and energy balance in young women distance runners. *Int J Sports Med* 1998;19:104-8.
79. Zanker C, Swaine I. Bone turnover in amenorrheic eumenorrheic women distance runners. *Scand J Med Sci Sport* 1998;8:20-26.
80. Hergenroeder AC. Bone mineralization, hypothalamic amenorrhea, and sex steroid therapy in female adolescents and young adults. *J Pediatr* 1995;126:683-689.
81. Manolagas SC, Bellido T, Jilka RL. Sex steroids, cytokines and the bone marrow: new concepts on the pathogenesis of osteoporosis. *Ciba Found Symp* 1995;191:187-196.
82. Heaney RP. Nutritional factors in osteoporosis. *Annu Rev Nutr* 1993;13:287-316.
83. Grinspoon SK, Baum HB, Peterson S, et al. Effects of rhIGF-I administration on bone turnover during short-term fasting. *J Clin Invest* 1995;96:900-6.
84. Grinspoon S, Baum H, Lee K, et al. Effects of short-term recombinant human insulin-like growth factor I administration on bone turnover in osteopenic women with anorexia nervosa. *J Clin Endocrinol Metab* 1996;81:3864-3870.
85. Gibson JH, Mitchell A, Reeve J, et al. Treatment of reduced bone mineral density in athletic amenorrhea: a pilot study. *Osteoporos Int* 1999;10:284-289.
86. DeCree C, Lewin R, Ostyn M. Suitability of cyproterone acetate in the treatment of osteoporosis associated with athletic amenorrhea. *Int J Sports Med* 1988;9:187-192.
87. Fisher EC, Nelson ME, Frontera WR, et al. Bone mineral content and levels of gonadotropins and estrogens in amenorrheic running women. *J Clin Endocrinol Metab* 1986;62:1232-6.
88. Snyder AC, Wenderoth MP, Johnston CC, Jr., et al. Bone mineral content of elite lightweight amenorrheic oarswomen. *Hum Biol* 1986;58:863-9.
89. Ding J-H, Sheckter CB, Drinkwater BL, et al. High serum cortisol levels in exercise-associated amenorrhoea. *Ann Int Med* 1988;108:530-534.
90. Harber VJ, Webber CE, Sutton JR, et al. The effect of amenorrhea on calcaneal bone density and total bone turnover in runners. *Int J Sports Med* 1991;12:505-8.
91. Baer JT, Taper LJ, Gwazdauskas FG, et al. Diet, hormonal, and metabolic factors affecting bone mineral density in adolescent amenorrheic and eumenorrheic female runners. *Journal of Sports Medicine and Physical Fitness* 1992;32:51-58.
92. Myerson M, Gutin B, Warren MP, et al. Total body bone density in amenorrheic runners. *Obstet Gynecol* 1992;79:973-978.
93. Wilmore JH, Wambsgans KC, Brenner M, et al. Is there energy conservation in amenorrheic compared with eumenorrheic distance runners? *J Appl Physiol* 1992;72:15-22.
94. Hetland ML, Haarbo J, Christiansen C, et al. Running induces menstrual disturbances but bone mass is unaffected, except in amenorrheic women. *The American Journal of Medicine* 1993;95:53-60.
95. Pearce G, Bass S, Young N, et al. Does weight-bearing exercise protect against the effects of exercise-

induced oligomenorrhea on bone density? *Osteoporos Int* 1996;6:448-452.

96. Moen SM, Sanborn CF, DiMarco NM, et al. Lumbar bone mineral density in adolescent female runners. *J Sports Med Phys Fitness* 1998;38:234-239.
97. Gibson JH, Harries M, Mitchell A, et al. Determinants of bone density and prevalence of osteopenia among female runners in their second to seventh decades of age. *Bone* 2000;26:591-598.
98. Snead DB, Stubbs CC, Weltman JY, et al. Dietary patterns, eating behaviors, and bone mineral density in women runners. *Am J Clin Nutr* 1992;56:705-711.
99. Cook SD, Harding AF, Thomas KA, et al. Trabecular bone density and menstrual function in women runners. *Am J Sports Med* 1987;15:503-507.
100. Lloyd T, Buchanan JR, Bitzer S, et al. Interrelationship of diet, athletic activity, menstrual status and bone density in collegiate women. *Am J Clin Nutr* 1987;46:681-684.
101. Baker E, Demers L. Menstrual status in female athletes: correlation with reproductive hormones and bone density. *Obstet Gynecol* 1988;72:683-687.
102. Buchanan JR, Myers C, Lloyd T, et al. Determinants of peak trabecular bone density in women: The role of androgens, estrogen, and exercise. *J Bone Miner Res* 1988;3:673-680.

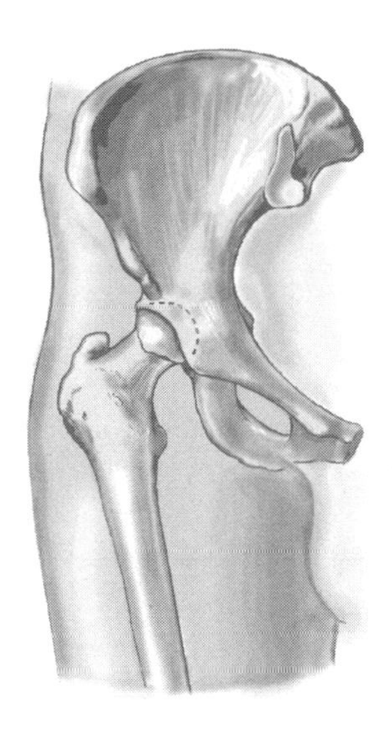

Chapter 17

Stress Fractures

Stress fractures occur relatively commonly in competitive athletes, with incidence rates in prospective studies averaging around 20% per year [1]. Recent research has improved our understanding of how these injuries occur, and how they may be prevented [2].

The purpose of this chapter is threefold. First, we provide background information regarding bone stress injuries. This includes a discussion of the concept that fractures exist on a continuum of bone overuse injury that begins with accelerated bone turnover at one end and extends to complete fractures. Necessary background also includes a summary of the prevalence of stress fractures and an overview of the clinical aspects of stress fracture diagnosis and treatment.

Our second aim in this chapter is to discuss why stress fractures occur. Of course we know that the material and structural properties of bone itself, the forces attenuated by muscle, and the amount of mechanical load on the bone all combine to cause stress fractures. (The biomechanical properties of bone were discussed in chapter 3.) However, because little is known about the precise role of muscle in stress fracture etiology, we focus on the mechanism whereby athletic activity induces stress fracture.

The third aim of the chapter concerns recent advances in knowledge. We summarize what is known about the key risk factors for stress fracture so that exercise specialists can incorporate injury prevention strategies into their training programs.

Continuum of Bone Overuse Injury

Stress fractures were first reported in military recruits in the 19th century and have become increasingly evident among athletes in the last two decades [3]. A stress fracture is a microfracture in bone that results from repetitive physical loading below the single loading cycle failure threshold. Overload stress can be applied to bone through two mechanisms: redistribution of impact forces resulting in increased stress at focal points in bone, and muscle pull across bone. The second of these is often ignored, but its role is evident in the stress fractures of the playing arms of tennis players and in rib fractures of rowers. Stress fracture is not a sudden, all-or-nothing phenomenon, but rather a process that takes time, and that can be stopped along the continuum.

Radiological methods can detect the process of injury in bone at various stages. Radioisotopic imaging ("bone scan") and magnetic resonance (MR) imaging, which can detect changes in bone at the phase of accelerated remodeling, are early detectors of overuse abnormalities in bone. Because CT scanning can reveal small cracks, this test becomes positive for stress fracture a little later in the evolution of a fracture. X-ray is the least sensitive imaging modality for stress fracture and does not always reveal fractures that might appear quite obvious on bone scans, MR imaging, or CT scans.

To facilitate discussion of the various processes along the stress fracture continuum, and because patients at different stages may require different treatments, clinicians have developed nomenclature to refer to the various stages of bone stress injury. The least degree of injury, which represents bony remodeling at a subclinical level, is termed *bone strain* or silent stress reaction, because it is pain free. Note that the term *strain* is used in this clinical context to describe the early phase of bone overuse injury; it has a different connotation when used in a biomechanical setting (see chapter 3). Slightly more significant injury to bone results in a *stress reaction*, which is characterized by bony tenderness clinically with mildly increased uptake of radioisotope on bone scan but without evidence of a defect in bone structure itself. This differs from a *stress fracture*, which is defined by significant focal uptake of radioisotope on bone scanning and evidence of a fracture on CT scan, MR imaging, or plain X-ray. The continuum of overuse injury in bone is illustrated in figure 17.1, and the clinical features of these injuries are summarized in table 17.1.

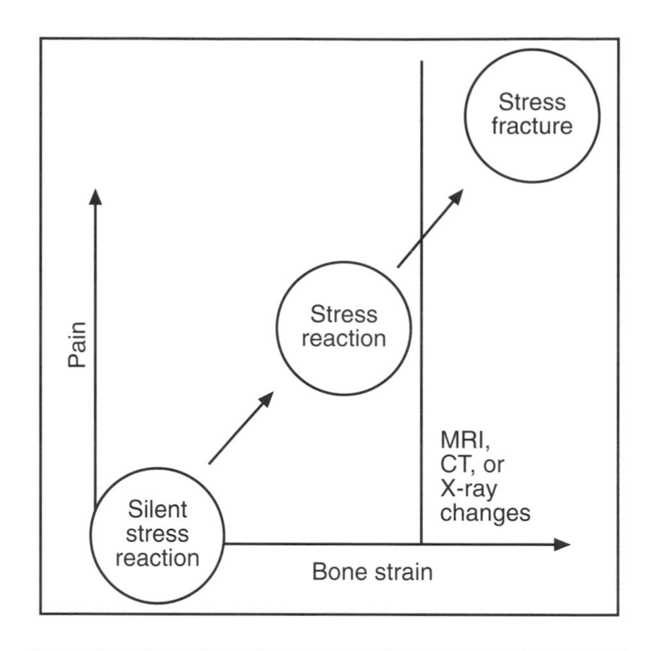

FIGURE 17.1 Continuum of overuse injuries in bone. Note that MR imaging can provide evidence of stress reaction (high signal) or stress fracture (e.g., a cortical defect).

Why Stress Fractures Occur—Pathophysiology

Athletic activity results in repetitive strains that are essential for the maintenance of normal bone strength and can also increase bone mass (see chapters 10 to 13). Intense activity, however, can overwhelm the normal repair process of bone (see the discussion of remodeling in chapter 2) and cause microscopic cracks to form and propagate. Microdamage has been demonstrated following repetitive loading in vivo [4-6]. Further, there is a threshold level of strain above which microdamage accumulates. This threshold is approximately 2000 microstrain, which represents the upper range of physiological values (see chapter 3, figure 3.8). If the load is applied continually, "microcracks" can coalesce into "macrocracks," and a stress fracture may eventually result. Thus, prolonged, repetitive small strains can cause a fracture, even if they are well below the level that would be needed to fracture a bone with a single event. This is similar to a nail being broken by continually bending it back and forth when there is insufficient force to snap it cleanly in one attempt.

As discussed in chapter 2, bone has the capacity to remodel and to repair fatigue-damaged regions; it adapts to the mechanical loads imposed upon it and becomes stronger. Since there is a lag between the increased osteoclastic activity and osteoblastic activity, however, bone is weakened during this time [7, 8]. Microdamage may occur at preexisting sites of accelerated remodeling when the bone is in this weakened state [9].

TABLE 17.1

Continuum of bony changes with overuse

Clinical features	Bone strain[1]	Stress reaction	Stress fracture
Local pain	Nil	Present	Present
Local tenderness	Nil	Present	Present
X-ray appearance	Normal	Normal	Abnormal (periosteal reaction or cortical defect in cortical bone, sclerosis in trabecular bone)
Radioisotopic bone scan appearance	Increased uptake (mild)	Increased uptake (moderate)	Increased uptake (strong)
CT scan appearance	Normal	Normal	Features of stress fracture (as for X-ray)
MR appearance	May show increased high signal	Increased high signal	Increased high signal ± cortical defect

[1]This is a descriptive clinical term and it does not refer to strain in the bioengineering sense of that term.

The processes of microdamage accumulation and bone remodeling both play an important part in the development of a stress fracture. If microdamage accumulates while repetitive loading continues, and remodeling cannot maintain the integrity of the bone, a stress fracture may result [10]. This may occur because the microdamage is too extensive to be repaired by normal remodeling, because depressed remodeling processes cannot adequately repair microdamage that occurs at a physiological rate, or because of a combination of these factors. Some stress fractures may also result from high-magnitude repetitive loads that damage bone integrity independent of the remodeling process. Preexisting decreased bone mass or bone quality may also make an area susceptible to microdamage.

Although athletic training influences bone loading, the relationship between the two is not linear. Remember that the volume of training includes both the total number of strain cycles (see chapter 3) and intensity (load per unit of time, or pace), which also affects the amplitude of strain cycles applied to bone. The magnitude and duration of each strain cycle depend on body weight, impact attenuation, and lower extremity biomechanical alignment. Muscular strength and conditioning can attenuate impact, as can extrinsic factors such as equipment and training surfaces. Lower extremity biomechanical alignment, including foot type, may affect gait mechanics, but fatigue, disease, and injury can also compromise gait. Clearly, many factors influence how training loads bone. The resilience of bone to bear this load depends on bone health—a factor determined by nutrition, genetics, endocrine and hormonal status, previous exercise, and the presence of bone disease.

Clinical Aspects

Stress fractures can occur in virtually any bone in the body. The most commonly affected bones are the tibia, metatarsals, fibula, tarsal navicular, femur, and pelvis [11-13]. A list of sites of stress fractures and the activities they may be associated with is shown in table 17.2.

Patients with stress fractures usually complain of pain and tenderness localized to the fracture site. The X-ray appearance of a stress fracture is often quite subtle. The most frequent finding is a periosteal reaction (figure 17.2). However, X-rays may not reveal a stress fracture until after it has been present for months, and some fractures are notoriously difficult to detect at all using plain X-ray [14]. Thus, further investigation using sophisticated imaging modalities such as radioisotopic bone scan, CT scan, or MR is indicated when stress fracture is suspected but the X-ray is normal.

Radioisotopic bone scan (scintigraphy) is used to reveal the site of an overuse bony lesion. A localized area of increased uptake or "hot spot" indicates a stress fracture (figure 17.3). Until the 1990s, a triple-phase bone scan was considered perfectly sensitive, so a negative bone scan essentially ruled out a stress fracture [15]. In recent

TABLE 17.2

Stress fractures: Site and commonly associated activity

Site of the stress fracture and the activity associated with this stress fracture
Coracoid process of scapula—Trapshooting
Scapula—Running with hand-held weights
Humerus—Throwing, racket sports
Olecranon—Throwing, pitching
Ulna—Racket sports (especially tennis), gymnastics, volleyball, swimming, softball, wheelchair sports
Ribs (first)—Throwing, pitching, ballet
Ribs (second to tenth)—Rowing, kayaking
Pars interarticularis—Gymnastics, ballet, cricket, fast bowling, volleyball, springboard diving
Public ramus—Distance running, ballet
Femur (neck)—Distance running, ballet
Femur (shaft)—Distance running
Patella—Running, hurdling
Tibia (plateau)—Running
Tibia (shaft)—Running, ballet
Fibula—Running, aerobics, race-walking, ballet
Medial malleolus—Basketball, running
Calcaneus—Long-distance military marching
Talus—Pole-vaulting
Navicular—Sprinting, middle distance running, hurdling, long/triple jumping, football
Metatarsal (general)—Running, ballet, marching
Metatarsal (base second)—Ballet
Metatarsal (fifth)—Tennis, ballet
Sesamoid bones of the foot—Running, ballet, basketball, skating

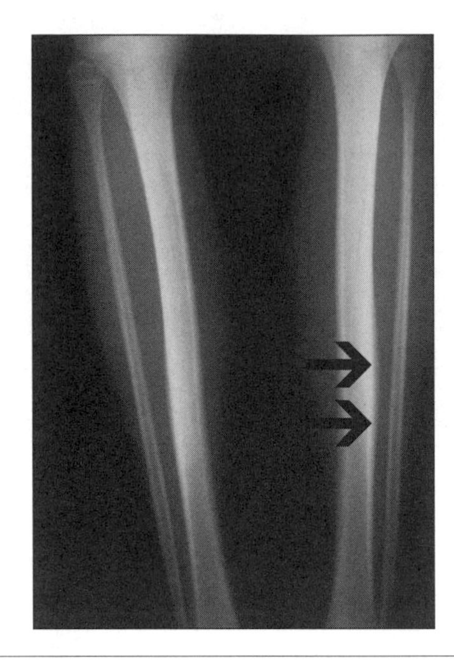

FIGURE 17.2 X-ray appearance of a recent stress fracture showing the characteristic periosteal new bone formation.

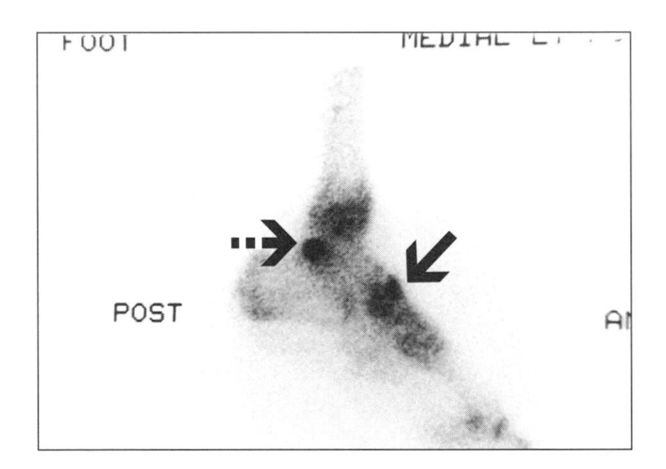

FIGURE 17.3 Bone scan showing focal uptake of radioisotope consistent with a stress fracture of the navicular bone (solid arrow). The increased uptake posteriorly in the talus (dotted arrow) illustrates that the bone scan is not a sensitive test—the athlete had no symptoms at that site.

years, a few authors have reported MR positive stress fractures with negative bone scans [16]. The clinical implication of these cases is that a negative bone scan only represents a 99% probability that there is no stress fracture, rather than a 100% probability. Bone scans remain a very valuable investigation.

In most cases of stress fracture, a radioisotopic bone scan is sufficient to confirm the diagnosis, and no further investigations are required. However, in a few sites that are known to present problems with treatment, such as the tarsal navicular, further information regarding the site and extent of the fracture is required. In the past, this was done with CT scanning (figure 17.4). MR imaging is increasingly advocated as the investigation of choice for stress fractures. While MR does not image fractures as clearly as CT scan does, it is of comparable sensitivity to isotope bone scan in assessing bony damage. For this reason, in practice, and depending on availability, MR is replacing the combination of bone scan and CT scan to investigate suspected stress fractures.

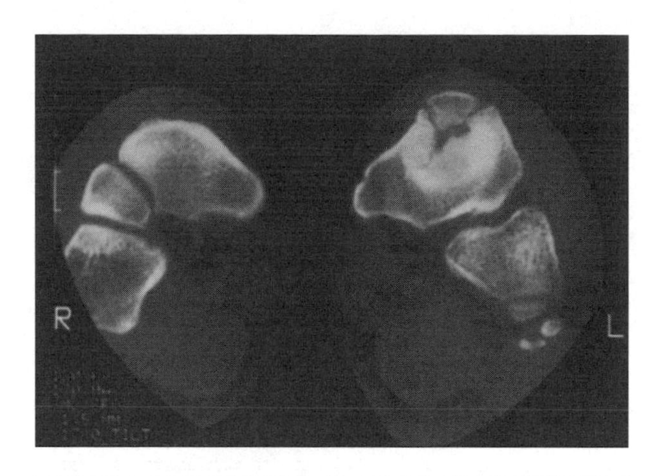

FIGURE 17.4 CT scan of the navicular of a 25-year-old triple jumper.

Treatment of Stress Fractures

Treatment of stress fractures generally requires avoidance of the precipitating activity. Although the majority of stress fractures heal within six weeks of beginning relative rest, recovery time varies according to the bone involved (table 17.3). Healing is assessed clinically by the absence of local tenderness and functionally by the ability to perform the precipitating activity without pain. In most cases it is not useful to attempt to monitor healing with X-ray or radioisotopic bone scan [3]. CT scan appearances of healing stress fractures can be deceptive because in some cases the fracture is still visible well after clinical healing has occurred [14].

In a number of sites of stress fractures, however, delayed union or nonunion of the fracture commonly occurs. These fractures need to be treated more aggressively such as with cast immobilization or even with surgery. The sites of these fractures are listed in table 17.4.

TABLE 17.3

Percentage of stress fractures healed at different times in a case series of 368 stress fractures in athletes

Stress fracture site	Healing period 2-4 wks %	1-2 mo %	> 2 mo %
Tibia			
Proximal third	0	43	57
Middle third	0	48	52
Distal third	0	53	47
Fibula	7	75	18
Metatarsals	20	57	23
Sesamoids	0	0	100
Femur			
Shaft	7	7	86
Neck	0	0	100
Pelvis	0	29	71
Olecranon	0	0	100

TABLE 17.4

Stress fractures that need treatment other than rest for optimal healing

Stress fracture site
Femoral neck
Talus (lateral process)
Navicular
Second metatarsal base
Sesamoid bone of the foot
Fifth metatarsal base
Anterior tibial cortex

Risk Factors

Risk factors for stress fractures are listed in table 17.5, and their interrelationship is shown schematically in figure 17.5

In the following sections we discuss in detail several risk factors that have been the subject of much recent research interest, including bone density, bone geometry, endocrine factors, and training. We then provide a brief summary of the role of the other risk factors listed in table 17.5. Readers interested in a more detailed review of stress fracture risks should refer to the work by Bennell and colleagues [2].

The Role of BMD

Some experts proposed that low bone density may contribute to the development of stress fractures by reducing bone strength and permitting the microdamage of repetitive loading to accumulate [17] (see chapter 15).

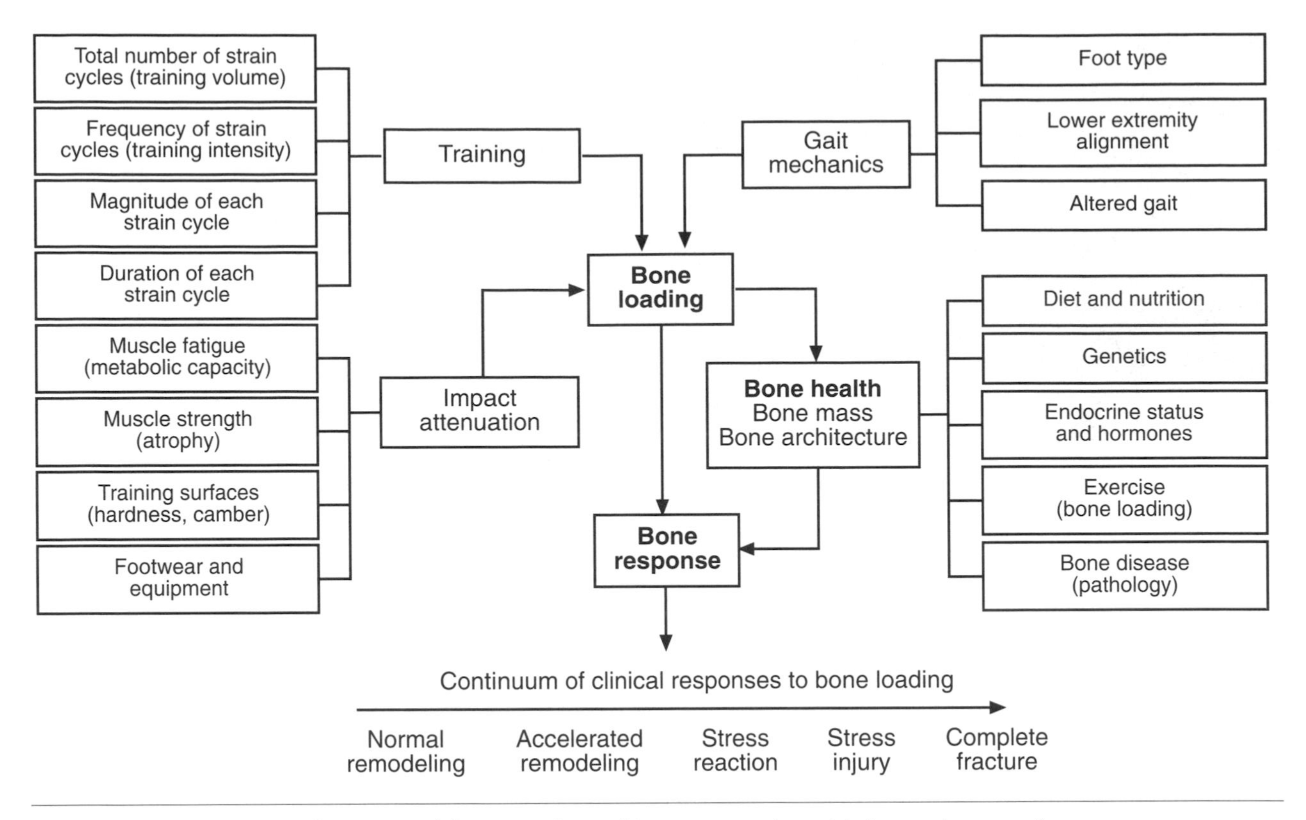

FIGURE 17.5 Schematic illustration of the interrelationship among various risk factors for stress fracture.
Reprinted, by permission, from Bennell KL, Matheson G, Meeuwisse W, et al., 1999, "Risk factors for stress fractures," *Sports Med* (28):96.

TABLE 17.5

Risk factors reported for stress fracture

Category	Risk factors
Biochemical	Bone density, bone geometry, skeletal alignment, body size, and composition
Physiological	Bone turnover, muscle flexibility and joint range of motion, muscular strength and endurance
Hormonal	Sex hormones, other hormones
Nutritional	Eating disorders, nutrient deficiencies
Physical training	Physical fitness, training regimen
External loading	Surface, footwear, insoles, orthotics
Other	Age, psychological factors

From [2].

Studies that investigated the relationship between bone density and stress fracture risk, however, have demonstrated varying results [2]. This may reflect differences in populations (military versus athletic, males versus females), type of sport (running, dancing, track and field), measurement techniques (single or dual photon absorptiometry, dual energy X-ray absorptiometry), and the bone regions studied.

In women, in the only prospective cohort study to date, track and field athletes who sustained stress fractures had significantly lower total body bone mineral content and lower bone density at the lumbar spine and foot than those without a fracture [1]. The subgroup of women who developed tibial stress fractures had 8% lower bone density at the tibia/fibula than those who did not. Although bone density was lower in the athletes with stress fractures, it nevertheless remained, as a group, higher than or similar to bone density of less active nonathletes. Thus, the women who eventually suffered stress fracture in this study would not have been identified as being at risk based on normative DXA values.

In men, on the other hand, very little prospective evidence supports a clear causal relationship between BMD and the risk of stress fractures. There was no difference in tibial bone density in 91 military recruits who developed stress fractures compared with 198 controls [18], nor was there a difference in BMD between male runners with and without a history of tibial stress fracture [19]. In a 12-month prospective study, BMD did not differ between male track and field athletes who sustained stress fractures and those who did not [20].

In a cross-sectional study, 41 male military recruits who suffered stress fractures had lower BMD at the femoral neck than their 28 counterparts who had no stress fractures [21]. When the group was subdivided by fracture site, femoral bone density was still lower in recruits with femoral and calcaneal stress fractures, but not in recruits with tibial, fibular, or metatarsal fractures. This site difference may be due to differences in the proportion of cortical to trabecular bone by region and highlights the problem of the specificity of the measurement site. In a study of 27 active duty army women with documented stress fractures, researchers found a strong negative association between femoral neck BMD and the probability of stress fractures (lower BMD, higher risk) [22]. As in other studies, women with stress fractures were more likely to be entry-level enlisted soldiers and exercising at a higher intensity than those without stress fractures.

The relationship between BMD and stress fracture risk remains unclear. Conflicting results from studies could indicate that the population of athletes with stress fractures is heterogeneous in terms of bone density and that other factors independent of bone mass contribute to the risk of fracture, particularly in men. The important clinical implication of these data is that bone densitometry does not appear to have a place as a general screening tool to predict the risk of stress fracture in otherwise healthy individuals. Bone densitometry may be warranted in athletes with multiple stress fracture episodes, however, or in females with menstrual disturbances.

The Role of Bone Geometry

Stress fracture provides a clinical application of the principles of the biomechanics of bone (see chapter 3). As discussed, bone strength is related not only to BMD but also to bone geometry. For bones loaded in tension or compression, the amount of load the bone can withstand prior to failure is proportional to the cross-sectional area of bone. The larger the area, the stronger and stiffer the bone. For bending and torsional loads, both the cross-sectional area and the cross-sectional moment of inertia determine bone strength. Thus, bones with a larger cross-sectional area and with bone tissue distributed farther from the neutral axis are stronger and hence less likely to fracture (see figure 17.6 and chapter 3).

Although bone geometry is largely dependent on body size [23], great variation exists even among individuals of similar age and build. In fact, much greater variation exists in structural geometry than in bone material properties, including bone mineral density [24]. Differences in bone geometry might therefore partly explain differences in stress fracture predisposition (figure 17.7).

A prospective observational cohort study of 295 male Israeli military recruits found that those who developed stress fractures had narrower tibias in the mediolateral plane at three different levels than those without stress fractures [25]. This result was found for total stress fractures as well as for stress fractures in the tibia and in the femur alone. The authors speculated that the dimension results generalized to femoral stress fractures because the size of the tibia may be proportional to the size of the femur.

The same authors found that the cross-sectional moment of inertia about the antero-

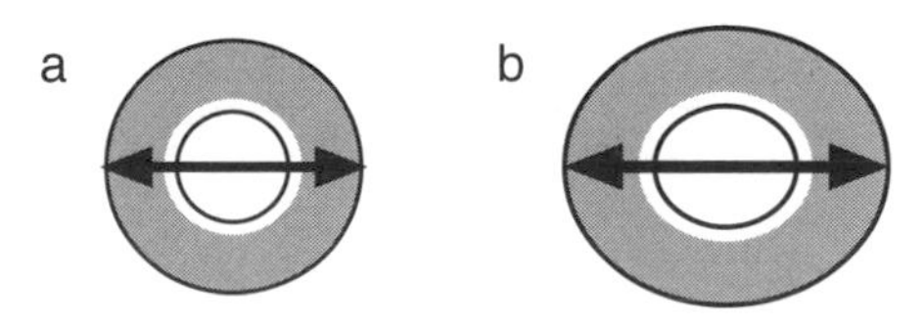

FIGURE 17.6 Schematic showing cross-sections of two bones. Section A has a smaller cross-sectional area than section B and is therefore at a greater risk for fracture.

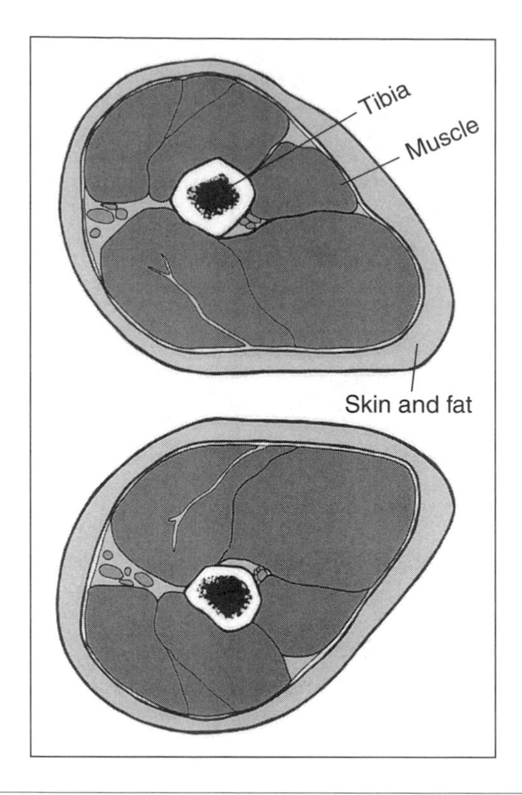

FIGURE 17.7 Cross-sectional views of distal tibia in triple jumper (top) and control subject (bottom), showing marked difference in cortical bone geometry. Measurement also reveals a difference in trabecular density, although that is not visible to the naked eye [70].
Drawn by Vicki Earle.

posterior axis (CSMI-AP), an estimate of the ability of bone to resist bending, was an even better indicator of the risk of stress fracture than tibial width [26]. Thirty-one percent of recruits with a low CSMI-AP developed tibial stress fractures compared with only 14% of recruits with a high CSMI-AP [27]. However, tibial geometry is complex and changes continuously along the length of the tibia. The researchers derived tibial widths from standard radiographs and based their calculations on the assumption that the tibia is an elliptical ring with an eccentric hole. This assumption is not necessarily valid. From our experience, tibial cross-sectional shape varies widely among individuals, and in many cases resembles a triangle more closely than an ellipse [19].

Nevertheless, the results showing differences in tibial geometry between recruits with stress fractures and those without were supported by a prospective study of more than 600 military recruits undergoing 12 weeks of basic training. DXA scan results were used to derive cross-sectional geometric properties of the tibia, fibula, and femur [28], a method that does not entail assumptions of cross-sectional shape. Recruits with stress fractures had smaller tibial width, cross-sectional area, and moment of inertia than those who remained fracture free.

There is also evidence in track athletes that tibial size is a risk factor for stress fractures. In a cross-sectional study of 46 male runners, CT scanning demonstrated tibial geometry at the level of the middle and distal third [19]. Athletes with a history of tibial stress fracture had a significantly smaller tibial cross-sectional area than those without fracture even after adjustment for height and weight.

Even if bone geometry plays a role in stress fracture development, this knowledge has limited clinical utility, as large-scale screening of tibial geometry is impractical. If, however, surrogate indicators of tibial geometry were derived from surface anthropometric measurements, screening of athletes would become feasible. Athletes could be examined prior to each season and those who were at risk could be monitored very closely for early symptoms of stress fracture, or they could have their training modified slightly (e.g., training surfaces, amount of plyometric training).

Amenorrhea, Oligomenorrhea, and Stress Fracture

As discussed in the previous chapter, athletic women have a higher prevalence of menstrual disturbance, including delayed menarche, anovulation, abnormal luteal phase, oligomenorrhea, and amenorrhea, than the general female population [29, 30]. These women may be at increased risk of stress fracture compared with regularly menstruating athletes.

Several retrospective surveys of runners and ballet dancers evaluated amenorrhea or oligomenorrhea and stress fracture risk [31-35]. These studies are characterized by small sample sizes and low questionnaire response rates. In other studies, ascertainment bias (a form of selection bias) may be present because subjects were specifically recruited according to certain criteria—either stress fracture history or menstrual status [36-41]. Studies generally categorized menstrual status by number of menses per year rather than by hormone levels, and definitions of menstrual status varied among studies. Studies that assessed hormone levels generally did so with single measurements, often

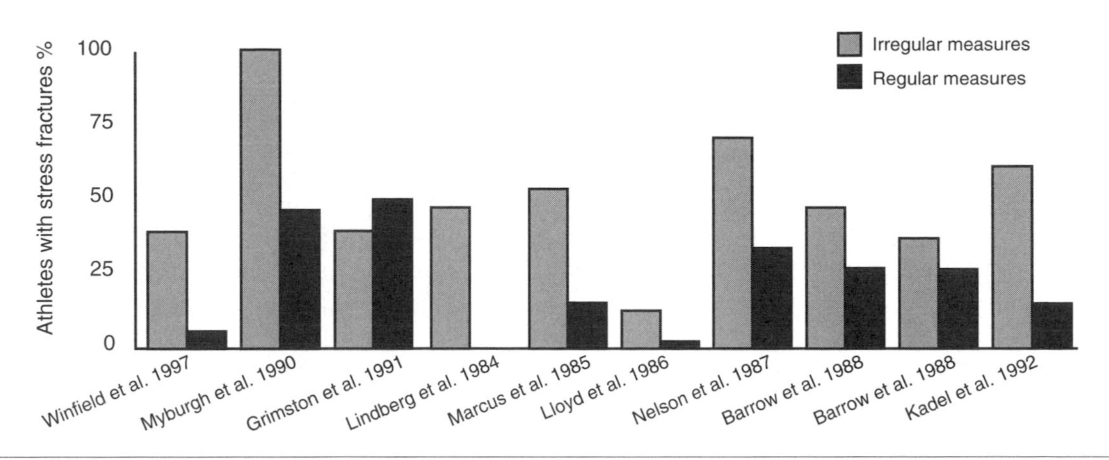

FIGURE 17.8 Summary of numerous studies showing that stress fractures appear to be more common in athletes with menstrual disturbance.

Reprinted, by permission, from Bennell KL, Matheson G, Meeuwisse, W, et al., 1999, "Risk factors for stress fractures," *Sports Med* (28):107.

nonstandardized with respect to menstrual cycle phase. The length of exposure to amenorrhea also differed within and among studies, a fact that may influence the risk of stress fracture.

Bearing in mind the methodology limitations, stress fractures are more common in athletes with menstrual disturbance (figure 17.8) [42]. Such athletes have a relative risk for stress fracture that is between two and four times greater than that of their eumenorrheic counterparts. However, the rate of stress fracture in ballet dancers with amenorrhea for longer than six months' duration was 93 times that of dancers with regular menses [34]. While this risk seems extraordinarily high, there were only six dancers with regular menses in this sample of 54 dancers, which may have affected the statistical analyses.

Menstrual disturbance appears to increase the risk of multiple stress fractures [32, 33]. In runners, amenorrheic and eumenorrheic groups reported a similar prevalence of single stress fractures, but 50% of the amenorrheic runners reported multiple stress fractures compared with only 9% of those menstruating regularly [33]. Another study reported that some amenorrheic athletes sustained as many as four to six stress fractures, whereas the 120 eumenorrheic runners sustained three or fewer stress fractures [32].

Grimston and colleagues [43] developed a menstrual index that summarized previous and present menstrual status. The index quantified the average number of menses per year since menarche. The researchers found no relationship between this menstrual index and the incidence of stress fractures in 16 female runners [43]. In contrast, others showed that athletes with fewer menses per year since menarche were at greater risk of stress fracture than those with a higher index [1, 32].

Keep in mind that athletes with menstrual disturbance generally also have greater training loads as well as differences in soft tissue composition and dietary intake. Since these were generally not controlled for in the studies discussed, it is impossible to be absolutely certain about the contributory factors.

Physical Training and Stress Fracture Risk

Repetitive bone loading arising from athletic training contributes to stress fracture development. How each training component (frequency, intensity, type, time, and rate of change) contributes to the risk of stress fracture remains unclear. Training may also influence bone indirectly through changes in levels of circulating hormones, through effects on soft tissue composition, and through associations with menstrual disturbance (see also chapters 7 and 16).

Physical Fitness. Whether lack of prior physical activity and poor physical conditioning, as markers of mechanical loading history, predispose a person to stress fracture is unclear. Many studies have relied on self-reporting rather than on standardized fitness tests to assess activity levels before beginning the exercise program. Most of the literature focuses on military recruits who are subjected to a short burst of intense, unaccustomed activity and are often unfit.

Some military studies have reported a correlation between self-reported previous physical activity levels and rate of stress fracture during basic training (table 17.6). On the other hand, several large studies found no relationship [18, 44-45]. For example, neither aerobic fitness (predicted $\dot{V}O_2max$) nor self-reported pretraining participation in sport activities was related to stress fracture incidence in 295 male recruits aged 18-20 years [44]. Although there are conflicting results, the majority of studies tend to suggest that poor physical fitness or prior physical inactivity may increase stress fracture risk in individuals undergoing basic military training. Poor physical conditioning does not seem to be relevant in athletes since they are, by definition, well-conditioned people.

Training Regimen. Aspects of the training regimen appear to predispose to stress fracture. Military studies have shown that various training modifications can decrease the incidence of stress fractures in recruits. These interventions include rest periods, the elimination of running and marching on concrete [46], the use of running shoes rather than combat boots [46], and a reduction of high-impact activity [47, 48]. These may reduce stress fracture risk by allowing time for bone microdamage to be repaired and by decreasing the load applied to bone.

In contrast, little controlled research has been conducted in athletes. Most are anecdotal observations or case series in which training parameters are examined only in athletes with stress fractures. For example, surveys reporting that up to 86% of athletes can identify some change in their training prior to the onset of the stress fracture [49] do not provide a similar comparison in uninjured athletes. Other researchers have blamed training errors in a varying proportion of cases without adequately defining these errors [50, 51]. Brunet and colleagues [52] surveyed 1505 runners and found that increasing mileage correlated with an increase in stress fractures in women but not in men. A reason for the apparent gender difference is unclear. In a study of ballet dancers, a dancer who trained for more than five hours per day had an estimated risk for stress fracture that was 16 times greater than that of a dancer who trained for less than five hours per day [34]. These studies suggest that training volume is a risk factor for stress fracture.

Bone remodeling will repair microdamage if given adequate time, but if a load is applied repeatedly, a stress fracture results. Cyclic training is therefore preferable to progressive training, as it allows both bone and soft tissue to rest from repetitive loading (this is also called periodization of training).

Muscle Strength and Endurance. Muscle strength and endurance strongly influence stress fracture risk. Some investigators believe that muscles act dynamically to cause stress fractures by increasing bone strain at sites of muscle attachment [53, 54]. Under normal circumstances, however, muscles exert a protective effect by contracting to reduce bending strains on cortical bone surfaces [55]. Following fatiguing exercise, bone strain, particularly strain rate, is increased [56, 57], a situation that applies especially in younger versus older persons [56].

TABLE 17.6

Studies in military recruits that found an association between physical fitness and risk of stress fractures

Author	Main finding
Montgomery 1989 [64]	Only 3% of trainees with running history (> 25 mi/wk) sustained stress fractures, compared with 12% of those without a running history (< 4 mi/wk) (men).
Gardner 1988 [65]	Previously inactive recruits had 24 times greater rate of stress fracture than previously very active recruits (men).
Winfield 1997 [66]	Only 4% of trainees with running history (> 2.8 mi/session) sustained stress fractures, compared with 16% of those who ran < 2.8 mi/session (women).
Cline 1998 [67]	Higher leisure activity energy expenditure was associated with lower risk of stress fracture (women).
Shaffer 1999 [68]	Recruits with "high risk" of stress fracture by algorithm of 5 physical activity questions and a 2.4-km run time suffered 3 times as many stress fractures as those of "low risk" by those criteria (women).

Studies measuring muscle strength or endurance [18, 58-60] have generally failed to find an association with stress fracture occurrence. Some indirect evidence for muscle fatigue as a risk factor comes from a study by Grimston and colleagues. During the latter stages of a 45-minute run, women with a history of stress fracture recorded increased ground reaction forces, whereas in the control group ground reaction forces did not vary during the run [61].

Measurements of muscle size may indicate the ability of a muscle to generate force. Male recruits with a larger calf circumference developed significantly fewer femoral and tibial stress fractures than those with smaller calf circumference [27]. This finding was also evident in female athletes, but not male athletes [1]. To establish a causal relationship, the effectiveness of a calf-strengthening program in reducing the incidence of stress fractures should be evaluated in a randomized, controlled trial.

Other Risk Factors

As stress fracture etiology is multifactorial, other variables related to bone, muscle, nutrition, and training have been examined as potential risk factors [2]. Because these can be difficult to examine in isolation, current data remain imperfect. We summarize current thinking, but warn the reader that future research may contradict what is accepted at present!

- Bone turnover markers (see chapter 3) have been evaluated both cross-sectionally and prospectively [62] in an attempt to predict stress fracture. To date, levels of these markers do not appear to differ between athletes who suffer from stress fractures and those who do not (see figure 4.11).
- Skeletal alignment has been investigated on numerous occasions, and leg length discrepancy appears the only factor that can be implicated in stress fracture risk [1, 52, 63].
- Conversely, abnormal and restrictive eating habits are associated with increased risk of fracture in physically active women [20, 30, 40]. Evidence does not show that dietary deficiency (in particular, low calcium intake) is a risk factor for stress fracture in otherwise healthy athletes.
- Large epidemiological studies of overall running injuries failed to show an association between training surfaces or terrain after controlling for running distance. Wearing aged, worn-out shoes increases the risk of stress fractures. Insoles reduced stress fracture risk in military populations, but this has not been demonstrated in athletic populations.

SUMMARY

- The histological and clinical continuum of bone overuse injury begins with accelerated bone remodeling (bone strain), continues through painful stress reaction, and, if the milieu is not changed and microcracks coalesce, proceeds to stress fracture. Stress fracture is evident using modern imaging techniques well before it is evident on plain X-ray.
- Stress fractures are seen clinically at specific sites in particular bones, and they are commonly associated with certain sporting activities. Treatment generally requires relative rest from the offending activity followed by a graduated return.
- Whether an athlete develops a stress fracture depends on properties of the bone (discussed in chapter 3), the ability of surrounding muscle to attenuate load (an area that requires a great deal more research), and the nature of the athletic activity itself. Repetitive activity can cause bone to respond by increasing BMD, but there appears to be a threshold of loading above which normal remodeling can no longer keep pace with microtrauma.
- Of the many potential risk factors for stress fracture, few have been proven unequivocally, as most studies have had design flaws. Nevertheless, it appears that BMD below that normally seen in athletes (but still above that of the non-athletic population) may increase female, but not male, athletes' risk of stress

fracture. Bone geometry probably plays a significant role; small bones are more susceptible to higher strains (per unit volume) and thus to fracture. Menstrual disturbance is a risk factor for stress fracture, but the mechanism of this risk remains under investigation. Abnormal and restrictive eating habits are associated with stress fractures, but there is little evidence to suggest calcium nutrition affects stress fracture risk.

• A background of physical activity appears to protect both female and male military recruits from sustaining stress fractures, compared with their previously inactive counterparts. Graduated training rather than continuous, hard training reduces stress fracture risk.

References

1. Bennell KL, Malcolm SA, Thomas SA, et al. Risk factors for stress fractures in track and field athletes. A twelve-month prospective study. *Am J Sports Med* 1996;24:810-818.
2. Bennell KL, Matheson G, Meeuwisse W, et al. Risk factors for stress fractures. *Sports Med* 1999;28:91-122.
3. Brukner P, Bennell K, Matheson G. *Stress fractures*. Melbourne: Blackwell Scientific, Asia, 1999.
4. Mori Y, Okumo H, Iketani H, et al. Efficacy of lateral retinacular release for painful bipartite patella. *Am J Sports Med* 1995;23:13-18.
5. Forwood MR, Parker AW. Microdamage in response to repetitive torsional loading in the rat tibia. *Calcif Tissue Int* 1989;45:47-53.
6. Burr DB, Martin RB, Shaffler MB, et al. Bone remodeling in response to in vivo fatigue microdamage. *J Biomech* 1985;18:189-200.
7. Roub LW, Gumerman LW, Hanley Jr EN, et al. Bone stress: A radionuclide imaging perspective. *Radiology* 1979;132:431-438.
8. Li G, Zhang S, Chen G, et al. Radiographic and histologic analyses of stress fracture in rabbit tibias. *Am J Sports Med* 1985;13:285-294.
9. Michael RH, Holder LE. The soleus syndrome: a cause of medial tibial stress (shin splints). *Am J Sports Med* 1985;13:87-94.
10. Schaffler MB, Radin EL, Burr DB. Long-term fatigue behaviour of compact bone at low strain magnitude and rate. *Bone* 1990;11:321-326.
11. Brukner P, Bradshaw C, Khan KM, et al. Stress fractures: A review of 180 cases. *Clin J Sport Med* 1996;6:85-89.
12. Baquie P, Brukner P. Injuries presenting to an Australian sports medicine centre: a 12 month study. *Clin J Sport Med* 1997;7:28-31.
13. Matheson GO, Clement DB, McKenzie DC, et al. Stress fractures in athletes. A study of 320 cases. *Am J Sports Med* 1987;15:46-58.
14. Khan KM, Fuller PJ, Brukner PD, et al. Outcome of conservative and surgical management of navicular stress fracture in athletes. Eighty-six cases proven with computerized tomography. *Am J Sports Med* 1992;20:657-666.
15. Matheson GO, Clement DB, McKenzie DC, et al. Scintigraphic uptake of 99mTc at non-painful sites in athletes with stress fractures. The concept of bone strain. *Sports Med* 1987;4:65-75.
16. Keene JS, Lash EG. Negative bone scan in a femoral neck stress fracture. A case report. *Am J Sports Med* 1992;20:234-236.
17. Carter DR, Caler WE, Spengler DM, et al. Uniaxial fatigue of human cortical bone: the influence of tissue physical characteristics. *J Biomech* 1981;14:461-470.
18. Giladi M, Milgrom C, Simkin A, et al. Stress fractures: identifiable risk factors. *Am J Sports Med* 1991;19:647-652.
19. Crossley K, Bennell KL, Wrigley T, et al. Ground reaction forces, bone characteristics, and tibial stress fracture in male runners. *Med Sci Sports Exerc* 1999;31:1088-1093.
20. Bennell KL, Malcolm SA, Thomas SA, et al. Risk factors for stress fractures in female track-and-field athletes: a retrospective analysis. *Clin J Sport Med* 1995;5:229-235.
21. Pouilles JM, Bernard J, Tremollieres F, et al. Femoral bone density in young male adults with stress fractures. *Bone* 1989;10:105-108.
22. Lauder TD, Dixit S, Pezzin LE, et al. The relation between stress fractures and bone mineral density: evidence from active-duty Army women. *Arch Phys Med Rehabil* 2000;81:73-9.
23. Miller GJ, Purkey WW. The geometric properties of paired human tibiae. *J Biomech* 1980;13:1-8.
24. Martens M, Van Auderkerke R, de Meester P, et al. The geometrical properties of human femur and tibia and their importance for the mechanical behaviour of these bone structures. *Acta Orthop Traumatic Surg* 1981;98:113-120.
25. Giladi M, Milgrom C, Stein M, et al. External rotation of the hip, a predictor of risk for stress fractures. *Clin Orthop & Rel Research* 1987;216:131-134.
26. Milgrom C, Giladi M, Simkin A, et al. An analysis of the biomechanical mechanisms of tibial stress fractures among Israeli infantry recruits. *Clin Orthop* 1988;231:216-221.

27. Milgrom C, Giladi M, Simkin A, et al. The area moment of inertia of the tibia: a risk factor for stress fractures. *J Biomech* 1989;22:1243-1248.
28. Beck TJ, Ruff CB, Mourtada FA, et al. Dual-energy X-ray absorptiometry derived structural geometry for stress fracture prediction in male US marine corps recruits. *J Bone Miner Res* 1996;11:645-653.
29. Malina RM, Spirduso WW, Tate C, et al. Age at menarche and selected menstrual characteristics in athletes at different competitive levels and in different sports. *Med Sci Sports Exerc* 1978;10:218-222.
30. Nattiv A, Puffer JC, Green GA. Lifestyles and health risks of collegiate athletes: a multi-center study. *Clin J Sport Med* 1997;7:262-272.
31. Lloyd T, Triantafyllou SJ, Baker ER, et al. Women athletes with menstrual irregularity have increased musculoskeletal injuries. *Med Sci Sports Exerc* 1986;18:374-379.
32. Barrow GW, Saha S. Menstrual irregularity and stress fractures in collegiate female distance runners. *American J Sports Med* 1988;16:209-216.
33. Clark N, Nelson M, Evans W. Nutrition education for elite female runners. *Phys Sportsmed* 1988; 16:124-136.
34. Kadel NJ, Tietz CC, Kronmal RA. Stress fractures in ballet dancers. *Am J Sports Med* 1992;20:445-449.
35. Warren MP, Brooks-Gunn J, Hamilton LH, et al. Scoliosis and fractures in young ballet dancers. *N Engl J Med* 1986;314:1348-1353.
36. Lindberg JS, Fears WB, Hunt MM, et al. Exercise-induced amenorrhea and bone density. *Ann Int Med* 1984;101:647-648.
37. Marcus R, Cann R, Madvig P. Menstrual function and bone mass in elite women distance runners: endocrine and metabolic features. *Ann Int Med* 1985;102:158-163.
38. Cook SD, Harding AF, Thomas KA, et al. Trabecular bone density and menstrual function in women runners. *Am J Sports Med* 1987;15:503-507.
39. Carbon R, Sambrook PN, Deakin V, et al. Bone density of élite female athletes with stress fractures. *Med J Aust* 1990;153:373-376.
40. Frusztajer NT, Dhuper S, Warren MP, et al. Nutrition and the incidence of stress fractures in ballet dancers. *Am J Clin Nutr* 1990;51:779-783.
41. Rutherford OM. Spine and total body bone mineral density in amenorrheic endurance athletes. *J App Physiol* 1993;74:2904-2908.
42. Bennell KL, Malcolm SA, Wark JD, et al. Skeletal effects of menstrual disturbances in athletes. *Scandinavian Journal of Medicine & Science in Sports* 1997;7:261-273.
43. Grimston SK, Sanborn CF, Miller PD, et al. The application of historical data for evaluation of osteopenia in female runners: the menstrual index. *Clin Sports Med* 1990;2:108-118.
44. Swissa A, Milgrom C, Giladi M, et al. The effect of pretraining sports activity on the incidence of stress fractures among military recruits. *Clin Orthop* 1989;245:256-260.
45. Hoffman JR, Chapnik L, Shamis A, et al. The effect of leg strength on the incidence of lower extremity overuse injuries during military training. *Mil Med* 1999;164:153-156.
46. Greaney RB, Gerber FH, Laughlin RL, et al. Distribution and natural history of stress fractures in US marine recruits. *J Bone Joint Surg* 1983;26-A:751-757.
47. Pester S, Smith PC. Stress fractures in the lower extremities of soldiers in basic training. *Orthop Rev* 1992;21:297-303.
48. Pope R. Prevention of pelvic stress fractures in female army recruits. *Mil Med* 1999;164:370-373.
49. Sullivan D, Warren RF, Pavlov H, et al. Stress fractures in 51 runners. *Clin Orthop* 1984;187:188-192.
50. Taunton JE, Clement DB, Webber D. Lower extremity stress fractures in athletes. *Phys Sportsmed* 1981;9:77-86.
51. McBryde Jr AM. Stress fractures in runners. *Clin Sports Med* 1985;4:737-752.
52. Brunet ME, Cook SD, Brinker MR, et al. A survey of running injuries in 1505 competitive and recreational runners. *J Sports Phys Fitness* 1990;30:307-315.
53. Meyer S, Saltzman C, Albright J. Stress fractures of the foot and leg. *Clin Sports Med* 1993;12:395-413.
54. Stanitski CL, McMaster JH, Scranton PE. On the nature of stress fractures. *Am J Sports Med* 1978;6:391-6.
55. Scott S, Winter D. Internal forces at chronic running injury sites. *Med Sci Sports Exerc* 1990;22:357-369.
56. Fyhrie DP, Milgrom C, Hoshaw SJ, et al. Effect of fatiguing exercise on longitudinal bone strain as related to stress fracture in humans. *Ann Biomed Eng* 1998;26:660-5.
57. Yoshikawa T, Mori S, Santiesban A, et al. The effects of muscle fatigue on bone strain. *J Exp Biol* 1994;188:217-233.
58. Milgrom C, Finestone A, Shlamkovitch N, et al. Youth is a risk factor for stress fracture; a study of 783 infantry recruits. *J Bone Joint Surg Br* 1994;76:20-22.
59. Ekenman I, Tsai-Fellander L, Westblad P, et al. A study of intrinsic factors in patients with stress fractures of the tibia. *Foot Ankle Int* 1996;17:477-482.
60. Hoffman J, Chapnik L, Shamis A, et al. The effect of leg strength on the incidence of lower extremity overuse injuries during military training. *Mil Med* 1999;164:153-156.
61. Grimston S, Nigg B, Fisher V, et al. External loads throughout a 45 minute run in stress fracture and

non-stress fracture runners [abstract]. *J Biomech* 1994;227:668.

62. Bennell KL, Malcolm SA, Brukner PD, et al. A 12-month prospective study of the relationship between stress fractures and bone turnover in athletes. *Calcif Tissue Int* 1998;63:80-85.

63. Friberg O. Leg length asymmetry in stress fractures. A clinical and radiological study. *J Sports Phys Fit* 1982;22:485-488.

64. Montgomery LC, Nelson FR, Norton JP, et al. Orthopedic history and examination in the etiology of overuse injuries. *Med Sci Sports Exerc* 1989; 21:237-243.

65. Gardner LI, Jr., Dziados JE, Jones BH, et al. Prevention of lower extremity stress fractures: a controlled trial of shock absorbent insole. *Am J Public Health* 1988;78:1563-1567.

66. Winfield AC, Moore J, Bracker M, et al. Risk factors associated with stress reactions in female Marines. *Mil Med* 1997;162:698-702.

67. Cline AD, Jansen GR, Melby Cl. Stress fracturs in female army recruits: implications of bone density, calcium intake, and exercise. *J Am Coll Nutr* 1998;17;128-135.

68. Shaffer RA, Brodine SK, Almeida SA, et al. Use of simple measures of physical activity to predict stress fractures in young men undergoing a rigorous physical training program. *Am J Epidemiol* 1999;149:236-242.

69. Brukner P, Khan KM. *Clinical Sports Medicine* 2nd Edition. McGraw-Hill, 2001.

70. Hulkko A, Orava S. Stress fractures in athletes. *Int J Sports Med* 1987;8:221-226.

71. Orava S. Stress fractures. *Br J Sports Med* 1980; 14:40-44.

72. Heinonen A, Sievänen H, Kyröläinen H, et al. Mineral mass, bone size and estimated mechanical strength of lower limb in triple jumpers. *Bone* 2001 (in press).

part V

Research Opportunities: Physical Activity and Bone Health

The aim of this part of the book is to provide an impetus to the graduate or postgraduate student who is interested in physical activity and bone health, but who may not be in the fortunate position of working with a large team of mentors. Let us be your mentors! Clearly we cannot prescribe a course of study for you, but we can outline areas of research that may coincide with your interests, thereby allowing you to focus your reading and begin the exciting task of formulating an original study.

We have not seen a book part like this elsewhere—and we know why. Material of this nature is rapidly outdated and can never truly represent the field. Despite this, we elected to include it, as we feel that ideas breed ideas. We consider the novel chapters in this part more as a template than a recipe—to be updated appropriately as the playing field of bone research changes.

Chapter 18 outlines the breadth of physical activity research being done in the field of bone health. It outlines research conducted at present and in the recent past and describes the different settings that permit bone research. This chapter does not attempt to outline new research ideas.

Specific research opportunities for short and long periods of study are suggested in chapters 19 and 20. Although the length of time required to complete a research project varies depending on many things, we outline those topics that would be suitable for a master's thesis (chapter 19) and those that would warrant a doctoral thesis (chapter 20). Because research demands original thought and creativity, we do not expect our suggestions to be followed like a recipe—in fact, it would be tragic if they were. Furthermore, this field of research is undergoing very rapid change. Our aim is to reveal the broad canvas of inquiry that has resulted from centuries of investigation by countless people who, like you, were interested in bone health. We encourage you, if you are so inclined, to make your own distinctive contribution to that field.

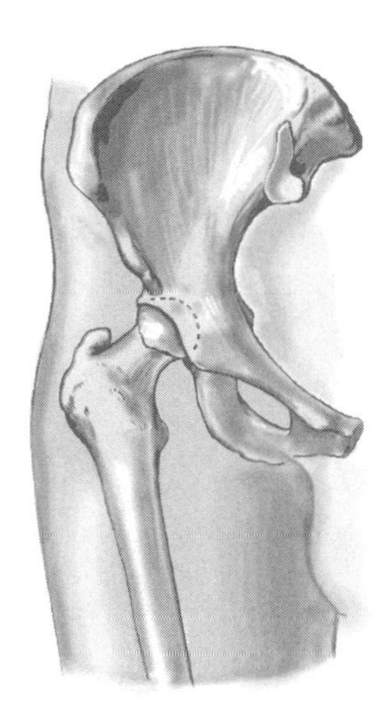

Chapter 18

Getting Involved in Bone Research

This chapter outlines the various settings in which bone research is being performed. The chapter is divided into eight sections, each of which defines an area of study within the larger field of bone research. Three sections concern healthy bone scrutinized at the tissue level, the organ level, and the level of the whole skeleton. Another section focuses on bone pathology. The next discusses physical activity research in the setting of pharmaceutical studies. The remaining sections concern bone research for nutritionists, research with special populations, and research opportunities in the field of measurement of physical activity with regard to bone health. These divisions are not mutually exclusive. In life, the study of bone often requires multidisciplinary collaboration, so one commonly finds teams of people from seemingly disparate disciplines working together.

An excellent way to supplement the listing of subject areas in bone research in this chapter is to check the index pages of journals that report advances in bone health and exercise. These include *Medicine and Science in Sport and Exercise* (the official journal of the American College of Sports Medicine), the *British Journal of Sports Medicine*, the *Clinical Journal of Sport Medicine*, the *Journal of Bone and Mineral Research* (the official journal of the American Society for Bone and Mineral Research), *Calcified Tissue International*, *Osteoporosis International* (a joint initiative between the International Osteoporosis Foundation and the National Osteoporosis Foundation of the USA), and *Bone* (the official journal of the International Bone and Mineral Society). The undergraduate or new graduate student, or the less experienced researcher, may find some of these journals a little daunting at first, but when they are read together with background material (such as is found in this book), things will quickly become clear.

Studies: Bone at the Molecular, Cellular, and Tissue Levels

An active branch of bone research focuses on the structure and function of bone at the molecular level, cellular level, and tissue level. Microscopic evaluation of bone is called histology, and biochemical analysis at this microscopic (light or electron) level is called histochemistry. Researchers interested in physical activity who work in this field address questions such as the following:

- What is the relationship between certain genes and the response of bone to mechanical loading?
- Which cells in bone respond to strain and influence bone gain and loss?
- How do bone cells respond to mechanical loading?
- What substances are the messengers that transmit the messages that bone is being loaded?
- What are the relationships among bone material properties, structural properties, and bone strength?

Histological and histochemical studies may be performed on excised bone samples (in vitro research), with human DNA, or in animals (in vivo research) in the laboratory. Sometimes, but not often, bone biopsies (e.g., iliac crest) from live humans are used to answer questions at this level.

Because genetic factors play a major role in bone health (see chapter 5), geneticists and molecular biologists can be found on many bone research teams. Some are looking for the "gene for osteoporosis," which may prove to be a series of genes [1]. This is an exciting area of research for the graduate student with a strong biochemistry and genetics background.

Bone as an Organ

Researchers interested in bone at a slightly more macroscopic level may study whole bones (e.g., the turkey ulna, the sheep calcaneum). They may load an entire bone in a live animal for a few minutes or a few hours a day to test the response of bone to strictly defined loading. These researchers may collaborate with others or perform their own histological or histochemical evaluations of bone specimens at the end of the research period. Tests may also be performed with excised bone (in vitro) to examine bone breaking strength [2].

Studies of bone physiology have become much more feasible and popular with the advent of biochemical markers of bone turnover (see chapter 4). This technology has allowed researchers to view whole bone changes using markers of bone resorption and bone formation, which can provide useful information about interventions. Many researchers study the interaction among hormones and factors, genes, or pharmaceutical agents, and bone loading.

An area of increasing interest combines the science of biomechanics with aspects of bone biology. As outlined in chapter 3, material and structural properties play a major role in bone strength as a tissue and as whole bone, respectively. Bone biomechanics is the field of research that measures forces on bone. Researchers from these fields work with bone as an organ and compare the strength, the geometric structure, and the kinetic and kinematic response of bone strength and structure to bone loading.

The Whole Skeleton

An enormous number of studies can be undertaken to evaluate bone in humans. For those interested in this field, training in various branches of the health and biological sciences provides an excellent background. Students of human movement (also called kinesiology, human kinetics, or a number of other names) are well placed to work in this field. Similarly, epidemiologists, nurses, physiotherapists, occupational therapists, and athletic trainers may all work directly on a bone research team or collaborate with bone researchers to study a physical activity intervention.

In humans, the most common outcome measure in bone studies over the past decade has been bone mineral density (BMD) by DXA (see chapter 4). In the early 1990s these instruments were generally located in hospital radiology departments, and physical activity researchers collaborated with medical doctors to use them. Although this is still often the case, an increasing number of nonmedical institutions and bone research laboratories house DXA scanners.

Bone researchers investigate humans at all stages of the life span. Studies have been undertaken in infants and small children, and research in pre-, peri-, and postpubertal children has increased dramatically in the past five years. Studies in adults are commonplace, and older populations are also attracting increased research attention. Researchers are well advised to collaborate with appropriate partners who understand human biology at the various ages and stages of development. The pediatric bone researcher may have a background in growth biology or may cooperate with those who understand this complex field. Similarly, gerontologists can provide useful information about age-appropriate functional tests and norms in elderly persons. Those who want to study biochemical markers can collaborate with biochemists who understand the strengths and weaknesses of laboratory tests.

Abnormal Bone: Stress Fractures and Osteoporosis

Stress fractures, traumatic fractures, and osteoporotic fractures provide fertile ground for interesting, clinically useful research.

Stress Fractures. Stress fractures are not easy to study, as there is not a good animal model for this pathology. In the past, cross-sectional and retrospective stress fracture studies were the norm, with researchers identifying a subject population and comparing them with controls. In recent years, however, prospective studies have provided very valuable information. These studies were the result of physical activity researchers collaborating with coaches and athletes to recruit a healthy, but at-risk cohort of subjects. Military boot camps have provided an ideal setting for studies such as this [3]. Once this cohort was assembled, collaborating clinicians performed necessary baseline measures and then performed clinical diagnosis and investigation as symptoms warranted.

Traumatic Fractures. Researchers have undertaken many studies of the prevalence of fractures due to low-energy trauma (e.g., falling). The large, high-quality studies of such phenomena are in the field of epidemiology—the study of populations. Hip fractures (figure 18.1) are of particular interest to researchers for two reasons. First, they are common and they cause the greatest morbidity and mortality of osteoporosis-related fractures. Second, they cause a patient to present to a hospital, and surgery is usually required. This differs from vertebral fractures or rib fractures, which may be asymptomatic and can be treated by a family doctor. The hospital serves as a catchment site for studying all fractures (epidemiological studies), as well as being the place the patient receives treatment for the painful hip fracture.

Regarding hip fractures, physical activity researchers have asked questions such as the following:

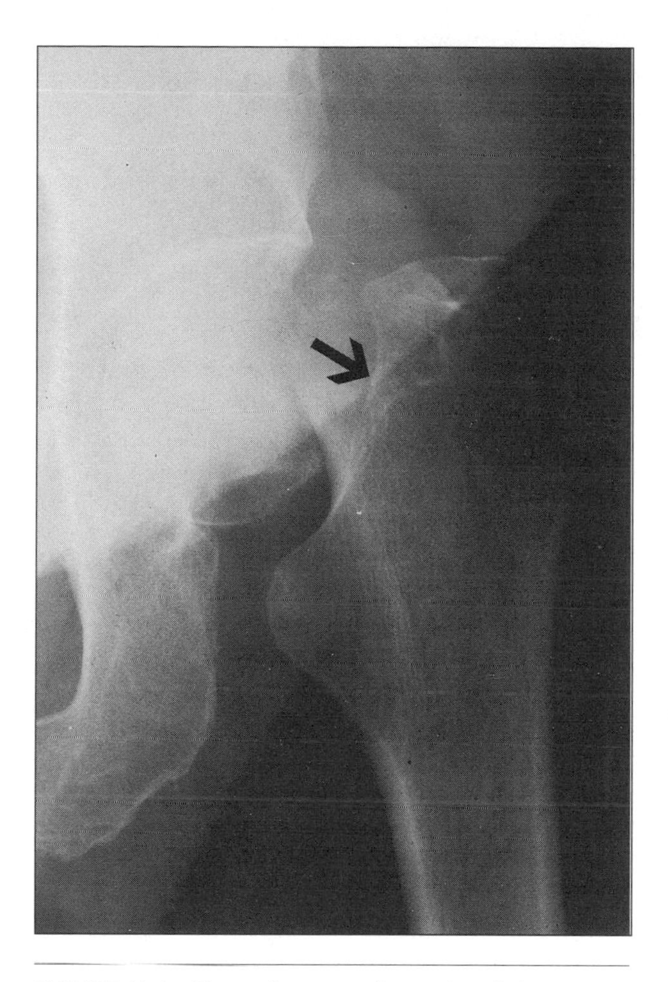

FIGURE 18.1 X-ray of a severe femoral neck fracture in a 77-year-old woman with osteoporosis who fell while stepping out of the bathtub.

- Do fractures result predominantly from falls in a certain direction?
- Does a first hip fracture predispose to a subsequent contralateral (opposite-side) hip fracture?
- What is the long-term physical and emotional health outcome after hip fracture?
- What is the cost of hip fracture to the community?
- What is the efficacy and effectiveness of hip protectors for preventing hip fractures due to falls?
- Can DXA scan scores, quadriceps strength, calf strength, or measures of balance predict hip fracture?
- Can increased physical activity improve muscle strength, balance, and coordination and reduce falls and related injuries?

These questions have been addressed, at least to some degree, in certain populations, but much remains unknown about fracture etiology and pathogenesis.

Fractures can occur in individuals with normal bone mineral, but clearly women and men with osteoporosis (low bone density, see chapter 15) are a population that is at greater risk of fracture. Osteoporosis has traditionally been studied in a medical environment with a large emphasis on testing the efficacy of pharmaceutical interventions, with little attention paid to the role of physical activity. Today, a wide range of researchers are increasingly studying various aspects of this condition, including the role of physical activity on factors such as BMD, bone strength, muscle strength, balance, and fear of falling. Because of the large number of subjects required to assess the efficacy of any intervention on fractures, no physical activity intervention has yet focused on fracture as an endpoint. Because such a study would need many thousands of patients, a multicenter study design would be needed.

Pharmaceutical Issues

Many pharmaceutical agents affect bone—either for better or for worse. Physical activity researchers have played a key role in studies that examined the effect of exercise combined with a drug on osteopenia or osteoporosis. For example, Dr. Wendy Kohrt compared the effect of estrogen therapy alone and estrogen therapy combined with an exercise program in older women [4].

Recent therapies for osteoporosis include the bisphosphonate family of drugs, among others; there is a need for physical activity research to be incorporated into studies of the efficacy of these drugs. Drs. Heinonen and Kannus from the UKK Institute in Tampere, Finland, are currently conducting a randomized, controlled trial that will study the effects of the bisphosphonate alendronate and exercise (alone and in combination) on bone loss in postmenopausal women.

Outcome measures from these bone and exercise studies might include ultrasound and MR imaging of bone, pQCT, and DXA (see chapter 4). Other key outcomes, especially in studies with elderly people, are strength, balance, falls, and quality of life. Graduate students trained in the physical activity sciences can play a key role in pharmaceutical research.

Nutrition and Bone

Graduate students with a nutrition background can be valuable members of the bone research team. Traditionally, nutritionists have been involved in designing instruments that determine various nutrient intakes or accurately assess calcium intake. They have also initiated important calcium supplementation studies in various age groups. The major studies of calcium in the postmenopausal years come from the Boston laboratory of the nutritionist Dr. Bess Dawson-Hughes (see chapter 8). Although there have been some attempts to evaluate the role of the interaction of physical activity and calcium nutrition [5, 6], further studies are needed (see chapter 20). Dietitians have also been pivotal in studies of vitamin D in the elderly [7, 8]. Nutritionists are crucial to clinical studies that evaluate bone health in subjects with various eating disorders.

Bone in Special Populations

As the understanding of bone in the normal population grows, researchers are attempting to apply that knowledge to individuals who may be at risk for bone disorders, such as osteoporosis, because of another medical condition. For example, individuals who take glucocorticoids—a powerful anti-inflammatory drug used to control conditions such as rheumatoid arthritis—inevitably suffer profound, and rapid, bone loss (figure 18.2). Exercise specialists can collaborate with care providers to examine whether physical activity can reduce the amount of bone loss in these populations. Also, targeted exercise prescription may alleviate some of the additional morbidity associated with inactivity as a result of numerous medical conditions that prevent patients from undertaking normal physical activity (e.g., cerebral palsy, blindness, spinal cord injuries, amputation).

Young and older adults are also often considered special populations. Further research is particularly relevant in these populations, since much less is generally known about bone in the extremes of age than is known about bone in the middle-aged adult.

Measurement of Physical Activity

Measurement of physical activity is often the domain of epidemiologists or human movement researchers. While measuring physical activity as

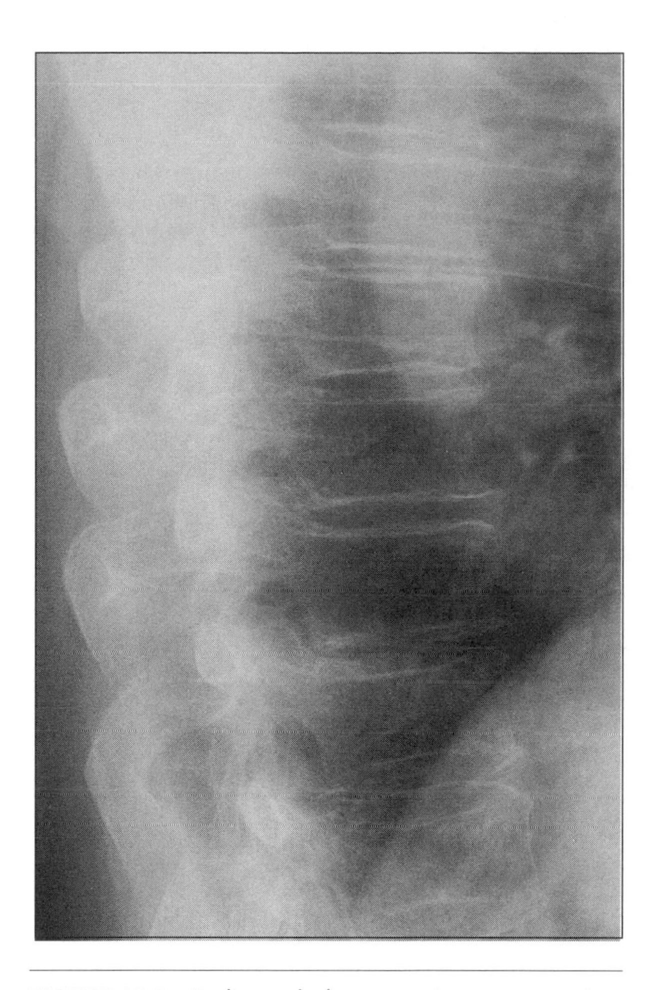

FIGURE 18.2 Radiograph demonstrating osteoporosis of the lumbar spine in a 24-year-old man who had begun taking corticosteroids for inflammatory bowel disease six months earlier. The vertebrae appear almost translucent because BMD is very low.

it benefits cardiovascular health is difficult, measuring physical activity that influences bone is even more challenging (see chapter 9). The graduate wishing to enter this potentially challenging area may consider studies that do the following:

- Evaluate measurement tools that discriminate between weight-bearing and non-weight-bearing physical activity, such as various questionnaires, and motion sensors
- Quantify the various forces on bone related to specific physical activity interventions (in collaboration with biomechanists)
- Measure the efficacy and compliance with exercise prescription in "normal" populations and their relationship to various dimensions of bone health (BMD, strength, balance)
- Measure the efficacy and compliance with exercise prescription in special populations for bone health (e.g., children with cystic fibrosis, adults on glucocorticoid therapy)
- Measure the feasibility of bone exercise programs in a community setting

SUMMARY

- In this chapter we introduced ways to approach the study of bone and physical activity. Our colleagues have coined the phrase "from cell to society" to describe the broad-ranging fields that contribute to unraveling the mysteries of bone and promoting bone health in our community.
- Students from a range of undergraduate backgrounds who choose to go on to graduate school may be well placed to contribute to a research team. The next two chapters provide some examples of the types of graduate research that might be suitable for a master's thesis (chapter 19) or a PhD thesis (chapter 20).

References

1. Sowers MF. Expanding the repertoire: The future of genetic studies (Editorial). *J Bone Miner Res* 1998;11:1657-1659.
2. Jarvinen TLN, Kannus P, Sievänen H, et al. Randomized controlled study of effects of sudden impact loading on rat femur. *J Bone Miner Res* 1998;1475-1482.
3. Milgrom C, Giladi M, Simkin A, et al. An analysis of the biomechanical mechanisms of tibial stress fractures among Israeli infantry recruits. *Clin Orthop* 1988;231:216-221.
4. Kohrt WM, Ehsani AA, Birge SJ. HRT preserves increases in bone mineral density and reductions in body fat after a supervised exercise program. *J Appl Physiol* 1998;84:1506-1512.
5. Specker BL. Evidence for an interaction between calcium intake and physical activity on changes in bone mineral density. *J Bone Miner Res* 1996; 11:1539-1544.
6. Friedlander AL, Genant HK, Sadowsky S, et al. A two-year program of aerobics and weight training enhances bone mineral density of young women. *J Bone Miner Res* 1995;10:574-585.
7. Dawson-Hughes B, Harris SS, Krall EA, et al. Effect of calcium and vitamin D supplementation on bone density in men and women 65 years of age or older. *N Engl J Med* 1997;337:670-676.
8. Chapuy MC, Arlot ME, Duboef F, et al. Vitamin D3 and calcium to prevent hip fracture in elderly women. *N Engl J Med* 1992;327:1637-1642.

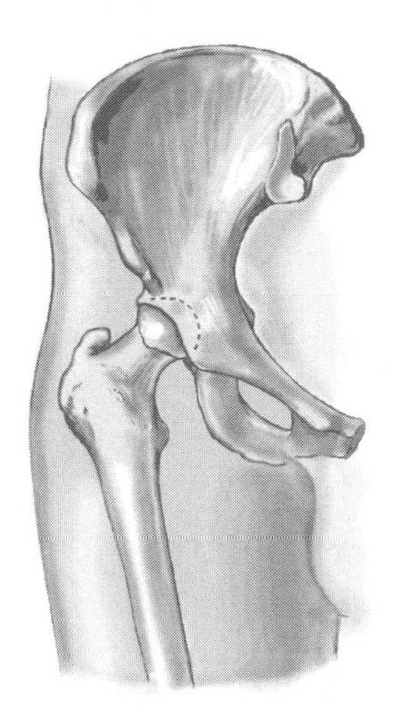

Chapter 19

Research Projects Suitable for a Master's Thesis

The aim of this chapter is to give the interested student a feel for the types of questions related to bone and physical activity that can be answered in a master's thesis. We use the general North American criteria for a master's degree: the student must do course work and a research project presented in a thesis. We will begin by suggesting some courses that may prove useful for the student who has a passion for bone research before outlining potential research projects.

When advising master's students about their research topics, we generally suggest they embark on a project that will permit them to write a manuscript suitable for a peer-reviewed journal. While publishing is not a strict criterion for passing a master's thesis, it is our way of deciding what is, and what is not, a research question that warrants the student's dedication.

Course Work

In addition to any necessary courses to update general communication skills, we recommend that students become familiar with the fields of anatomy, physiology (general and exercise), biomechanics, nutrition, and biochemistry as they relate to bone. Courses in molecular biology and principles of genetics appear to be increasingly valuable.

Students planning to work with animal subjects should take appropriate animal biology courses. Those planning to work with humans at various stages of the life span should take additional courses such as growth (if working with children), adolescent health (if working with teenagers), or gerontology (if working with seniors).

Bone research requires an understanding of upper-level multivariate statistics, linear regression, and the use of statistical software packages. We also recommend that master's students become familiar with a reference manager software; referencing software (e.g., EndNote) is particularly user-friendly and affordable for students. As literature searches, writing, data management, and statistical analysis all require computer literacy, we encourage every student to also hone these important information technology skills.

Studies Appropriate for a Master's Thesis

In this section, we group studies according to the student's area of specialization—for example, human movement studies, pediatrics/growth, gerontology, groups with special needs, nutrition, biomechanics, biochemistry, various health sciences, as well as basic science. In the real world, of course, bone research often requires a multidisciplinary collaboration.

Master's theses must be rather limited in scope because as a new graduate student you only have a short time in which to do your coursework, become familiar with a research area, learn techniques, and then start (and finish) a study! Remember that you have the rest of your life to win the Nobel prize—after your master's—if you are still that way inclined.

Human Movement Studies

Studies that provide good training for further bone research, and a perspective of current issues in the bone field, include the following:

- Evaluating the association among various components of body composition as measured by DXA and other methods as available, e.g., bioelectrical impedance, conventional anthropometry.
- Evaluating the difference in BMD when different amounts of fatty substances (e.g., lard) are placed on the abdomen, simulating real-life differences in obesity among subjects.
- Comparing the difference between resistance and endurance training protocols on biomarkers of bone turnover.
- Measuring endocrine function and BMD in chronically trained, ultraendurance runners and controls to evaluate whether hormone levels are suppressed.
- Measuring BMD in various special populations, both athletic and nonathletic, and examining the determinants of BMD in these populations.
- Measuring the possible acute biochemical response of bone to exercise.
- Assessing BMD between athletic and nonathletic populations at different ages.

Although athlete-control studies have been numerous, there may still be a role for a carefully thought-out, well-controlled bone study in certain athlete groups. This is particularly the case if outcome measures include modalities such as pQCT or MRI (rather than merely DXA).

There is a great deal of interest in studies evaluating side-to-side differences (technically known as the unilateral control model) in sports, in which athletes prefer to perform the movement or activity on one side. The Bone Research Group from the UKK Institute in Tampere, Finland, has performed wonderful studies in racket sport players (figure 19.1) [1], but these studies bear repeating in other centers and with the new instruments for measuring bone, such as pQCT, MRI, and bone biomarkers.

FIGURE 19.1 The racket sport model of bone loading has provided much useful information because it controls for genetics, hormones, medications, nutrition, and other lifestyle factors. Many further studies could be undertaken using this model.

Pediatrics and Growth

The study of bone growth is by its very nature longitudinal if it is to be assessed appropriately. Students would gain insight into this field by assessing physical characteristics (e.g., muscle mass, fat mass, bone mass) of children of the same age who are at different stages of maturity. They might also explore the lifestyle determinants (nutrition, exercise) of bone mass in children at different ages and/or maturity levels.

There is scope for the study of children at risk of osteoporosis because of the burden of disease (e.g., cystic fibrosis) or because of the medication required to treat the disease (particularly corticosteroids for pediatric rheumatological conditions). This is an ideal field for collaboration between clinicians and bone scientists.

Men

Because of the current focus on women's bones, research into aspects of men's bone physiology, pathology, and treatment of osteopenia or osteoporosis is fertile ground. Osteoporosis and related fractures is increasing in men and has been greatly underdiagnosed.

Gerontology

Measuring fall risk factors accurately and reliably is a major research challenge. Candidates for master's degrees can test the reliability of the various methods used to measure fall risk factors and compare different laboratory assessment tools. A master's student could compare various questionnaires that are commonly used to measure quality of life in the aged and in those with osteoporosis. For example, one might compare the various methods used to test static and dynamic balance (fall risk factors) [4]. A master's thesis could also explore the determinants of BMD by DXA in this population.

Groups With Special Needs

Research is needed to investigate the bone health of special populations who are undernourished or prone to suffer the effects of immobilization in addition to their medical condition (e.g., those with cerebral palsy or blindness).

Biomechanics

There is a great need to describe the biomechanical forces of various exercises that are being, or might be, used in exercise interventions for bone health. Also, a master's thesis could be devoted to measuring the breaking strength of bone samples of various sizes and orientations and correlating these with anatomical properties such as proportions of cortical and trabecular bone. Physical therapists could collaborate with laboratory-based biomechanists to measure the forces on vertebrae when performing manual therapy such as posteroanterior mobilizations in both normal and osteoporotic vertebrae.

Biochemistry

A student with a background in biochemistry could develop expertise in the use of bone biomarkers and the measurement of hormone levels. Testing reliability and stability of these biological substances would be good training in research methodology and is of clinical value. Ideally this work could be carried out with researchers who were measuring bone mass by DXA or another technique.

Radiology

Radiology provides many potential opportunities for research. The various technologies that measure bone require reliability studies and interinstrument as well as intertechnique correlative studies (e.g., DXA, QCT, pQCT, QUS—as outlined in chapter 4).

Physical Therapy

A physical therapist or neurophysiologist could use electromyography (EMG) to identify the site of muscle activity in specific exercises that are commonly prescribed for bone health or fall prevention, allowing them to test the effectiveness of an exercise prescription. As physical therapists will be working in clinical settings after graduation, it would be of interest to assess changes in BMD following an accident or injury or during rehabilitation. A physical therapist might be interested in assessing the reliability of activity questionnaires.

Sports Medicine

Studies are still required in stress fracture diagnosis and outcome. Studies of the female athlete are needed to describe the physiological mechanisms in bone (including hormones and factors) that are associated with undernutrition.

The effect of immobilization using a model such as a patient with an anterior cruciate ligament reconstruction warrants investigation. If this study could be extended to monitor the rate of bone recovery after a return to full physical activity, perhaps by another student, it would provide two very appropriate master's theses.

Basic Science

A student could spend a very worthwhile year developing experiments with an animal model such as a rat or mouse. Various forms of exercise such as treadmill running, voluntary wheel running, jumping, and swimming have been used as interventions in animal studies. Researchers have even used innovative methods to develop a form of resistance training for animals [2, 3].

The master's student who gained experience in molecular biology would be well placed for further studies and employment in studies of cell mechanisms or genetics (see chapter 5).

Physical Activity Questionnaires

Questionnaires that are specifically designed to measure loading or degrees of loading are very limited at present. A master's in this field could easily be developed into an applied PhD thesis.

SUMMARY

- In this chapter we outlined a range of subjects that might form a springboard for the reader to discuss bone research ideas with a suitable adviser. Many different undergraduate backgrounds provide good training for bone health research. Even without expensive equipment, such as DXA, the master's student can make an excellent contribution to this area and learn a great deal about bone.

References

1. Kannus P, Haapasalo H, Sankelo M, et al. Effect of starting age of physical activity on bone mass in the dominant arm of tennis and squash players. *Ann Int Med* 1995;123:27-31.
2. Bennell K, Page C, Khan K, et al. Effects of resistance training on bone parameters in young and mature rats. *Clin Exp Pharmacol Physiol* 2000; 27:88-94.
3. Jarvinen TLN, Kannus P, Sievanen H, et al. Randomized controlled study of effects of sudden impact loading on rat femur. *J Bone Miner Res* 1998;1475-1482.
4. Hill K, Schwartz J, Flicker L, et al. Falls among healthy, community-dwelling, older women: A prospective study of frequency, circumstances, consequences and prediction accuracy. *Aust N Z J Public Health* 1999;23:41-48.

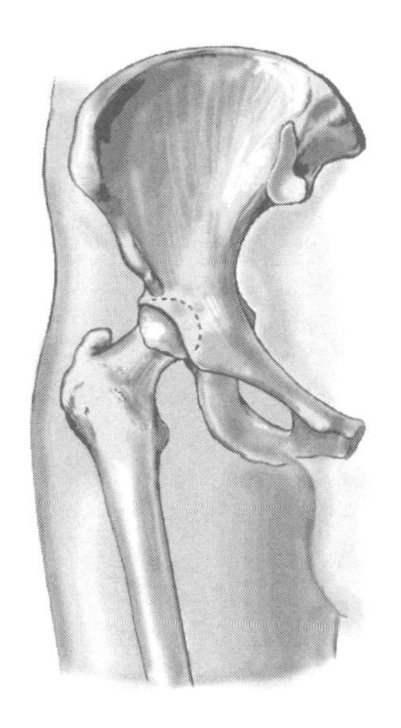

Chapter 20

Research Projects Suitable for a PhD Thesis

Because the PhD degree demands original research, we are not in a position to suggest a specific PhD thesis for any student. We can, however, illustrate the broad range of subjects that have made up past PhDs and draw attention to questions that remain to be answered. As most doctoral students know, of course, posing questions is easy—the hard part is designing appropriate studies to answer those questions.

When planning a PhD, the student should aim to generate a minimum of three publishable papers from the work. In many cases, one paper relates to some form of methodology or basic science related to the study and another two or more studies report the main research. Alternatively, a study may have cross-sectional and longitudinal outcomes that provide the basis for at least two papers.

In our experience, the student who attempts to tackle a thesis in toto can quickly become daunted, discouraged, or thoroughly disenchanted with research. For this reason we advocate breaking the process down into manageable chunks, that is, individual papers. In planning a PhD thesis we suggest you focus on single questions that are appropriate for individual papers rather than on entire theses. These questions can generate a sizable yet manageable amount of data, which, when combined with papers focused on related questions, can form a substantive thesis.

Recent Bone and Physical Activity Doctoral Theses

Some of the excellent studies you read in the literature, and that we have cited throughout this book, come from PhD theses. In this section we provide examples of theses in the bone and physical activity field. In each case we provide a descriptive title of the thesis that is generally *not* the same as the formal thesis topic. We then provide a one-line summary of the aims and the methods and cite just one of the numerous papers that were published from the thesis. The theses summarized here are only a few of the large number that investigate bone and physical

Tips for Graduate Students

Hit the ground running. Do your homework by researching your topic before approaching a potential adviser.

Know your research team. Very few studies are performed in isolation. Know something about the people you are thinking of working with; their academic, research, and publishing histories; their basic philosophies; and their level of funding for you.

Measure the fit. Consider whether you can contribute meaningfully to the research team and whether your personality is compatible.

Be realistic. Design a project or program that can be completed in the time allotted.

Delimit your topic. Be directed in the design of your thesis topic and stay on course.

activity. We included theses familiar to us and want to emphasize that inclusion here is not meant to imply any relative merit compared with those performed in other centers by other excellent researchers.

Studies in Athletic Populations

A longitudinal study of young competitive gymnasts evaluating their bone mineral density, their longitudinal growth, and their endocrine function. The study investigated whether gymnastics was beneficial to bone mineral accrual, whether it had any negative effects on growth, and if the hormonal milieu in these subjects could explain the bone changes. Normal size- and maturity-matched control subjects came from a conveniently located school [1].

A study of the effect of physical activity on growing bone, with particular attention to the development of peak bone mass in racket sport players. This study series asked four different questions regarding physical activity and bone. One question related to the effect of starting age of playing on the side-to-side differences in BMC, BMD, bone dimension, and estimated bone strength variables. A cross-sectional study of tennis players at various ages compared side-to-side differences by DXA [2]. The issue of starting age was discussed in detail in chapter 10 and requires prospective evidence from intervention trials to identify the optimal "critical window" (if one exists) when loading might optimize bone mineral accrual.

A study of the bone mineral density of young (15- to 18-year-old) elite ballet dancers. The aim of the study was to determine whether menstrual disturbance was associated with diminished bone mass at either non-weight-bearing sites, weight-bearing sites, or both. Control data were obtained from healthy subjects and from patients with anorexia nervosa. Determinants of BMD were also examined [3].

A study of the bone mineral density of female former elite ballet dancers aged from 21 to 78 years. The researcher addressed whether retired ballet dancers have greater or lesser BMD than their nondancing age-, height-, and weight-matched counterpart (figure 20.1.). The thesis also examined the determinants of dancers' BMD, particularly the role of childhood ballet training [4].

Studies of Physical Activity in Nonsporting Populations

Although many studies have focused on bone mass in active athletic populations such as tennis players, gymnasts, and ballet dancers, a large number of studies have examined exercise intervention in normally active individuals.

A study of the contribution of physical activity and calcium intake to bone mass, bone size, and strength in women ranging in age from 8 to 72 years. In total, the six published studies examined 919 women using cross-sectional study designs in various populations [5].

A unilateral exercise study in postmenopausal women comparing two strength

FIGURE 20.1 Ballet dancers have been studied in numerous PhD theses, as they have risk factors for low bone mass but also undertake a great deal of bone loading.

training regimes that differed only in the number of repetitions. The study examined whether endurance resistance training (20 repetitions) or high-intensity resistance training (12 repetitions) had more effect on BMD in postmenopausal women [6].

Studies of the effect on bone of various exercise regimens in young pre- and perimenopausal women. The researcher was seeking to measure the effect of exercise on bone strength in young pre- and perimenopausal healthy women. He performed six different studies that evaluated over 400 women in two cross-sectional studies, two longitudinal studies, and two very demanding randomized, controlled trials [7].

Studies of the effect of an exercise intervention in 9- to 11-year-old Australian girls [8] and boys [9]. The exercise intervention was provided two to three times a week in addition to school physical education. Exercise prescription included jumping and resistance training. The main outcome measure was BMD by DXA.

A study of the effect of a school-based exercise intervention on elementary school children in Canada. This study included Asian and Caucasian children (figure 20.2). The exercise intervention was a carefully designed modification of school physical education that was administered by classroom teachers (not physical education specialists). The exercise intervention was randomized by schools. The main outcome measure was BMD by DXA [22].

A cross-sectional study of 115 healthy Caucasian women aged 18 years and their mothers. The aim of the thesis was to examine potential determinants of BMD such as genetics, physical activity, muscle strength, and nutrition. The main outcome measure was BMD by DXA. The researchers measured blood and urine biochemistry, serum sex hormones, nutrient intakes, aerobic fitness, trunk strength, and habitual levels of physical activity [10].

Animal Studies

PhD research may include studies with animals, humans, or both. An example of exclusively

FIGURE 20.2 Studies of physical activity in children of different ethnicities can provide useful new information regarding ethnic-specific response to loading the skeleton.

animal work comes from the UKK Institute in Tampere, Finland.

Studies of the effect of loading and unloading on various characteristics of the bones of a male rat. The investigators examined the effect of various states of loading and unloading on bone. They utilized a male rat model using DXA, bone dimension measurements, and mechanical testing of bone as an outcome measure [11].

Current Questions in the Field

In this section we list the types of studies that could form the basis of PhD theses. Given today's emphasis on evidence-based medicine, we strongly suggest that researchers consider controlled, randomized trials in which investigators are blinded to the groups the subjects are in. While this is not always feasible, young researchers would do well to aspire to perform this study type wherever appropriate since it generally provides the best quality clinical information.

FIGURE 20.3 The loads in many exercise interventions remain to be quantified. When they are, it will be easier to prescribe exercise precisely for bone health.

Human Movement Studies

As the optimal osteogenic activities still have not been defined, there is a need for biomechanical studies that measure the forces generated by certain physical activities. It would be particularly useful to measure the loading components (kinetic and kinematic) in exercise intervention studies (figure 20.3). This might explain why certain exercise regimens are osteogenic and others are not. An assessment of the attenuation of forces at different sites along the lower limbs to the proximal femur would also be of interest.

The effects of detraining on bone also requires clarification. This question is currently being addressed at the UKK Institute in Tampere, Finland.

Because of the importance of falling as a risk factor for fracture, studies of the mechanism and impact of various types and directions of falls would be very useful. What are the mechanisms that underpin falling?

Health Sciences

The borders between the health sciences are often blurred in the multidisciplinary study of bone. A nutrition graduate may be working in the lab analyzing bone turnover markers together with a physical therapy graduate. A genetics student may be using PCR (polymerase chain reaction) technology to determine the vitamin D receptor gene haplotype of subjects in an exercise intervention study.

Human studies of exercise and bone are naturally limited to noninvasive outcome measures such as DXA, QCT, pQCT, quantitative ultrasound, MRI, and so on (see chapter 4). Although DXA has been used in a large number of studies, other recently devised modalities offer novel information to the bone researcher in many settings. All modalities provide distinct information on the changes in bone with different types of exercise regimes (see chapters 3 and 4).

There is excellent scope for collaboration between clinicians who prescribe ovarian hormone therapy (ERT, HRT) and bone researchers. While the mechanisms of how estrogen reduces bone loss are well documented, the effects of exercise on bone remain to be described. Dr. Wendy Kohrt has published elegant papers evaluating the outcome of exercise and hormone therapy on bone, but she states that further studies in this population, using bone markers, are needed to test the hypothesis she has expressed [12].

Bone Physiology and Biochemistry. At the cellular level, there is a need for further investigation of the mechanism whereby a mechanical loading signal is transmitted to bone (mechanotransduction; see chapter 2). Mechanical bone loading is the laboratory equivalent of physical activity and may provide answers to questions that cannot be answered in vivo.

To our knowledge, only one study has compared the effect of training or detraining at various ages on measures of bone strength [11]. As discussed in chapter 19 numerous modalities exist for exercising rats, each with advantages and disadvantages.

Therapeutic ultrasound therapy (as opposed to QUS, the bone measurement technique) has been shown to stimulate fracture repair and osteoblasts in vivo. There may be scope for more studies evaluating this modality in a cell culture model and in an animal model. If preliminary data suggest its efficacy, such a study could be indicated in humans.

Animal training laboratories are quite common in physiology departments, but they are generally the domain of the muscle physiologists. We suggest that bone researchers collaborate with these scientists and use the same animals to study both muscle and bone and their interaction (see chapter 6). The bone researcher may choose to compare various exercise modalities, such as resistance training, endurance training, jumping, weight-vested training, and swimming. Outcome measures should include bone mineral using various protocols, if possible, including an assessment of bone size and strength.

A longitudinal study of bone turnover markers, as well as other measures of bone strength, both in vivo and then ex vivo, in exercising and control sheep would be of great value. Large animal models are relatively expensive but provide unique research benefits [14].

Nutrition. A study using both exercise and calcium intervention to determine if there are any direct effects on bone could be very informative. Such a study could also investigate the interactions between exercise and calcium intervention. Separate studies need to be carried out in the young, in adults, and in postmenopausal women. Outcome measures need to include determinants of bone strength, not BMD alone.

A study of the role of various micronutrients on bone could also provide important information. This could be combined with laboratory work investigating the mechanisms of the action of these nutrients on bone.

Weight loss is thought to influence BMD, but the data remains so conflicting that a recent review suggested that "the consequences of weight reduction on BMD in obese subjects remain unknown" [13]. Further longitudinal studies on changes in BMD in the postmenopausal years are needed to determine whether changes in BMD are sensitive to changes in body composition and, furthermore, to delineate the mechanism of any such association [13].

Students of nutrition may want to evaluate the role of exercise in patients with eating disorders. This clinical population is difficult to study, and exercise may be counterproductive to therapy aimed at weight gain. Other populations that would benefit from research advances would be those with celiac disease, lactase intolerance, or milk allergy. Such studies would need to be performed with great care and prudence and in close collaboration with the appropriate clinicians.

Veterinary Medicine

There is a large scope for veterinarians and animal scientists to study bone. One only has to look at the seminal bone papers [15, 16] to realize how animal studies underpin bone science. Valuable studies would combine animal and human studies (e.g., one animal project repeated in humans). As this book is aimed at the reader who works in the laboratory or with humans, this section will not be expanded beyond the various references to animal models in other parts of this chapter.

Medicine

The following topics relating to stress fractures, osteoporosis, exercise in special populations, and pharmaceutical agents may provide material for doctoral theses.

Stress Fractures. A prospective study of the effect of risk factors for stress fractures in young men and women in various sports could make a novel contribution. This study should examine whether more recently developed serum bone biomarkers can predict the onset of stress fractures. Urinary bone biomarkers could not do this [17], but more recent serum biomarkers [18, 19] may have greater predictive capacity, given their smaller intra-individual variation.

The role of muscle fatigue in attenuating forces and in the development of stress fractures would also be of interest. Researchers might look at specific stress fracture sites (tibia versus metatarsals or navicular) and investigate possible determinants (foot and lower limb biomechanics and training environment) of fracture between sites.

Oral Contraceptive Pill and Bone. The effect of the oral contraceptive pill on BMD in normally menstruating women remains unclear [20, 23]. Because the oral contraceptive pill is often prescribed for treatment of menstrual disturbances in female athletes, whether BMD improves in these situations warrants investigation. Furthermore, whether estrogen in the oral contraceptive pill helps prevent stress fractures in female athletes is also unknown. The dearth of good-quality data about important issues regarding the oral contraceptive pill highlights this as an excellent area for future research studies.

Osteoporosis. At present a new instrument is needed to measure quality of life in women and men with osteoporosis. This questionnaire could be compared to other commonly used assessments.

Researchers interested in osteoporosis may also want to conduct a study comparing the effect of low-dose bisphosphonate medication (currently prescribed to "prevent" osteoporosis) together with exercise to determine whether exercise can provide additional benefit to the subjects. Outcome measures should include measures of bone strength, not BMD alone.

We know that the first hip fracture predisposes an individual to a second hip fracture [21]. Students may undertake an intervention study targeting women with one hip fracture using exercise, osteoporosis medication, and scientifically tested hip protectors [24] as the intervention.

Men with osteoporosis are rarely the subject of research study. A study of an exercise intervention program in men with this condition could be enlightening to both health workers and patients.

Vertebral fractures in patients with osteoporosis have received much less research attention than hip fractures, and many questions need answers. Researchers would do well to consider the prevalence of symptomatic and asymptomatic vertebral fractures in this population (figure 20.4).

Short, medium, and long-term outcomes of exercise intervention in women with osteoporosis would be an excellent subject for a doctoral thesis. Although a few studies have evaluated exercise in the laboratory setting, none have examined the efficacy of exercise intervention in osteoporotic patients using a wide-scale community-based program.

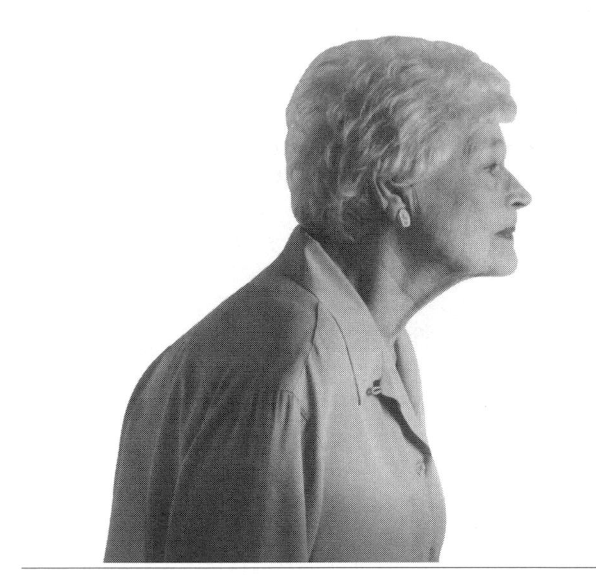

FIGURE 20.4 Thoracic kyphosis can be a sign of osteoporotic vertebral fracture.

Although studies of exercise intervention programs designed to prevent falls and fall-induced injuries have been performed in the healthy elderly, few have addressed people with osteoporosis. This would be another worthy area of research.

Bone and Exercise in Special Populations. There is a great deal of scope for exercise-related research in people who are at particular risk of osteoporosis. These include patients on medication that causes bone loss (glucocorticoids), the frail elderly, and those whose activity is limited for medical reasons (e.g., blindness). Also, there is still a great deal to be learned about influences on prenatal, neo-natal, and infant bone.

Role of Exercise in Pharmaceutical Interventions. Studies can evaluate the efficacy, safety, and mechanisms of exercise interventions combined with pharmaceutical agents prescribed to treat established osteoporosis. The pharmaceutical agents are often tested in large clinical trials that do not include an exercise arm. The protocols used for such studies, however, can easily be adapted to perform appropriate studies combining exercise with the drug in question.

Studies Measuring Physical Activity

Studies that define bone-relevant physical activity are urgently needed. Once this has been achieved, a questionnaire will be needed that accurately evaluates this type of activity. The majority of physical activity questionnaires are designed to measure cardiovascular response, which is not necessarily relevant to bone health. This task would need to be undertaken for different age groups, as bone-relevant activity is likely to be different across the life span.

A longitudinal observational study of bone-relevant physical activity from childhood through the attainment of peak bone mass is needed. Researchers could explore how physical activity patterns influence bone change throughout childhood, throughout the adult years, and with aging. They may also research whether the benefits gained from childhood physical activity are maintained into adulthood.

Although there have been some studies in this field, there remains a need for studies of the acute versus the chronic response of bone to loading. Bone-specific (cortical or trabecular) response related to size, shape, or dimensions of the target bone should be assessed.

A balance and a fitness test battery for different age groups that identifies those at risk of poor bone health and falling would be very beneficial. Particularly in elderly people, the tests would ideally measure risk for falls and osteoporotic fracture.

Another research direction could be to focus on different outcomes that affect the likelihood of fracture. Researchers may explore, for example, what type of exercise provides the most effective improvement in balance—an important risk factor for falling. A challenge to exercise researchers is to define the dose-response of exercise, as people often want to do the least amount possible.

SUMMARY

- There are more research studies on physical activity and bone health to be done than there are "dot-coms" on the Web. You are only limited by your own personal limitations and the research budget—more commonly the latter than the former!
- The major physical-activity-related questions that need answering are related to optimizing bone strength (not necessarily just mass by BMD) in children; defining the interaction among physical activity, calcium, and bone strength across the ages; and refining exercise interventions for bone health so that it becomes possible to truly prescribe effective exercise in the same way that one prescribes insulin precisely according to serum glucose level.
- These are not small tasks, but we fully encourage all those interested in physical activity and bone health research to collaborate vigorously and lobby decision makers vociferously so that physical activity becomes an even greater contributor to the bone health of our communities.

References

1. Bass S, Pearce G, Bradney M, et al. Exercise before puberty may confer residual benefits in bone density in adulthood: studies in active prepubertal and retired female gymnasts. *J Bone Miner Res* 1998;13:500-507.
2. Haapasalo H, Kannus P, Sievänen H, et al. Effect of long-term unilateral activity on bone mineral density of female junior tennis players. *J Bone Miner Res* 1998;13:310-319.
3. Young N, Formica C, Szmukler G, et al. Bone density at weight-bearing and non weight-bearing sites in ballet dancers: The effects of exercise, hypogonadism and body weight. *J Clin Endocrinol Metab* 1994;78:449-454.
4. Khan KM, Bennell KL, Hopper JL, et al. Self-reported ballet classes undertaken at age 10-12 years and hip bone mineral density in later life. *Osteoporos Int* 1998;8:165-173.
5. Uusi-Rasi K, Sievanen H, Vuori I, et al. Associations of physical activity and calcium intake with bone mass and size in healthy women at different ages. *J Bone Miner Res* 1999;13:133-142.
6. Kerr D, Morton A, Dick I, et al. Exercise effects on bone mass in postmenopausal women are site-specific and load dependent. *J Bone Miner Res* 1996;11:218-225.

7. Heinonen A, Oja P, Sievänen H, et al. Effect of two training regimes on bone mineral density in healthy perimenopausal women: a randomised control trial. *J Bone Miner Res* 1998;13:483-490.
8. Morris FL, Naughton GA, Gibbs JL, et al. Prospective 10-month exercise intervention in premenarcheal girls: positive effects on bone and lean mass. *J Bone Miner Res* 1997;12:1453-1462.
9. Bradney M, Pearce G, Naughton G, et al. Moderate exercise during growth in prepubertal boys: changes in bone mass, size, volumetric density and bone strength. A controlled study. *J Bone Miner Res* 1998;13:1814-1821.
10. Henderson NK, Price RI, Cole JH, et al. Bone density in young women is associated with body weight and muscle strength but not dietary intakes. *J Bone Miner Res* 1995;10:384-393.
11. Jarvinen TLN, Kannus P, Sievanen H, et al. Randomized controlled study of effects of sudden impact loading on rat femur. *J Bone Miner Res* 1998;1475-1482.
12. Kohrt WM, Snead DB, Slatopolsky E, et al. Additive effects of weight-bearing exercise and estrogen on bone mineral density in older women. *J Bone Miner Res* 1995;10:1303-1311.
13. Skerry TM, Lanyon LE. Immobilisation induced bone loss in the sheep is not modulated by calcitonin treatment. *Bone* 1993;14:511-6.
14. Cauley J, Salamone LM, Lucas FL. *Postmenopausal endogenous and exogenous hormones, degree of obesity, thiazide diuretics and risk of osteoporosis*. In: Marcus R, Feldman D, Kelsey J, ed. Osteoporosis. San Diego, CA, USA: Academic Press, 1996: 551-576.
15. Lanyon LE. Analysis of surface bone strain in the calcaneus of sheep during normal locomotion. *J Biomechanics* 1973;6:41-49.
16. Lanyon LE, Rubin CT. Static versus dynamic loads as an influence on bone remodelling. *J Biomech* 1984;17:897-905.
17. Bennell KL, Malcolm SA, Brukner PD, et al. A 12-month prospective study of the relationship between stress fractures and bone turnover in athletes. *Calcif Tissue Int* 1998;63:80-85.
18. Szulc P, Seeman E, Delmas PD. Biochemical measurements of bone turnover in children and adolescents. *Osteoporos Int* 2000;11:281-94.
19. Fink E, Cormier C, Steinmetz P, et al. Differences in the capacity of several biochemical bone markers to assess high bone turnover in early menopause and response to alendronate therapy. *Osteoporos Int* 2000;11:295-303.
20. Bennell KL, White S, Crossley K. The oral contraceptive pill: a revolution for sportswomen? *Br J Sports Med* 1999;33:231-238.
21. Stewart A, Walker LG, Porter RW, et al. Predicting a second hip fracture. *J Clin Densitom* 1999;2:363-370.
22. McKay HA, Petit MA, Schutz RW, et al. Augmented trochanteric bone mineral density after modified physical education classes: A randomized school-based exercise intervention in prepubertal and early-pubertal children. *J Pediatr* 2000;136:156-162
23. Burr DB, Yoshikawa T, Teegarden D, et al. Exercise and oral contraceptive use suppress the normal age-related increase in bone mass and strength of the femoral neck in women 18-31 years of age. *Bone* 2000;27:855-863.
24. Kannus P, Parkkari J, Niemi S, et al. Prevention of hip fractures in elderly people with use of hip protector. *N Engl J Med* 2000;343:1506-1513.

Appendix A Tables

As new information becomes available, the following tables will be updated on-line at **www.bonebook.net**.

TABLE A.1

Retrospective studies of childhood activity and adult BMD

Authors (no. in exercise group)	Mean age (SD)	Type of exercise	Outcome
Fehily et al. 1992 [63] (182)	20–23	Sport participation at age 12 evaluated by recall questionnaire	Positive correlation between physical activity (age 12) and radial BMC
Halioua & Anderson 1984 [63] (181)	20–50	Activity questionnaire	High lifetime physical activity (> 45 min/wk) associated with greater distal radial BMC
Kannus et al. 1995 [45] (105)	27	Tennis	3–4 times greater humeral BMC side-to-side difference in those who started playing at or before puberty
Khan et al. 1998 [57] (99)	51 (14)	Classical ballet	Weekly ballet class at each age between 10 and 12 was positively associated with difference in BMD between dancers and controls at both the femoral neck and total hip sites.
Kriska et al. 1988 [74] (223)	58	Past physical activity by questionnaire at age 14–21	Difference in bone area but no difference in BMD according to physical activity levels
McCulloch et al. 1990 [75] (101)	20–35	Adolescent and childhood activity evaluated by recall questionnaire	Calcaneal BMD (by QCT) greater in high activity group
Talmage & Anderson 1984 [65] (1200 with subset < 25 yrs)	> 25	Adolescent activity evaluated by recall questionnaire	Higher radial BMC at age 25 in those who did secondary school athletics or heavy farm labor
Tylavsky et al. 1992 (705)	17–23	Self-administered activity questionnaire during high school and college	Radial BMC greater in those with greater physical activity
Valimaki et al. 1994 [71] (153)	9–29	Activity questionnaire	Physical activity (> 30 min/wk) predicted femoral neck BMD
Welten et al. 1994 [71] (98)	13–28	Activity questionnaire 4–6×/yr from age 13 to 28	Physical activity not a predictor of lumbar spine BMD

Table taken from chapter 10.

TABLE A.2

Physical activity as a determinant of premenopausal BMD

Authors (no. of subjects)	Determinant	Site and measure of bone mass	Outcome
Alekel et al. 1995 [55] (28)	Walking by physical activity questionnaire	Lumbar spine and proximal femoral BMD	Walking associated with increased BMD at both sites
Aloia et al. 1988 [56] (24)	Physical activity (sensor)	Lumbar spine BMD (DPA)	Significant correlation that explains 16% of the variance in BMD
Davee et al. 1990 [57] (9)	Hours per week of physical activity	Lumbar spine BMD	Not correlated
Halioua et al. 1989 [58] (181)	Lifetime physical activity by questionnaire	Radial BMC/BMD	High lifetime physical activity associated with increased forearm bone mineral
Henderson et al. 1995 [37] (115)	Physical activity score	Hip BMD including subregions	No effect independent of trunk flexor strength or $\dot{V}O_2max$
Kanders et al. 1988 [59] (60)	Physical activity measured by Minnesota Leisure Time Physical Activity Questionnaire	Lumbar spine and radial BMD	Lumbar spine but not radial BMD related to physical activity
Kirk et al. 1989 [60] (19)	Physical fitness ($\dot{V}O_2max$)	Lumbar spine BMD	Correlation between $\dot{V}O_2max$ and lumbar BMD ($r = 0.51$)
Mazess et al. 1991 [61] (300)	Activity measured by accelerometer and pedometer	Lumbar spine (DPA) and forearm (SPA)	Activity not correlated with BMD
Sowers et al. 1985 [62] (86)	Physical activity questionnaire	Forearm bone mineral by SPA	No correlation
Stevenson et al. 1989 [63] (112)	Lack of regular exercise	Lumbar spine, proximal femoral BMD	No effect of current regular exercise
Uusi-Rasi et al. 1994 [64] (31)	Daily walking distance measured	Lumbar spine, proximal femur, distal radial BMC	No difference between subjects and controls
Young et al. 1995 [65] (450)	Physical activity measured by questionnaire	Lumbar spine and proximal femoral sites	Not correlated
Zhang et al. 1992 [32] (264)	Physical activity by Caltrac personal activity computer	Lumbar spine, midradius, distal radius—perimenopausal women	Significant correlation between physical activity and BMD at spine and forearm sites

Table taken from chapter 11.

TABLE A.3

Prospective studies of spinal bone mineral in postmenopausal women: Effect of physical activity with minimal mechanical loading

Authors	Comments about the degree of mechanical loading at the spine	Effect size at the spine (%)
Bloomfield et al. 1992 [24]	Stationary cycling (60–80% max. heart rate) would not be expected to generate high-impact, fast strain rate forces on the lumbar spine, although the authors postulated that psoas may be contracting on its lumbar spine attachments	Ex: +3.6 Controls: –2.4
Cavanaugh & Cann 1988 [49]	Walking at 60–85% of max. heart rate	Ex: 0.0 Controls: –6.1
Grove & Londeree 1992 [20]	Low impact group: exercises below 1.5 × body weight in peak ground reaction force would be attenuated at the lumbar spine	Ex: +0.6 Controls: 0.0
Hatori et al. 1993 [39]	Walking above the anaerobic threshold	Ex: +1.1 Controls: –1.7
Hatori et al. 1993 [39]	Walking below the anaerobic threshold	Ex: –1.0 Controls: –1.7
Krølner et al. 1983 [25]	Running, walking, exercises on all fours, exercises in sitting and in standing, ball games: all of these could have loaded the lumbar spine	Spine BMC by DPA: +6.2 (NS)
Lau et al. 1992 [47]	Climbing up and down a 23-cm bench 100 times (no change to protocol over the year) and 15 min of submaximal trunk exercises	Controls: –2.5
Martin et al. 1993 [47]	Walking for 30 min at 70–85% max. heart rate	Spine by DPA Ex: –0.5 Controls: –0.6
Martin et al. 1993 [50]	Walking for 45 min at 70-85% max. heart rate	Spine by DPA Ex: +0.8 Controls: –0.6
Revel et al. 1993 [51]	Psoas exercises: seated hip flexion against 5-kg resistance	Total lumbar spine BMD (by QCT) Ex: 0.0 Controls: –7
Tsukahara et al. 1994 [52]	All exercises were done partially non-weight-bearing in water	Ex: +2.2 Controls; –0.9
White 1984 [53]	Walking program	Ex: –1.7 Controls: –1.6

Table taken from chapter 12.

TABLE A.4

Summary of cross-sectional studies showing the percentage difference in bone density at different sites of athletes with menstrual disturbance compared with controls with normal menses

First author, year [reference]	Sport	Number with menstrual disturbance	Number with normal menses	Assessment of estradiol	Ovulation documented in subjects with normal menses	Exclusion of eating disorders	Assessment of bone mass	% Difference in lumbar spine
(1) Menstrual disturbance was amenorrhea								
Drinkwater 1984 [50]	Mixed[a]	14	14	Y	Y	Y	DPA, SPA	–13.82
Lindberg 1984 [52]	Runners	11	15	Y	N	N	SPA	*
Linnell 1984 [74]	Runners	10	12	N	N	N	SPA	*
Marcus 1985 [54]	Runners	11	6	Y	Y	N	CT, SPA	–17.0
Fisher 1986 [87]	Runners	11	17	Y	N	Y	DPA, SPA	–8.3
Nelson 1986 [53]	Mixed[a]	11	17	Y	N	Y	DPA, SPA	–8.1
Snyder 1986 [88]	Rowers	4	7	N	N	N	DPA, SPA	–8.4
Ding 1988 [89]	Mixed[b]	19	35	Y	N	N	DPA	–17.5
Grimston1990 [40]	Runners	3	9	N	N	N	DPA, SPA	–6.9
Wolman1990 [68]	Mixed[c]	25	21	N	N	N	CT	–20.0
Harber 1991 [90]	Runners	11	17	Y	N	N	CS	*
Warren 1991 [2]	Dancers	22	29	Y	N	Y	DPA	–12.9
Baer 1992 [91]	Runners	7	6	Y	N	Y	DP	–0.9
Myerson 1992 [92]	Runners	13	13	Y	Y†	YD	PA	–5.4
Sncad 1992 [62]	Runners	11	24	Y	Y	Y	DP	–11.2
Wilmore 1992 [93]	Runners	8	5	Y	Y	N	PA	–0.2
Wolman 1992 [76]	Mixed[c]	25	27	N	N	N	CT	–20.4
Hetland 1993 [94]	Runners	13	93	Y	Y	Y	DXA, SPA	–9.0
Jonnavithula 1993 [34]	Dancers	5	12	Y	N	N	DPA SPA	–16.0
Myburgh 1993 [11]	Mixed[b]	9	12	N	N	N	DXA	–11.6
Rutherford 1993 [13]	Mixed[d]	15	15	N	N	N	DXA	–9.2
Pearce 1996 [95]	Dancers[f]	41	46	N	N	Y	DXA	–2.6
Rencken 1996 [44]	Mixed[a]	29	20	Y	Y	Y	DXA	–12.0
Kcay 1997 [43]	Dancers	38	12	N	N	N	DXA	**
Moen1998 [96]	Runners	10	10	Y	N	N	DXA	–2.7
Tomten 1998 [47]	Runners	13	15	Y	Y	Y	DXA	–12.0
Petterson 1999 [45]	Runners	10	10	Y	Y	N	DXA	–16.0
Gibson 2000 [97]	Runners[g]	33	18	Y	N	N	DXA	***

(continued)

TABLE A.4 ***(continued)***

First author, year [reference]	Sport	Number with menstrual disturbance	Number with normal menses	Assessment of estradiol	Ovulation documented in subjects with normal menses	Exclusion of eating disorders	Assessment of bone mass	% Difference in lumbar spine
(2) Menstrual disturbance was oligomenorrhea or amenorrhea								
Snead 1992 [98]	Runners	13	19	Y	Y	Y	DPA	–11.5
Micklesfield 1995 [45]	Runners	10	15	Y	N	N	DXA	–13.1
Robinson 1995 [10]	Mixed[c]	16	26	N	N	N	DXA	–5.7
(3) Menstrual disturbance was oligomenorrhea								
Snyder 1986 [88]	Rowers	5	7	N	N	N	DPA, SPA	–2.4
Cook 1987 [99]	Runners	19	17	N	N	N	DPA	–7.7
Lloyd 1987 [100]	Mixed[b]	6	10	N	N	Y	CT	–15.2
Baker 1988 [101]	Mixed[b]	6	10	Y	Y	Y	CT	–15.2
Buchanan 1988 [102]	Mixed[b]	9	10	Y	Y	Y	CT	–12.0
Snead 1992 [62]	Runners	8	24	Y	Y	Y	DPA	–12.2
Hetland 1993 [94]	Runners	17	93	Y	Y	Y	DXA, SPA	–2.7

Subjects are grouped according to whether subjects (1) all had amenorrhea, (2) had either amenorrhea or oligomenorrhea, or (3) had oligomenorrhea.

*Statistically significant; **Amenorrheic/Oligomenorrheic group significantly lower than normal menses group, Z-scores only presented; ***Significantly different from zero (Oligomenorrheic < zero and normal menses group > zero; T-scores reported using European reference population.

[a]Mostly runners; [b]Collegiate athletes; [c]Rowers, runners, and dancers; [d]Runners and triathletes; [e]Runners and gymnasts; [f]Dancers and non-dancing controls; [g]16-35 years of age

NS = not stated but non-significant; Y = yes; N = no; DXA = dual energy X-ray absorptiometry; DPA = dual photon absorptiometry; SPA = single photon absorptiometry; CT = computer tomography

†Measurements at proximal femur not included. Figures given are an average of all other lower limb sites.

††Figures given are an average of all upper limb sites.

‡Figures given are an average of 0 < 40 months and > 40 months.

Table taken from chapter 16.

Appendix B Questionnaires

QUESTIONNAIRE B.1

Lifestyle Questionnaire

Name of participant: ______________________________ **Date:** ____________

1. Indicate the physical activities that you participated in over the last month during your leisure time.

	Number of occasions over the last month	Average number of activity minutes spent on each occasion				Slight change from normal stage	Some perspiration, faster than normal breathing	Heavy perspiration, heavy breathing
	Frequency	**Duration**				**Intensity**		
		1-15	**16-30**	**31-60**	**60+**	**Light**	**Medium**	**Heavy**
Walking for exercise	______	❑	❑	❑	❑	❑	❑	❑
Bicycling	______	❑	❑	❑	❑	❑	❑	❑
Swimming	______	❑	❑	❑	❑	❑	❑	❑
Jogging/running	______	❑	❑	❑	❑	❑	❑	❑
Home exercises	______	❑	❑	❑	❑	❑	❑	❑
Ice skating	______	❑	❑	❑	❑	❑	❑	❑
Cross-country skiing	______	❑	❑	❑	❑	❑	❑	❑
Tennis	______	❑	❑	❑	❑	❑	❑	❑
Golf	______	❑	❑	❑	❑	❑	❑	❑
Popular dance	______	❑	❑	❑	❑	❑	❑	❑
Baseball/softball	______	❑	❑	❑	❑	❑	❑	❑
Alpine skiing	______	❑	❑	❑	❑	❑	❑	❑
Ice hockey	______	❑	❑	❑	❑	❑	❑	❑
Bowling	______	❑	❑	❑	❑	❑	❑	❑
Exercise classes	______	❑	❑	❑	❑	❑	❑	❑
Racquetball	______	❑	❑	❑	❑	❑	❑	❑
Curling	______	❑	❑	❑	❑	❑	❑	❑
Others (please specify):								
______________	______	❑	❑	❑	❑	❑	❑	❑
______________	______	❑	❑	❑	❑	❑	❑	❑
______________	______	❑	❑	❑	❑	❑	❑	❑

2. How long have you been doing some physical activity in your leisure time at least once a week?

   ❑ I don't do an activity each week

   ❑ For less than 3 months

   ❑ From 3 months to just under 6 months

   ❑ From 6 months to just under 1 year

3. If you want to participate more in physical activities than you do now, why aren't you able to? (Check at most three reasons.)

   ❑ I don't want to participate more

   ❑ Ill health

   ❑ Injury or handicap

   ❑ Lack of energy

   ❑ Lack of time because of work/school

   ❑ Lack of time because of other leisure activities

   ❑ Costs too much

   ❑ No facilities nearby

   ❑ Available facilities are inadequate

   ❑ No leaders available

   ❑ Requires too much self-discipline

   ❑ Lack necessary skills

   ❑ Other ______________________________

4. If you wanted to participate more in physical activities, which of the following would increase the amount of physical activity you do? (Check at most three.)

   ❑ Nothing

   ❑ Better or closer facilities

   ❑ Different facilities

   ❑ Less expensive facilities

   ❑ More information on the benefits

   ❑ Organized fitness classes available

   ❑ Employer or union sponsored activities available

   ❑ Organized sports available

   ❑ Organized fitness classes available

   ❑ Fitness test with personal activity program available

   ❑ People with whom to participate

   ❑ Common interest of family

   ❑ Having necessary equipment

   ❑ Common interest of friends

   ❑ More leisure time

   ❑ More energy

   ❑ More self-discipline

   ❑ Better health

5. Here is a list of reasons why some people do physical activities during their leisure time. How important is each of these to you?

	Very important	Of some importance	Of little importance	Of no importance
To feel better mentally and physically	❑	❑	❑	❑
To be with other people	❑	❑	❑	❑
For pleasure, fun, or excitement	❑	❑	❑	❑
To control weight or to look better	❑	❑	❑	❑
To move better or to improve flexibility	❑	❑	❑	❑
As a challenge to my abilities	❑	❑	❑	❑
To relax or reduce stress	❑	❑	❑	❑
To learn new things	❑	❑	❑	❑
Because of fitness specialist's advice for improving health in general	❑	❑	❑	❑
Because of doctor's order for therapy or rehabilitation	❑	❑	❑	❑
Other ________________________	❑	❑	❑	❑

6. How important are each of the following to you in gaining a feeling of well-being?

	Very important	Of some importance	Of little importance	Of no importance
Adequate rest and sleep	❑	❑	❑	❑
A good diet	❑	❑	❑	❑
Low calorie snacks between meals	❑	❑	❑	❑
Maintenance of proper weight	❑	❑	❑	❑
Participation in social and cultural activities	❑	❑	❑	❑
Control of stress	❑	❑	❑	❑
Regular physical activity such as exercise, sports, or games	❑	❑	❑	❑
Being a nonsmoker	❑	❑	❑	❑
Adequate medical and dental care	❑	❑	❑	❑
Positive thinking/meditation	❑	❑	❑	❑

7. How would you compare yourself to others of your own age and sex?

❑ More fit

❑ Less fit

❑ As fit

8. In the past year, what physical activities have you stopped doing and why? (Do not include those stopped due to a change in the season.)

____

____

____

9. What physical activities would you like to start in order to improve your fitness and health? Why haven't you started yet?

____

____

____

____

____

____

10. With whom do you/would you *usually* do your physical activities in your leisure time?

❑ No one

❑ Co-workers

❑ Friends

❑ Classmates at school

❑ Immediate family or relatives

❑ Others

11. At what time do you/would you usually do your physical activities? (Indicate more than one if you usually do activities more than once a day.)

❑ In the morning

❑ In the evening

❑ At lunchtime

❑ In the afternoon

❑ At no special time

12. (a) How would you describe your state of emotional well-being?

❑ Very good

❑ Good

❑ Adequate

❑ Poor

❑ Very poor

(b) How do you think this might affect your physical activity/fitness goals?

❑ Aid

❑ Hinder

❑ No effect

Please explain: ____

____

____

____

____

____

____

13. What do you *usually* eat for breakfast? (*Usually* means at least four days a week.) Check all that apply.

❑ I don't eat breakfast

❑ Eggs

❑ Bacon or other meat, fish, or poultry

❑ Bread, danish, or donut

❑ Granola

❑ Other cereals

❑ Fruit or fruit juice

❑ At least 6 oz milk

❑ Cheese

❑ Yogurt

❑ Tea/coffee

14. In the last year, how often have you been eating the following?

	More	Less	Same as before
Sweet food and candies	❑	❑	❑
Fruit and vegetables	❑	❑	❑
Fats and fried foods	❑	❑	❑
Salt and salty foods	❑	❑	❑
Meals on a regular basis	❑	❑	❑
The same amount of food or calories	❑	❑	❑

15. (a) About how many hours of sleep do you *usually* get each day?

❑ 6 hours or less

❑ 7

❑ 8

❑ 9

❑ 10

❑ 11 hours or more

(b) Do you think you are getting enough sleep?

❑ Always

❑ Usually

❑ Seldom

❑ Never

16. (a) About how often do you *usually* drink alcohol?

❑ More than once a day

❑ 1-3 times a week

❑ 4-7 times a week

❑ Less than once a month

❑ 1-3 times a month

❑ I don't drink alcohol (go to question 17)

(b) About how many drinks do you usually have at a time?
Where one drink is: One pint of beer = 12 oz
One small glass of wine
One shot of liquor/spirits (i.e., 1-1½ oz with/without mix)

❑ 1 ❑ 2-3 ❑ 4-5 ❑ 6-7 ❑ 8+

17. Which of the following best describes your experience with tobacco? Check all that apply.

- ❑ I haven't smoked
- ❑ I currently smoke
  - ❑ Cigarettes occasionally
  - ❑ Less than ½ pack of cigarettes daily
  - ❑ About a pack of cigarettes daily
  - ❑ Two or more packs of cigarettes daily
  - ❑ A pipe, cigar, or cigarillo occasionally
  - ❑ A pipe, cigar, or cigarillo daily
- ❑ I stopped smoking
  - ❑ Cigarettes recently
  - ❑ Cigarettes over a year ago
  - ❑ A pipe, cigars, or cigarillos recently
  - ❑ A pipe, cigars, or cigarillos over a year ago

18. In general, how would you describe your state of health?

- ❑ Very good
- ❑ Good
- ❑ Average
- ❑ Poor
- ❑ Very poor

QUESTIONNAIRE B.2

Lifestyle Needs and Activity Preferences

Name of participant: ______________________________ **Date:** ______________

Lifestyle Need:
It is important to me to . . .

- ❑ like the people I'm with
- ❑ be in a group
- ❑ be independent
- ❑ get to know other people
- ❑ have other people like me
- ❑ be physically active
- ❑ use my imagination
- ❑ create something
- ❑ find activities challenging
- ❑ feel safe and secure
- ❑ try something new and different
- ❑ be myself
- ❑ use my talents
- ❑ improve myself and my skills
- ❑ accomplish something
- ❑ relax
- ❑ spend time with my family
- ❑ release energy
- ❑ have common interests with other people
- ❑ be able to contribute something to a group
- ❑ meet many new people
- ❑ be a leader
- ❑ feel confident
- ❑ learn something
- ❑ be in pleasant, attractive surroundings
- ❑ be alone
- ❑ have a structured activity
- ❑ be able to do things at the last minute
- ❑ follow rules
- ❑ be praised
- ❑ have fun and enjoy myself
- ❑ release frustration
- ❑ take risks

Once you have checked the lifestyle needs that are important to you, list the three most important and identify activities that would most probably satisfy those needs.

Lifestyle needs	Activity preferences
1. ____________	____________
2. ____________	____________
3. ____________	____________

QUESTIONNAIRE B.3

Action Plan Worksheet

Name of participant: ______________________________ **Date:** ____________

Goals and Action Steps	**Time Frame**
Goal #1 ______________________	__________
Action steps	
1. ______________________	__________
2. ______________________	__________
3. ______________________	__________
Goal #2 ______________________	__________
Action steps	
1. ______________________	__________
2. ______________________	__________
3. ______________________	__________
Goal #3 ______________________	__________
Action steps	
1. ______________________	__________
2. ______________________	__________
3. ______________________	__________

Prescription for Physical Activity

Frequency ______________________________

Intensity ______________________________

Time ______________________________

Type ______________________________

Comments

____

____

____

____

Success Indicators

1. ______________________________
2. ______________________________
3. ______________________________
4. ______________________________
5. ______________________________

Date for next appraisal: ______________________________

Index

Note: Page numbers followed by *t* or *f* refer to the table or figure on that page, respectively

U

V

W

X

Y

Z

About the Authors

Dr. Karim Khan, MD, PhD, is a clinician-researcher at the University of British Columbia in Vancouver, Canada. He is an assistant professor with a joint appointment in the Faculty of Medicine (Department of Family Practice—Sports Medicine) and the School of Human Kinetics. He divides his clinical time between the Osteoporosis Program at the Women's Health Centre, BC Women's and Children's Hospital, and UBC's Allan McGavin Sports Medicine Centre. Dr. Khan is also a visiting professor at the University of New South Wales and the University of Melbourne, both in Australia.

Dr. Khan's scientific work in bone research includes investigations into the treatment and etiology of stress fractures, studies of the role of activity in bone mineral density in both young and older athletes, and exercise intervention for bone health in schoolchildren and older women with osteoporosis. He is currently developing a falls and fracture prevention program at UBC and BC Women's Hospital. The common thread in all of these studies is the role of exercise prescription as a therapeutic modality.

Dr. Khan is a renowned speaker on the international medical conference circuit, and he may be found anywhere between his native Australia, North America, Asia, Europe, or Africa at various times of the year. His clinical textbook of sports medicine (*Clinical Sports Medicine,* coauthored with Peter Brukner) is a best-seller and in its second edition. He is on the editorial board of the *British Journal of Sports Medicine,* the *Journal of Medicine and Science in Sport, Physician and Sportsmedicine,* and the *International Sports Med Journal.*

Dr. Heather McKay, PhD, is an associate professor in the School of Human Kinetics at the University of British Columbia, Vancouver, Canada. She is presently a Michael Smith Foundation for Health Research Scholar. She is deputy director of the Vancouver site of the Canadian Multi-Center Osteoporosis Study and a key collaborator with the osteoporosis research program at BC Women's Hospital and Health Center, the bone health research program at BC Children's Hospital, and the BC Center of Excellence in Women's Health (BCCEWH).

Dr. McKay and colleagues at the University of Saskatchewan monitored bone growth and development and lifestyle in healthy children over seven years. This study provided pivotal insights into the normal patterns of bone gain in healthy children. Her current program of prospective research at UBC has contributed crucial insights into the key time during childhood for exercise intervention to optimize bone mass. Her research group was the first to intervene with exercise over two years in a multi-ethnic group of both boys and girls at different stages of maturity in a school-based setting. She is currently evaluating the feasibility of a wide-scale dissemination of a school-based bone health curriculum in the elementary school system. Dr. McKay has recently extended her bone health research program to include exercise intervention studies that assess bone and balance in women with osteoporosis.

Dr McKay is a regular reviewer for scholarly journals, including *Journal of Bone and Mineral Research, Bone,* and *Journal of Pediatrics* as well as for national and international granting agencies. Dr. McKay is a member of the Province of BC's Bone Health and Fracture Steering Committee, and the Osteoporosis Society of Canada Physical Activity and Fracture Sub-Committee.

Dr. Pekka Kannus, MD, PhD, is the chief physician and head of the Accident and Trauma Research Center at the UKK Institute in Tampere, Finland. He is a professor of injury prevention at the University of Tampere and associate professor of sports medicine at the University of Jyväskylä, Finland. Dr. Kannus is also a visiting professor with the Department of Orthopedics and Rehabilitation at the University of Vermont College of Medicine in Burlington, Vermont.

Dr. Kannus' scientific work has focused on basic and applied research of the musculoskeletal system of the human body. His primary interest in clinical practice is sports injuries. He has published more than 200 scientific articles in international peer-reviewed journals and has written a full book and 20 book chapters. Dr. Kannus has given keynote lectures more than 50 times at conferences around the world, including the Vince Higgins Lecture at the Australian Conference of Science and Medicine in Sport in 1995.

Dr. Kannus is section editor of the *Scandinavian Journal of Medicine & Science in Sports*; foreign consulting editor of *Clinical Exercise Physiology*; and is on the editorial board of *Physician and Sportsmedicine*, the *British Journal of Sports Medicine*, the *Clinical Journal of Sport Medicine*, and *Isokinetics and Exercise Science*. Dr. Kannus is a regular reviewer for scholarly journals including the *Lancet*, *JAMA*, the *British Medical Journal*, and the *Journal of Bone and Mineral Research*. He is a former president of the Scandinavian Foundation of Medicine and Science in Sports.

Dr. Don Bailey, PhD, is an emeritus professor and senior researcher in the College of Kinesiology at the University of Saskatchewan in Saskatoon, Canada. He is also a visiting professor in the School of Human Movement Studies at the University of Queensland in Brisbane, Australia.

Throughout his illustrious scientific career, Dr. Bailey has been concerned with studying child growth and development and, more recently, the relationship between bone mineral accrual and physical activity in the growing years. He directed the landmark Saskatchewan Growth Study, a 10-year longitudinal investigation of growth and physical fitness in school-aged children, which was followed by a 7-year longitudinal trial on the determinants of bone mineral accrual in healthy children. In this unique study, the research team annually measured growth parameters, physical activity, nutrition, and bone mineral density in more than 200 boys and girls aged 8 to 16 years. This has led to pivotal publications in key journals such as the *Journal of Bone and Mineral Research*, *Journal of Bone and Joint Surgery*, *Journal of Pediatrics*, *American Journal of Clinical Nutrition*, *International Journal of Sports Medicine*, *Medicine and Science in Sports and Exercise*, and *Calcified Tissue International*.

Dr. Bailey is much sought after as a speaker at international bone and growth conferences, having given more than 50 invited lectures around the world. Currently he is vice president of the International Council for Physical Activity and Fitness Research.

Dr. John Wark, MD, PhD, is professor of medicine at the University of Melbourne, Australia. He is head of the Bone and Mineral Service at The Royal Melbourne Hospital and director of the Centre for Osteoporosis and Bone Studies on the same leading teaching hospital campus. He also directs a community-based osteoporosis centre in Melbourne. Dr. Wark is an active contributor to scientific groups of the World Health Organization and an active member of the Australian and New Zealand Bone and Mineral Society. His current projects with WHO include a genetic-environmental life course model of health and functional capacity, policy for the prevention of falls, models of care in hip fracture, and a life course perspective on the functional capacity of postmenopausal women. Dr. Wark is a Specialist Editor for the journal *Clinical Science* and is a regular reviewer for all the leading international bone journals.

As both a specialist endocrinologist and an extremely well-credentialed scientist with more than 150 major international publications and many large grants to his credit, Dr. Wark's expertise covers a wide range of issues regarding bone, genetic, and environmental determinants of health, nutrition, and physical activity. He is the principal investigator of an internationally renowned, ongoing, longitudinal cohort study of genetic and environmental determinants of bone health in more than 2000 female twins over a wide age range. He also has key publications in stress fractures, osteoporosis, childhood calcium supplementation, athletes and bone mineral density, bone health across the menopause, and the role of vitamin D in bone metabolism. He was the principal investigator in the first controlled trial of physical activity intervention in schoolgirls, and this model has been successfully implemented in boys by his collaborators. Further evidence for Dr. Wark's passion for physical activity came in 1999 when he was found trekking through the Andes at 4000 meters, and he is often found supporting his favorite Australian Rules football team, the Collingwood Magpies.

Dr. Kim Bennell, PT, PhD, is associate professor and director of the Centre for Sports Medicine Research and Education in the School of Physiotherapy (faculty of medicine, dentistry and health sciences) at the University of Melbourne, Australia. She is also a director of a private physiotherapy clinic that specializes in exercise prescription in the prevention and treatment of osteoporosis. She has won the Young Investigator Award at the Australian Conference of Science and Medicine in Sport on two occasions and has received a National Young Tall Poppy Award in recognition of outstanding achievement and excellence in biomedical research.

Dr. Bennell's research spans many areas of physical activity and physiotherapy, but her reputation in the bone field derives from her pivotal stress fracture research and her subsequent work studying the relationship between physical activity and bone mineral density in active people. Dr. Bennell is the author of the first prospective study of the risk factors for stress fractures in athletes. She is currently undertaking research on the effect of ballet training in young girls, which has been funded by the National Health and Medical Research Council.

Dr. Bennell is on the editorial board of the *British Journal of Sports Medicine, Clinical Journal of Sports Medicine, Physical Therapy in Sport,* and *Manual Therapy*. She serves on a number of professional committees including the Sports Physiotherapy Group and lectures regularly to community groups on the benefits of exercise for bone health.